ACS SYMPOSIUM SERIES **1039**

Organic Thin Films for Photonic Applications

Warren N. Herman, Editor
Laboratory for Physical Sciences, University of Maryland

Steven R. Flom, Editor
Optical Sciences Division, Naval Research Laboratory

Stephen H. Foulger, Editor
School of Materials Science and Engineering, Clemson University

Sponsored by the
ACS Division of Polymer Chemistry

American Chemical Society, Washington, DC

Library of Congress Cataloging-in-Publication Data

Organic thin films for photonic applications / [edited by] Warren N. Herman, Steven R. Flom, Stephen H. Foulger ; sponsored by the ACS Division of Polymer Chemistry.
p. cm. -- (ACS symposium series ; 1039)
Includes bibliographical references and index.
ISBN 978-0-8412-2563-3 (alk. paper)
1. Thin film devices. 2. Optoelectronic devices. 3. Organic thin films. 4. Photonics--Materials. I. Herman, Warren N. II. Flom, Steven R. III. Foulger, Stephen H. IV. American Chemical Society. Division of Polymer Chemistry.
TK7872.T55O675 2010
621.36--dc22

2010012905

The paper used in this publication meets the minimum requirements of American National Standard for Information Sciences—Permanence of Paper for Printed Library Materials, ANSI Z39.48n1984.

Distributed by Oxford University Press

PRINTED IN THE UNITED STATES OF AMERICA

Foreword

The ACS Symposium Series was first published in 1974 to provide a mechanism for publishing symposia quickly in book form. The purpose of the series is to publish timely, comprehensive books developed from the ACS sponsored symposia based on current scientific research. Occasionally, books are developed from symposia sponsored by other organizations when the topic is of keen interest to the chemistry audience.

Before agreeing to publish a book, the proposed table of contents is reviewed for appropriate and comprehensive coverage and for interest to the audience. Some papers may be excluded to better focus the book; others may be added to provide comprehensiveness. When appropriate, overview or introductory chapters are added. Drafts of chapters are peer-reviewed prior to final acceptance or rejection, and manuscripts are prepared in camera-ready format.

As a rule, only original research papers and original review papers are included in the volumes. Verbatim reproductions of previous published papers are not accepted.

ACS Books Department

Contents

Part II. Organic Optoelectronic Materials and Devices

Indexes

Preface

Organics are an extraordinarily large class of materials with promising applications for photonics and electronics. This volume is based on a symposium entitled "Organic Thin Films for Photonics Applications(OTF)" held at the 236th Annual Meeting of the American Chemical Society in Philadelphia, PA August, 2008. The symposium was the 15th in a series of topical meetings hosted either alternately or jointly by the American Chemical Society and the Optical Society of America dating back to the first jointly sponsored meeting in Toronto in October, 1993.

These meetings originated from a recognition of the potential impact of organic thin films for applications in fields such as optical communications, optical signal processing, optical displays, and optical data storage. The meetings have evolved to also include discussions of three dimensional photolithography and solar power generation. Over the years, these interdisciplinary meetings have provided a unique opportunity for chemists, materials scientists, physicists and engineers to gather to present and discuss new theoretical and experimental results on organic thin film materials and photonic devices.

The 15th symposium included 18 invited talks and 21 contributed papers from internationally recognized scientists covering a wide variety of topics including: Electro-optic polymers for high speed data modulation and terahertz emission and sensing; Third order nonlinear optical materials for all-optical signal processing and optical limiting; Multiphoton absorption polymerization fabrication of optical devices; Organic photovoltaics; Terahertz spectroscopy of organic semiconducting materials; Polymer based plasmonic devices; Photorefractive polymers and devices for holographic displays; All-polymer lasers; Two-photon absorbing materials for 3D optical data storage; Electro-luminescent colloids; and Conducting and semiconducting polymers for photovoltaic and transistor applications.

Today, organic light emitting diodes for optical displays and polymer waveguides discussed in several of the early OTF meetings have been developed to the point of highly successful commercialization. The development of organic nonlinear optical and optoelectronic materials for optical communications, power generation, data storage, and other applications have not yet reached the same level of maturity but offer the similar levels of promise that the early OLED and polymer waveguides studies showed. This book describes recent research in materials development, characterization techniques, and device development related to organic thin films for photonic applications.

Acknowledgments

This symposium was sponsored by the American Chemical Society Division of Polymer Chemistry with additional generous support provided by the U.S. Air Force Office of Scientific Research and the U.S. Office of Naval Research.

Warren N. Herman
Editor

Laboratory for Physical Sciences, University of Maryland

Steven R. Flom
Editor

Optical Sciences Division, Naval Research Laboratory

Stephen H. Foulger
Editor

School of Materials Science and Engineering, Clemson University

Chapter 1

Introduction

Warren N. Herman,*[,1] Steven R. Flom,[2] and Stephen H. Foulger[3]

[1]Laboratory for Physical Sciences, University of Maryland, 8050 Greenmead Drive, College Park, MD 20740
[2]Optical Sciences Division, Naval Research Laboratory, Washington, DC 20375
[3]Center for Optical Materials Science and Engineering Technologies, School of Materials Science and Engineering, Clemson University, Clemson, SC 29634
***herman@lps.umd.edu**

The chapters of this book arose from a symposium that took place in Philadelphia at the annual ACS meeting in August 2008. They describe research progress in the development of materials, characterization techniques, and devices related to organic thin films for photonic applications. The topics covered can be divided into two general categories: the first involves organic nonlinear optics and the second involves organic semiconductors with emphasis on organic optoelectronics. Included are the development of chromophores with recording-setting electro-optic coefficients and poling of thick polymer slabs, demonstration of plasmonic waveguides in electro-optic materials, fabrication of the largest photorefractive 3D display to date, new photosensitive polymeric systems for optical data storage, 3D photonic structures fabricated by multiphoton absorption polymerization, a practical technique for measuring the phase of the complex third-order nonlinear susceptibility without the need for a reference material, electroluminescent colloidal particles that can be designed to produce a wide range of emission colors, the creation of thin-film lasers through the coextrusion of multilayered polymeric films, issues related to improving the power conversion efficiency of organic solar cells, various organic semiconductor materials and the use of terahertz spectroscopy techniques for determining carrier dynamics in organic semiconductors. In this introduction we provide some background information and a brief summary of the exciting developments reported in the remainder of the book.

Organic Nonlinear Optical Materials and Devices

Electro-Optic Organic Materials and Devices

Electro-optic materials are a special case of second-order nonlinear optical (NLO) materials (*1*). They possess the property than an applied electric-field (generated by a voltage applied to electrodes) modifies the refractive index of the material. This property permits the fabrication of optical waveguide modulator and switching devices for telecommunications applications and enables photorefractive effects for holographic information storage and displays when combined with molecular components that provide charge generation, transport, and trapping. The linear electro-optic effect (or Pockels effect) is related to the second order electric susceptibility $\chi^{(2)}$ appearing under the dipole approximation in an expansion of the macroscopic electric polarization vector **P** in powers of the electric field **E**. To third order in powers of **E**, this expansion can be written

$$P_i = P_{i0} + \varepsilon_0 \chi_{ij}^{(1)} E_j + \varepsilon_0 \chi_{ijk}^{(2)} E_j E_k + \varepsilon_0 \chi_{ijkl}^{(3)} E_j E_k E_l + \dots,$$

where ε_0 is the permittivity of free space and the coordinate indices take the values 1,2,3 with summation over repeated indices understood. Because **P** and **E** are polar vectors, the even-order susceptibilities must vanish for a material that possesses a center of inversion. The macroscopic second-order optical nonlinearity in organic materials originates from chromophores, organic molecules that lack inversion symmetry and exhibit optical nonlinearity on the microscopic scale (*2*). Macroscopic structures incorporating chromophores that are randomly oriented will possess inversion symmetry, and consequently a zero $\chi^{(2)}$. Inversion symmetry can be removed by crystal growth, by Langmuir-Blodgett deposition, or by orientationally ordering the chromphores with an applied electric field, referred to as electric field poling (*3–5*).

The existence of second order optical nonlinearities in organic molecules has been known since the late 1960s (*6, 7*). Experimental findings were primarily based on second-harmonic generation (SHG) in organic crystals or electric-field induced SHG in liquids until the first electric-field poled second-order nonlinear polymeric thin solid films were reported in 1982 (*8, 9*). Active research through late 1970 and the 1980s led to a number of professional symposia resulting in numerous symposium books (*10–13*) on both second-order and third-order nonlinear optical properties of organic thin films.

In 1992, electro-optic polymeric materials garnered serious attention with the demonstration of a wideband 40 GHz Mach-Zehnder optical modulator using a methacrylate polymer with side chain groups containing

N-methylaminonitrostilbene that was electric-field poled to give an electro-optic coefficient of 16 pm/V (*14*). This modulator had a half-wave voltage of 10 volts at an optical wavelength of 1.3 μm. However, organic electro-optic materials were not yet ready for commercialization, primarily because of the lack of temporal stability of the required orientational order. A overview of these early development years can be found in a review with extensive references by Burland, Miller, and Walsh that appeared in 1994 (*15*).

That year was also the occasion of the second ACS/OSA sponsored Organic Thin Films for Photonics Applications symposium which resulted in an ACS Symposium Series book on second order nonlinear optical polymers (*16*). This was followed in the next four years by other conference-based books (*17–20*), and in 1997 a polymer traveling wave electro-optic phase modulator operating at frequencies up to 113 GHz was demonstrated, again at an optical wavelength of 1.3 μm (*21*). In 2000, a polymeric Mach-Zehnder modulator based on the chromophore CLD-1with an electro-optic coefficient of 60 pm/V and a sub-1V drive voltage at 1.3 μm was demonstrated, although not at GHz frequencies due to electrode design (*22*). Then in 2007, a 1V-drive Mach-Zehnder modulator with a high electro-optic coefficient of 138 pm/V based on the chromophore AJC146 was demonstrated at the popular telecom wavelength of 1550 μm (*23*). Although again a low frequency demonstration, this result was the first to show in-device electro-optic coefficients greater than four times larger than those of the benchmark inorganic material lithium niobate. Recently, the first 40 GHz electro-optic polymer intensity modulators able to pass industry standard environmental stress, long-term stability, and other important telecom qualification tests were demonstrated (*24*). These modulators were based on the chromophore LPD-80 and had a drive voltage of 3 V. For telecom and RF photonic applications, further research and development is expected to lead to modulators with all the characteristics described above in a single device: sub-1V drive voltages, 3 dB bandwidths greater than 100 GHz, low optical loss, and environmental stress qualification.

Although this only scratches the surface of organic electro-optic material research, it sets the stage for the next six chapters of this book. Future fiber-optic communication systems will push beyond 100 Gb/s driving demand for very high bandwidth low drive power optical modulators (*25*). High speed electro-optic polymer modulators with large electro-optic coefficients have the potential to provide the needed low drive power devices. Chapters 2-4 deal with design and synthesis of organic chromophores with very high electro-optic activity. In Chapter 2 by Dalton, *et al.*, time-dependent Density Functional Theory and quantum mechanical and pseudo-atomistic Monte Carlo statistical mechanical computations are used to guide the systematic improvement of organic electro-optic materials with the goal of predicting electro-optic activity of even complex materials from first principles theory. The focus is on optimizing the electro-optic activity of dipolar organic chromophore-containing polymers and dendrimers processed by electric field poling near the glass transition temperature of the material.

In Chapter 3 by Lindsay and Chafin, quantum chemistry computations using Density Functional Theory on large (33-53 atom) organic asymmetric chromophores reveal the ability to increase electro-optic activity by altering the interior positions of electron-donating and electron-withdrawing substituents. Trade-off scenarios for improving the optical loss are also recommended.

In Chapter 4, Jen *et al.* explore arene-perfluoroarene stacking interactions in a new series of dendronized nonlinear optical chromophores to create extended network structures and overcome intermolecular electrostatic interactions for improving poling efficiency and alignment stability. These supramolecular self-assembled molecular glasses showed very high electro-optic activity and excellent thermal stability at 85 °C.

An interesting application of electro-optic polymers making use of thin metal stripes that support long range surface plasmon polaritons (LR-SPP) (*26*) is discussed by Lee *et al.* in Chapter 5. Although various SPP devices have been demonstrated, none previously have been demonstrated in an electro-optic material. Here the authors discuss the fabrication of LR-SPP structures in several different polymers as a first step toward active plasmonics whereby plasmonic structures can be used to create electro-optic information processing systems.

As mentioned above, electro-optic activity enables photorefractive effects when combined with molecular components that provide charge generation, transport, and trapping. In Chapter 6, Norwood *et al.* review the achievement of the first updatable holographic 3D display based on photorefractive polymers and report the fabrication of a 6 inch × 6 inch display, the largest photorefractive 3D holographic display to date. Photorefractive polymers are attractive because polymer systems are easier to process than inorganic crystals, flexible in formulation, readily tailored to obtain specific target properties and lower cost.

The last topic in the area of organic electro-optics deals with successful electric-field poling of polymer slabs that are nearly three-quarters of a millimeter thick. By contrast, electro-optic polymer films in devices for optical modulation and switching in telecommunication applications are necessarily thin films (on the order of 10 μm or less) because they must also perform as single-mode optical waveguides. In Chapter 7, Hoffman *et al.* describe for the first time the fabrication and poling of a 700 μm thick bulk piece of polymethylmethacrylate (PMMA) doped with various concentrations of Disperse Red 1 (DR1). Such poled thick films could find application in terahertz generation and detection (*27*).

Third-Order Nonlinear Optical Organic Materials, Characterization, and Device Fabrication

Two-photon absorption (*28*) is related to the imaginary part of $\chi^{(3)}$, the third-order electric susceptibility discussed in connection with expansion of the electric polarization in powers of the electric field in the first paragraph of the previous section. Two-photon writing, erasing, and reading was first achieved by the photochromic interconversion of spyrobenzopyrans in 1989 (*29*) and photo-polymerization using two-photon absorption in the near-infrared has proven to be an extraordinarily useful technique for producing 3D microstructures (*30*).

In Chapter 8, Yanez *et al.* report the use of new photosensitive polymeric systems comprised of novel two-photon absorbing dyes in different polymer matrices and go on to demonstrate a true three dimensional Write-Once-Read-Many (WORM) optical data storage system where the two-photon advantage is fully exploited. This improved ODS system is shown to allow the two-photon recording and readout of voxels in five consecutive layers within the polymer matrix.

Li *et al.* in Chapter 9 review their work with acrylic optical waveguide-based devices created with multiphoton absorption polymerization (MAP), a technique that is attracting increasing attention in the realm of micro- and nanofabrication. They also present new results obtained with epoxy-based polymers. The higher refractive index of the epoxy polymers is expected to lead to further enhancements in device performance. Interesting 3-D architectures that will provide improved performance and new functionalities in devices created using MAP are also discussed.

In contrast to two-photon absorption applications, which require a large imaginary component of $\chi^{(3)}$, highly nonlinear polymer waveguide devices based on materials with a large real part of $\chi^{(3)}$ offer the possibility of major advances in optical signal processing and development of parametric amplifiers. Fundamentally, the real part of $\chi^{(3)}$ enables the lossless exchange of energy between optical beams. For fiber optic communications, materials with a large real $\chi^{(3)}$ could be used for optical switching, gating and routing signals.

Development of new nonlinear materials requires a characterization set-up which is able to quickly measure complex $\chi^{(3)}$ without the many processing steps involved in device fabrication. In Chapter 10, Cao *et al.* present an elegant practical technique for absolute measurement of the phase φ of the complex third-order nonlinear optical susceptibility $\chi^{(3)}=|\chi^{(3)}|\exp(i\varphi)$ in thin films using forward Degenerate Four Wave Mixing (DFWM) in the BOXCARS configuration. A novel aspect of this technique is that determination of φ does not depend on knowledge of the magnitude of $\chi^{(3)}$ and is absolute in the sense that it does not require a separate reference material, such as fused silica.

Organic Optoelectronic Materials and Devices

Over the last two decades, academic and private industry researchers have actively pursued the development of all-organic devices as alternatives to inorganic based electronic and optoelectronic systems. The inherent advantage of organic small molecule and polymeric systems, relative to inorganic systems that are fabricated through established lithography routes, is the potential that low-cost, large area electronics may be realized when these materials are coupled with conventional commercial printing technologies (*31–33*). Specific applications envisioned for organic small molecule/polymeric semiconducting systems include, for example, light-emitting devices (*34*), lasers (*35, 36*), photovoltaic (*37*), and thin-film transistors (*38*). The following chapters cover these topics with contributions that highlight current points of interest in these specific applications.

Organic Light Emitting Devices

In Chapter 11, Huebner *et al.* describe a simple route to producing electroluminescent colloidal particles that can be designed to produce a wide range of emission colors. The full spectrum of colors could be realized with a simple mixture of red, blue, and green emitting particles. The additive color response of the colloids was achieved through a frustration of energy transfer between particles. These particles could be converted into printable inks, a useful characteristic if these systems are to be employed in roll-to-roll printing of optoelectronic devices.

Lott *et al.* describe the creation of thin-film lasers through the coextrusion of multilayered polymeric films in Chapter 12. These stacked films exhibited a photonic band gap whose rejection wavelength and peak shape can be controlled by changing layer thickness. Control of layer thickness was achieved by adjusting processing conditions or by post processing biaxial stretching. Doping the layered films with organic fluorescent dyes resulted in the creation of both distributed Bragg reflector (DFB) and distributed feedback (DFB) optically pumped lasers, with both architectures displaying low lasing thresholds and high efficiencies.

Organic Photovoltaic Devices (OPVs)

In Chapter 13, Lane describes the state-of-the-art and the historical design considerations of organic photovoltaic devices, with a specific focus on the role that interfaces and morphology play in defining the ultimate efficiency of an organic solar cell.

Ballarotto *et al.* describe in Chapter 14 the role that the conductivity of a hole-doped polymer transport layer has on the fill-factor of organic bulk heterojunction photovoltaic devices. Specifically, devices with a structure consisting of ITO/(hole-doped layer)/(intrinsic layer)/metal were investigated with PEDOT:PSS as the hole transport layer and a blend of regio-regular P3HT and PCBM-C60 as the intrinsic active layer. Photovoltaic device performance characteristics were analyzed as a function of varying conductivity of the hole-transport layer.

Organic Semiconductors: Devices and Characterization

In Chapter 15, Platt *et al.* review the optical, fluorescent, and photoconductive properties of functionalized anthradithiophene (ADT), pentacene, and dicyanomethylenedihydrofuran (DCDHF) derivatives. Specifically, the charge carrier photogeneration and photoconductive gain were studied in solution-deposited films. Different charge carrier dynamics could be achieved at ps-ms time-scales after 100-fs pulsed photoexcitation with mixtures of the derivatives.

Ma *et al.* in Chapter 16 describe the use of organophosphonic acid self-assembled monolayers (SAMs) on metal oxides (MOs) such as AlO_x and HfO_2 as ultrathin gate dielectrics and templates. These structures were utilized in the fabrication of low-voltage organic field-effect transistors with low leakage currents and small subthreshold slopes.

Terahertz time-domain spectroscopy as well as optical pump-terahertz probe time domain spectroscopy have proven to be very useful tools in exploring the carrier dynamics in organic semiconductors. In Chapter 17, Esenturk *et al.* summarize results from relative mobility measurements of the organic semiconductor polymers poly(3-hexylthiophene) (P3HT) and poly(2,5-bis(3-tetradecylthiophen-2yl)thieno[3,2-b]thiophene (PBTTT) as model active layers in transistor devices. They also discuss recent results from relative efficiency studies of blend and alternating multilayer films of zinc phthalocyanine (ZnPc) and buckminsterfullerene (C60) as model active nanolayers for efficient solar cells.

In Chapter 18, Peng *et al.* discuss issues relating to ambiguity in determination of carrier dynamics using terahertz time-domain spectroscopy when the carrier population varies abruptly during the time of THz probing, particularly at the early stage of the photoexcitation. The fast dynamics determined from measurement of the transient THz peak absorption may not necessarily reflect the true charge carrier dynamics in a sample for the same reason. They further experimentally validate a technique to remove this ambiguity and apply the technique to obtain the photoinduced carrier dynamics of P3HT/PCBM blends using optical-pump terahertz-probe time domain spectroscopy with subpicosecond resolution.

The editors are grateful to our colleagues whose inspired work and dedicated diligence has resulted both in the symposium presentations and the chapters of this book. We hope that a result of these efforts is to inspire others to follow the research directions outlined here and to continue to develop the exciting and still-growing field of organic thin films for photonic applications.

References

1. Boyd, R. W. *Nonlinear Optics*, 3rd ed.; Academic Press: Burlington, MA, 2008; Chapter 11.
2. Singer, K. D.; Kuzyk, M. G.; Sohn, J. E. *J. Opt. Soc. Am. B* **1987**, *4*, 968.
3. Hayden, L. M.; Anderson, B. L.; Lam, J. Y. S.; Higgins, B. G.; Stroeve, P.; Kowel, S. T. *Thin Solid Films* **1988**, *160*, 379.
4. Singer, K. D.; J. E. Sohn, J. E.; S. J. Lalama, S. J. *Appl. Phys. Lett.* **1986**, *49*, 248.
5. Mortazavi, M. A.; Knoesen, A.; Kowel, S. T.; Higgins, B. G.; Dienes, A. *J. Opt. Soc. Am. B* **1989**, *6*, 733.
6. Bass, M.; Bua, D.; Mozzi, R.; Monchamp, R. R. *Appl. Phys. Lett.* **1969**, *15*, 393.

7. Lalama, S. J.; Garito, A. F. *Phys. Rev. A* **1979**, *20*, 1179 and references therein.
8. Meredith, G.; VanDusen, J.; Williams, D. *Macromolecules* **1982**, *15*, 1385.
9. Garito, A. F.; Singer, K. D. *Laser Focus* **1982**, *18*, 59.
10. *Nonlinear Optical Properties of Organic and Polymeric Materials*; Williams, D. J., Ed.; ACS Symposium Series 233, American Chemical Society: Washington, DC, 1983.
11. *Nonlinear Optical Properties of Organic Materials and Crystals*; Chemla, D., Zyss, J., Eds.; Academic Press: Orlando, FL, 1987; Vols. 1, 2.
12. Prasad, P.; Williams, D. *Nonlinear Optical Effects in Molecules and Polymers*; John Wiley and Sons: New York; 1991.
13. *Materials for Nonlinear Optics-Chemical Perspectives*; Marder, S. R., Sohn, J. E., Stucky, G. D., Eds.; ACS Symposium Series 455, American Chemical Society: Washington, DC, 1991.
14. Teng, C. C. *Appl. Phys. Lett.* **1992**, *60*, 1538.
15. Burland, D. M.; Miller, R. D.; Walsh, C. A. *Chem. Rev.* **1994**, *94*, 31.
16. *Polymers for Second-Order Nonlinear Optics*; Lindsay, G. A., Singer, K. D., Eds.; ACS Symposium Series 601, American Chemical Society: Washington, DC, 1995.
17. *Organic Thin Films for Waveguiding Nonlinear Optics (Advances in Nonlinear Optics)*; Kajzar, F., Swalen, J. D., Eds.; Gordon & Breach: Amsterdam, 1996.
18. *Nonlinear Optics of Organic Molecular and Polymeric Materials*; Nalwa, H. S., Miyata, S., Eds.; CRC Press: Boston, MA, 1996.
19. *Poled Polymers and Their Applications to SHG and EO Devices (Advances in Nonlinear Optics)*; Miyata, S., Sasabe, H., Eds.; Gordon and Breach: Amsterdam, 1997.
20. *Photonic and Optoelectronic Polymers*; Jenekhe, S. A., Wynne, K. J., Eds.; ACS Symposium Series 672, American Chemical Society: Washington, DC, 1997.
21. Chen, D.; Fetterman, H.; Chen, A.; Steier, W. H.; Dalton, L. R.; Wang, W.; Shi, Y. *Appl. Phys. Lett.* **1997**, *70*, 3335.
22. Shi, Y.; Lin, W.; Olson, D. J.; Bechtel, J. H.; Zhang, H.; Steier, W. H.; Zhang, C.; Dalton, L. R. *Appl. Phys. Lett.* **2000**, *77*, 1.
23. Enami, Y.; DeRose, C. T.; Mathine, D.; Loychik, C.; Greenlee, C.; Norwood, R. A.; Kim, T. D.; Luo, J.; Tian, Y.; Jen, A. K.-Y.; Peyghambarian, N. *Nat. Photonics* **2007**, *1*, 180.
24. Dinu, R.; Jin, D.; Yu, G.; Chen, B.; Huang, D.; Chen, H.; Barklund, A.; Miller, E.; Wei, C.; Vemagiri, J. *J. Lightwave Technol.* **2009**, *27*, 1527.
25. Tkach, R. *Bell Labs Technol. J.* **2010**, *14*, 3.
26. Berini, P. *Phys. Rev. B* **2000**, *61*, 10484.
27. Sinyukov, A.; Hayden, L. M. *Opt. Lett.* **2002**, *27*, 55.
28. Boyd, R. W. *Nonlinear Optics*, 3rd ed.; Academic Press: Burlington, MA, 2008; Chapter 4.
29. Parthenopoulos, D. A.; Rentzepis, P. M. Three-dimensional optical storage memory. *Science* **1989**, *245* (4920), 843.

30. Cumpston, B. H.; Ananthavel, S. P.; Barlow, S.; Dyer, D. L.; Ehrlich, J. E.; Erskine, L. L.; Heikal, A. A.; Kuebler, S. M.; Sandy Lee, I.-Y.; McCord-Maughon, D.; Qin, J.; Röckel, H.; Rumil, M.; Wu, X.-L.; Marder, S. R.; Perry, J. W. *Nature* **1999**, *398*, 51.
31. Mitschke, U.; Bauerle, P. *J. Mater. Chem.* **2000**, *10*, 1471.
32. Patel, N. K.; Cinà, S.; Burroughes, J. H. *IEEE J. Sel. Top. Quantum Electron.* **2002**, *8*, 346.
33. Kukhto, A. V. J. *Appl. Spectrosc.* **2003**, *70*, 165.
34. Burroughes, J. H.; Bradley, D. D. C.; Brown, A. R.; Marks, R. N.; Mackay, K.; Friend, R. H.; Burns, P. L.; Holmes, A. B. *Nature* **1990**, *347*, 539.
35. Weinberger, M. R.; Langer, G.; Pogantsch, A.; Haase, A.; Zojer, E.; Kern, W. *Adv. Mater.* **2004**, *16*, 130.
36. Ubukata, T.; Isoshima, T.; Hara, M. *Adv. Mater.* **2005**, *17*, 1630.
37. Yu, G.; Heeger, A.J. *Synth. Met.* **1997**, *85*, 1183.
38. Tsumura, A.; Koezuka, H.; Ando, T. *Appl. Phys. Lett.* **1986**, *49*, 1210.

Part I. Organic Nonlinear Optical Materials and Devices

Chapter 2

Organic Electro-Optic Materials

Understanding Structure/Function Relationships Critical to the Optimization of Electro-Optic Activity

Larry R. Dalton,[*,1] Philip A. Sullivan,[1] Denise Bale,[1] Benjamin Olbricht,[1] Joshua Davies,[1] Stephanie Benight,[1] Ilya Kosilkin,[1] Bruce H. Robinson,[1] Bruce E. Eichinger,[1] and Alex K.-Y. Jen[2]

[1]Department of Chemistry, University of Washington, Seattle, WA 98195
[2]Department of Materials Science & Engineering, University of Washington, Seattle, WA 98195
[*]dalton@chem.washington.edu

Time-dependent density functional theory quantum mechanical and pseudo-atomistic Monte Carlo statistical mechanical computations are used to guide the systematic improvement of organic electro-optic materials. In particular, the dependence of molecular first hyperpolarizability upon dielectric permittivity and optical frequency is defined as is the dependence of electric field poling-induced acentric order on a variety of intermolecular electrostatic interactions. A number of measurement techniques are used to obtain independent test of theoretical predictions of chromophore order. The conclusion of these studies is that electro-optic activity of even complex materials (multi-chromophore-containing dendrimers and composite materials containing several types of chromophores, i.e., binary chromophore organic glasses) can be predicted from first principles theory. Laser-assisted electric field poling applied to a binary chromophore organic glass is used to provided an additional test of theory.

Introduction

The focus of this chapter is upon optimizing the electro-optic (EO) activity of dipolar organic chromophore-containing polymers and dendrimers processed by electric field poling near the glass transition temperature of the material. Organic electro-optic (OEO) materials can also be prepared by self-assembly, sequential synthesis methods including Langmuir-Blodgett (*1*) and Merrifield (*2*) type methods. While many of the theoretical methods discussed here are relevant to such materials, discussion of self-assembly, sequential synthesis preparation of materials is beyond the scope of this chapter as is the preparation of materials by crystal growth.

Electro-optic activity can be viewed as voltage control of the index of refraction of a material deriving from the effect of an applied electric field altering the charge distribution of a material and thus influencing an optical field transiting the material. For organic materials, the magnitude of electro-optic activity is defined by the low frequency component of the molecular first hyperpolarizability, $\beta(-\omega;0,\omega)$, of the constituent chromophores and the degree of acentric order, $\langle\cos^3\theta\rangle$, induced by the electric poling field. Molecular first hyperpolarizability depends upon optical frequency, ω, and dielectric permittivity, ε. The acentric order parameter will depend on chromophore dipole moment, μ, electric poling field strength, E_p (also denoted as E_{pol}), intermolecular electrostatic interactions, and dielectric permittivity. These latter two quantities will depend on chromophore number density, N. Indeed, the master equation for the principal element of electro-optic tensor, r_{33}, can be expressed as $r_{33} = N \beta(\omega,\varepsilon)\langle\cos^3\theta(N,\varepsilon)\rangle \text{constant}(n,\varepsilon)$ where n is the index of refraction. As is evident from a close consideration of this equation, the variables are not independent. This interdependence requires a simultaneous consideration of quantum and statistical mechanics to achieve a "first principles" understanding of the structure/function relationships that define electro-optic activity in organic materials.

While a variety of quantum mechanical methods provide good prediction of trends for the variation of molecular first hyperpolarizability with chromophore structure (*3*), quantitative prediction has been elusive. Time-dependent density functional theory (TD-DFT) has attracted considerable attention recently because time-dependent fields can be explicitly incorporated into computations permitting quantitative simulation of linear and nonlinear optical properties (*4–7*). More recently, TD-DFT and real-time TD-DFT (RTTD-DFT) methods have been extended to the treatment of dielectric permittivity effects. As shown in Figure 1, new TD-DFT methods permit essentially quantitative (within experimental error) simulation of the dependence of β on ε.

Quantum mechanical calculations are critical to providing input (μ, β, and electron distributions) to statistical mechanical calculations if "first principles" prediction of the electro-optic activity of materials is to be achieved. Our use of "first principles" prediction here is not meant to imply that macroscopic properties are calculated from a single fully-atomistic quantum mechanical treatment but rather that quantum and statistical mechanical calculations are correlated so that adjustable parameters are not used in calculations and that results ultimately relate back to quantum mechanical principles. Statistical

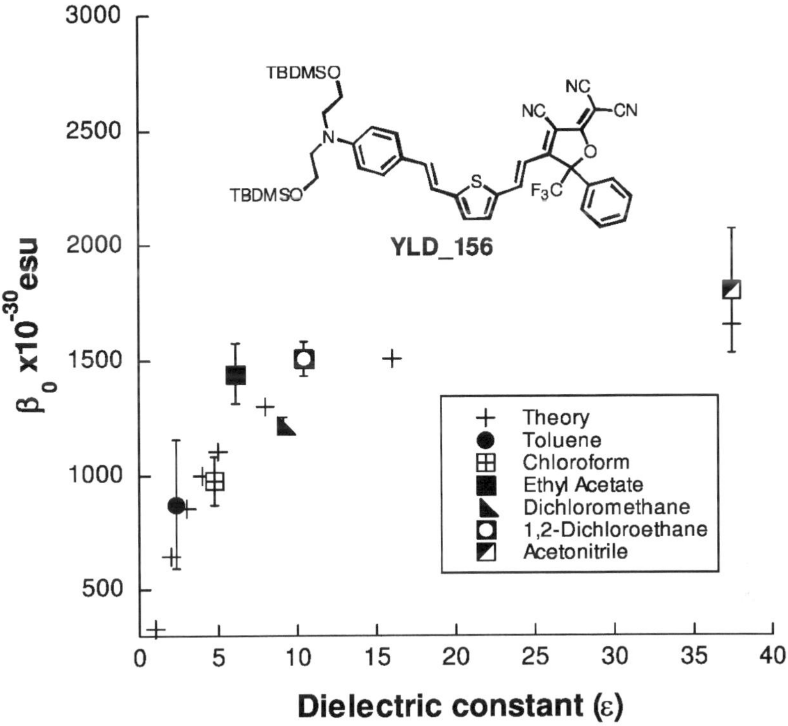

Figure 1. Comparison of experimental (HRS determined) and theoretical (TD-DFT calculated) values of molecular first hyperpolarizability is shown for the YLD156 chromophore (structure shown in the figure) as a function of solvent dielectric permittivity. In this figure, theory has been used to adjust experimental data taken at 6.361 x 10^{-15} Hz (1.907 microns wavelength) to zero frequency to facilitate comparison with a broader range of theoretical calculations. TD-DFT calculations (7) establish that the frequency dependence is also adequately described by the two-state model (a similar level of agreement is observed when data is compared at higher frequencies). HRS measurements have been carried out over the range from 700-3000 nm to evaluate the frequency dependence and resonance contributions to β.

mechanical methods (Monte Carlo and molecular dynamical methods) are clearly required as the poling-induced acentric order is dependent on the interaction of many chromophores. Fully atomistic Monte Carlo methods (*8, 9*) are attractive but the number of atoms required for treatment is far too large to make this approach generally viable for the range of materials to be considered here. We have improved computational efficiency by treating relatively-rigid extended π-electron molecular segments as rigid ellipsoids with dipole moments and electron distributions defined by quantum mechanical calculations. This treatment is referred to as the "United Atom" approximation. We, in turn, label this United Atom approximation-based approach as the pseudo-atomistic Monte Carlo method (*10–12*).

The pseudo-atomistic Monte Carlo approach has been employed to successfully simulate the dependence of electro-optic activity on chromophore number density and upon chromophore shape for chromophore/polymer composite materials. For these materials, the crucial observation is that chromophore shape is the parameter critical for optimizing electro-optic activity with the observation that spherically-shaped chromophores lead to the largest poling-induced EO activity. The dependence on chromophore shape reflects the competition of chromophore intermolecular dipolar interactions and nuclear repulsive (steric) forces, each of which have different spatial anisotropies, in influencing polar order. For such chromophore/polymer composite materials, the detailed structure of the polymers has little impact on realizable EO activity. Such materials represent a limited opportunity for the improvement of EO activity. This lack of significant dependence on polymer structure reflects weak coupling of the chromophores to the polymer structure near the glass transition temperature. A more important question is whether or not correlated quantum/statistical mechanical methods can lead to prediction of EO activity for more complicated materials such as chromophores covalently incorporated into polymers and dendrimers and such as multi-chromophore containing materials (e.g., chromophore guests doped into chromophore-containing host materials referred to as binary chromophore organic glasses). In addition to focus on the application of theoretical methods to these more complex materials, this chapter discusses experimental techniques used to measure molecular first hyperpolarizability, electro-optic activity, and poling-induced order. Moreover, laser-assisted electric field poling (*13*) of disperse red (DR) dye containing binary chromophore organic glasses is used to test conclusions regarding structure/function relationships that dominate electro-optic activity for organic materials.

Experimental

Synthesis and Processing

The synthesis of YLD124 (OR = H) type chromophores (see Scheme 1 below) illustrates the general sequence of reactions used to produce chromophores based on amine donors and tricyanovinylfuran (TCF) acceptors. These chromophore have been doped into a variety of polymers (polymethylmethacrylate, PMMA; amorphous polycarbonate, APC; polyimides, PI) to form composite materials. They are also covalently incorporated into dendrimers and polymers. DR1-co-PMMA was purchased from Aldrich Chemical. The YLD124 and YLD156 chromophores were also doped into chromophore-containing dendrimers and polymers to prepare binary chromophore organic glass composite materials (*13–17*).

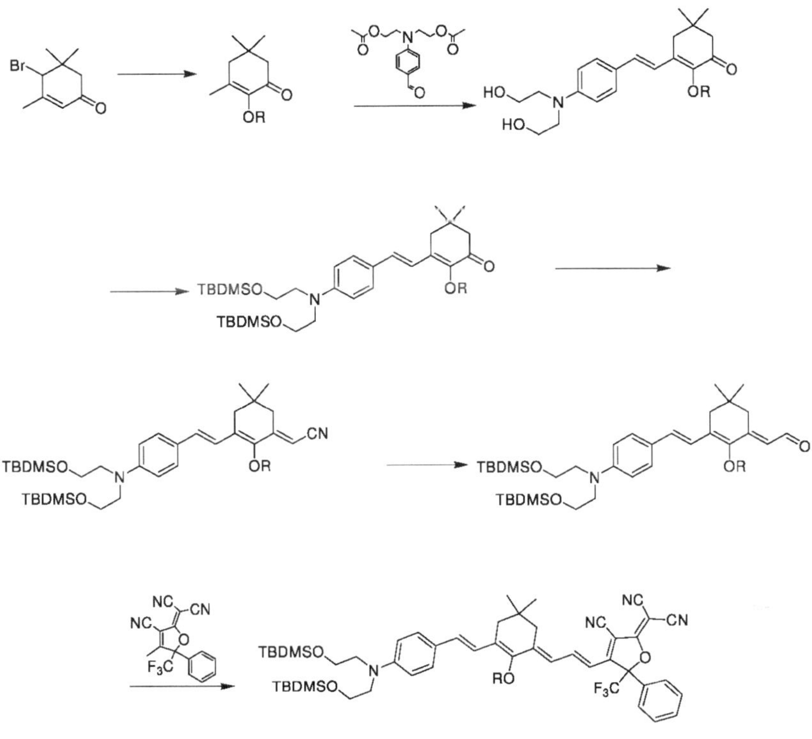

Scheme 1. The synthesis of YLD124-type chromophores is shown.

EO activity was induced into materials by contact (electrode) electric field poling of materials near their glass transition temperatures. Poling electrodes typically involved a gold electrode and an indium tin oxide (ITO) electrode. To reduce conductivity problems, a nanoscopic layer of TiO_2 was, in some cases, coated onto the ITO electrode. Such TiO_2 layers were not used for the results discussed here. For the materials studied here, a linear relationship between r_{33} and poling field, E_p, is predicted and experimentally observed. Thus, it is useful to obtain r_{33}/E_p by least squares fitting.

For device applications, chromophores and host materials typically must be modified to permit crosslinking to elevate the final glass transition temperature. Typically, fluorovinyl either (*18*, *19*) or Diels-Alder/Retro-Diels-Alder (*20*, *21*) cycloaddition reactions are employed. Such crosslinking chemistries permit final material glass transition temperatures in the range 200-250°C to be routinely achieved. Because the focus of this communication is definition of structure/function relationships, materials are not crosslinked.

Characterization of Molecular First Hyperpolarizability, $\beta(\omega,\varepsilon)$

Molecular first hyperpolarizability, $\beta(\omega,\varepsilon)$, was characterized employing wavelength-agile, femtosecond–time-resolution Hyper-Rayleigh Scattering

(HRS) as described elsewhere (*22*). Measurements were carried out as a function of chromophore concentration in solvents of various dielectric constants. To correct for laser pulse fluctuations, signals were referenced against chromophore samples of known hyperpolarizability and against solvent values. Measurements were carried out from 700 nm to 3000 nm. Even with the use of multiple reference samples and an integrating sphere to determine absolute β values, experimental errors remain large due to fundamental signal-to-noise limitations and due to uncertainty in absolute values of β determined for common reference solvents. Within experimental uncertainty, experimental values are reproduced by TD-DFT calculations. We are continuing efforts to improve the experimental reliability of HRS measurements; the present report is a progress report with the best available data.

The product of dipole moment and molecular first hyperpolarizability was measured employing electric field induced second harmonic generation (EFISH). A comparison of theoretical and experimental values was carried out as is illustrated in Table I.

Characterization of Electro-Optic Activity, r_{33}

Electro-optic activity was measured employing a variety of techniques including Teng-Man ellipsometry (*14, 23–25*) and attenuated total reflection, ATR (*26, 27*). Because of large potential error associated with the use of thin organic films and issues associated with the use of indium tin oxide (ITO) electrodes, Teng-Man ellipsometry was used primarily for the *in situ* monitoring of the induction and relaxation of electro-optic activity. The corrections suggested by Herman and coworkers (*24, 25*) were used to attempt to obtain improved electro-optic coefficients, which approximately agreed with values obtained with ATR. However, because of the larger potential error associated with Teng-Man measurements, this technique is not used for quantitative definition of electro-optic activity. As discussed elsewhere (*13*), the Teng-Man apparatus was modified to permit laser-assisted electric field poling experiments using polarized light to alter the order of disperse red (DR1) chromophores.

The EO coefficients reported in this communication were measured by ATR (apparatus shown in Figure 2), with measurements carried out at both 1.3 and 1.55 microns. To correlate with the bulk of literature data, the electro-optic data reported in this communication are for 1.3 microns wavelength. In recent research publications, e.g., (27), we report both 1.3 and 1.55 micron ATR measurements to permit better correlation with optical loss and photostability measurement data. The ATR technique is a modification of the widely used prism-coupling refractive index measurement. A change in the prism-coupling angle, due to the electro-optic effect, is measured by monitoring the change in intensity of the laser light reflected from the prism/sample interface as a function of applied voltage. The sample is mounted on the prism using a pneumatic plunger. The plunger, prism, and detector are mounted atop a stepper motor to allow precise control over the incidence angle of the measurement laser. The electro-optic coefficient, r_{33}, is calculated using the approximate relation: $r_{33} = [(2d\Delta R\delta n_{eff})/(n_{TM}^3 V\delta\theta)/[(\delta R\delta n_{eff})/(\delta\theta\delta n_{TM})]$ where *d*

Table I. Comparison of Theory and Experiment for $\mu\beta_0$ of the FTC chromophore* in the Solvents Indicated

Solvent	ε	λ_{max} *(nm)*	$\mu\beta_0 \times 10^{-44}$ *esu (experiment)*	$\mu\beta_0 \times 10^{-44}$ *esu (theory)*
Dioxane	2.21	622	3.41	4.9
CH_3Cl	4.81	675	8.52	8.9

* The FTC chromophore has the same structure as the YLD156 chromophore of Figure 1 except for the CF_3 and phenyl ring of the TCF acceptor being replaced by CH_3 and H respectively. Experimental measurements were carried out at 1.907 microns wavelength and values were adjusted using the two state model to the long wavelength limit.

is the thickness of the EO film and n_{TM} is the film refractive index. As the sample is rotated past the coupling angle, the intensity of the DC reflected signal, R, is recorded. Simultaneously, the change in R, related to the change in refractive index caused by application of an AC modulation voltage, V, is recorded using a lock-in amplifier. Here, $n_{eff} = n_p \sin\theta$ where n_p is the prism refractive index. Rutile, TiO_2, has a refractive index of n_0(1310 nm) = 2.462 and n_0(1550 nm) = 2.453.

Characterization of Order Parameters: Variable Angle Polarized Absorption Spectroscopy

Direct measurement of acentric order parameters, such as $\langle\cos^3\theta\rangle$, does not exist; however, techniques such as polarized absorption spectroscopy (PAS) permit direct measurement of the related centrosymmetric order parameter, $\langle P_2\rangle = (3\langle\cos^2\theta\rangle - 1)/2$. Despite its ease and simplicity, the most common form of PAS, the normal incidence method (*14*), suffers from the disadvantage of not permitting deconvolution of spectral changes associated with poling-induced order and molecular decomposition. Thermal relaxation of poling-induced order can also be a problem. To address the problems associated with the normal incident PAS method, Graf and coworkers (*28*) introduced the variable-angle PAS (VAPAS) method. With VAPAS, the absorbance is measured as a function of the incident angle of a polarized probe beam, thereby assessing the orientation distribution of the transition dipoles of the chromophores embedded in the electrically-poled film. The accuracy of the VAPAS method relies on theoretical modeling of beam propagation through a multiple-interface sample described by Snell's Law and Fresnel corrections. The modeling requires knowledge of the refractive index of each layer at each measurement wavelength. VAPAS is best suited for the measurement of order for corona-poled samples in which the film surface properties have not be perturbed by electrode deposition and removal. However, contract (electrode) poling permits measurement of electric poling field strength, E_p, and is more compatible with Teng-Man and ATR techniques. A serious problem with the method of Graf et al. is that the real and imaginary components of the refractive index are treated independently. Following the work of Manisuripur (*29*) using rigorous simulation through Jones Matrix analysis of raw experimental data, we have shown that the simple Grad analysis is not

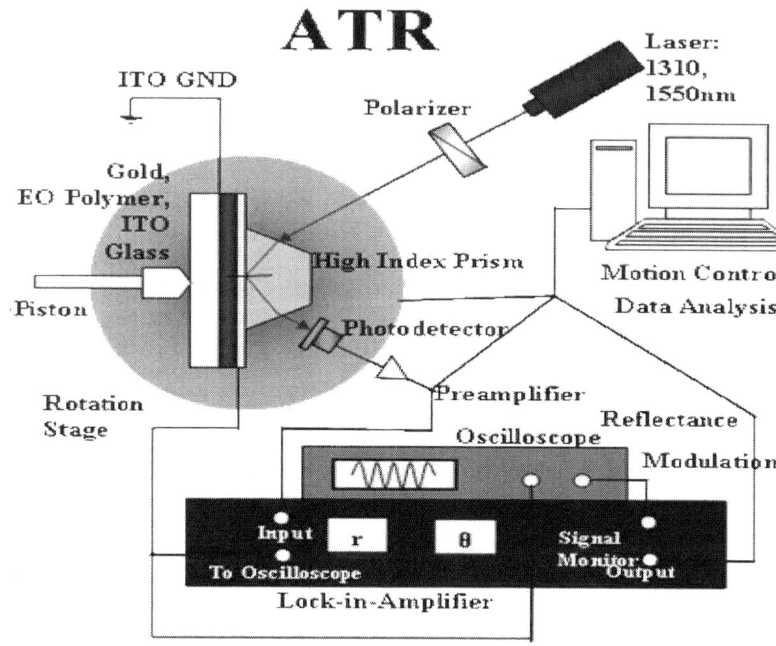

Figure 2. A schematic diagram of the ATR apparatus is shown.

sufficient in the case of highly absorptive media such as encountered here. The complex refractive index must be rigorously included in order to accurately determine $\langle P_2 \rangle$. However, we have identified a modification of the Graf method that yields numbers very close to the rigorous analysis.

We have modified the VAPAS technique to remove the above complications by using s-polarized light as a reference (see Figure 3). We refer to this modified technique as the RVAPAS method. For such light, the average absorption, $\langle \alpha_f(\nu) \rangle_s$, does not depend on the angle at which the light enters the sample and is given by $\langle \alpha_f(\nu) \rangle_s = \alpha_f[1 - r\langle P_2 \rangle]$. The anisotropy of the transition dipoles (for an axially-symmetric system), r, is defined as $r = (t_0 - t_\bullet)/(2t_\bullet + t_0)$ where the transition dipole moments are $t_0 = \langle 0|\mu_z|f\rangle\langle f|\mu_z|0\rangle$ and $t_\bullet = \langle 0|\mu_x|f\rangle\langle f|\mu_x|0\rangle$. For the OEO materials considered here, $r \approx 1$. For p-polarized light, the average absorbance does depend on the angle at which light enters the sample: $\langle \alpha_f(\nu) \rangle_p = \alpha_f[1 + r[3\cos^2\theta - 1]\langle P_2 \rangle]$. The ratio of absorbance for s and p polarized light is $[\langle \alpha_f(\nu) \rangle_\bullet / \langle \alpha_f(\nu) \rangle_0] = 1 + \sin^2\theta[3r\langle P_2 \rangle/(1 - r\langle P_2 \rangle)]$. If we denote the slope of the plot of the ratio versus $\sin^2\theta$ as m then $\langle P_2 \rangle \approx m/(3 + m)$. For the materials studied here, the following relationship approximately holds between order parameters $\langle \cos^3\theta \rangle \approx 2\sqrt{(\langle P_2 \rangle/15)}$. Of course, all order parameters ($\langle \cos^2\theta \rangle$, $\cos^3\theta\rangle$, etc.) are calculated with each pseudo-atomistic Monte Carlo trajectory calculation so measurement of any order parameter provides a test of theory.

In addition to use of the RVAPAS method to determine the order parameters for device relevant materials such as multi-chromophore-containing dendrimers discussed in the next section, a model system study was carried out on the YLD124 chromophore dissolved in PMMA to yield the concentrations shown in Table II.

These samples were poled using an applied electric field of 50 V/micron. Data were also analyzed by full Jones matrix analysis.

With the RVAPAS technique, $\langle P_2 \rangle$ is observed to be essentially constant as a function of wavelength (650 to 800 nm) over the resonance absorption and as a function of sample thickness (0.8 to 1.4 microns). For comparison, it can be noted that full Jones matrix analysis for the N = 1.2 x 10^{20} molecules/cc sample yielded a $\langle P_2 \rangle$ value of 0.034. Moreover, the scatter in data is substantially reduced relative to VAPAS data. More complete discussions of the RVAPAS technique will be presented elsewhere.

Results and Discussion

Model System Studies

In order to test the effectiveness of theoretical simulations for simple chromophore/polymer materials, the YLD124/PMMA material was chosen as a model system. As expected from many studies of the variation electro-optic activity of simple chromophore/polymer composite materials with chromophore number density, order decreases with increasing number density (see Table II).

The theoretical values of $\langle P_2 \rangle$ reproduce the experimental values reasonably well. The important point derivative from the data of Tables II and III is that the order parameter is low and thus there is the potential for further substantial improvement of order.

Before leaving a discussion of the YLD124/PMMA model system it is useful to carry out an additional correlation of data, namely, correlation of β valves calculated from electro-optic (r_{33}) and RVAPAS ($\langle P_2 \rangle$) measurements with direct measurement by HRS. In this comparison, theoretical guidance must be used in converting a $\langle \beta_{HRS}(-2\omega;\omega,\omega)$ measured in chloroform of 6900 x 10^{-30} esu to a PMMA equivalent $\beta_{zzz}(-\omega;0,\omega)$ of 7800 x 10^{-30} esu. This HRS-based value can be compared with the RVAPAS/electro-optic estimated value of 8767 x 10^{-30} esu. This level of agreement is within experimental error of and suggests an internal consistency in agreement between theory and experiment. A more detailed discussion will be presented elsewhere

Multi-Chromophore-Containing Dendrimers

The multi-chromophore-containing dendrimer (denoted PAS-33) shown in Figure 4 has been used as a test of the correlated TD-DFT and pseudo-atomistic methods. The flexible segments are treated via fully-atomistic calculations while the conjugated π-electron segments are treated in the United Atom Approximation. A full range of data has been collected for this material (see Figures 5, 6, and 7).

Theoretical calculations yield a value of 25 for the limiting dielectric permittivity compared to the experimental value of 27 (see Figure 7).

Pseudo-atomistic Monte Carlo calculations yield a poling efficiency of $r_{33}/E_{pol} = 1.5\ (nm/V)^2$ compared to the experimental value of 1.35 (at 1.3 microns). The chromophore number density in this dendrimer material is 6.5×10^{20} molecules/cc. We have also diluted PAS-33 in amorphous polycarbonate (APC) to a concentration level of 3.5×10^{20} molecules/cc. The experimentally measured poling efficiency for this composite material is $r_{33}/E_{pol} = 0.72\ (nm/V)^2$ compared to a theoretically computed value of 0.81. Further dilution of chromophore number density of the PAS-33 dendrimer in APC and preparation of higher generation dendrimers by addition of Frechet-type dendrons yield a linear relationship between poling efficiency and chromophore number density implying an order parameter $<\cos^3\theta>$ that is independent of chromophore concentration. A linear dependence is also predicted by pseudo-atomistic Monte Carlo calculations. The covalent bond potentials and density of the dendrimers considered here likely prevent interpenetration of dendrimers and the concentration-dependent centrosymmetric aggregation of chromophores (This is indicated by the linear dependence of r_{33} on concentration or equivalently of the independence of order parameter on concentration). It is also clear that multi-chromophore-containing dendrimers permit very high chromophore concentrations to be achieved without problems associated with phase separation and associated light scattering (optical loss).

The poling efficiency for the PAS-33 dendrimer can also be expressed in terms of the odd and even order parameters; namely, $<\cos^3\theta>/E_p = 0.0015\ (V/\mu m)^{-1}$ and $<P_2>/E_p = 0.00075\ (V/\mu m)^{-1}$.

The restrictions placed on chromophore organization by covalent bond potentials are evident in the dendrimers discussed here. When the length of flexible spacers in the dendrimers are increased quite different behavior is observed with the results approaching that expected for composite materials where chromophore movement (and association) is relatively unhindered.

Binary Chromophore Organic Glasses

Theory suggests that improved order can be achieved if two different chromophores are mixed to form a composite material (a binary chromophore organic glass). For example, if the lattice symmetry surrounding a chromophore is reduced from 3-D (Langevin lattice) to 2-D (Bessel lattice) to 1-D (Ising lattice), the order parameter increases significantly. If the two chromophores experience spatially-anisotropic interactions, then order will be increased since the electric poling field increases the order of both chromophores and both chromophores are coupled by spatially-anisotropic interactions.

Doping the YLD-124 chromophore into the pure dendrimer materials discussed in the preceding section results in a factor of 2.5 improvement in poling efficiency. Coarse-grained Monte Carlo calculations yielded a theoretically-predicted improvement of comparable magnitude (*16*).

A variety of binary chromophore organic glasses exploiting different spatially anisotropic interactions have been investigated. These include quadrupolar interactions between protonated and fluorinated Frechet-type dendrimers (*14*)

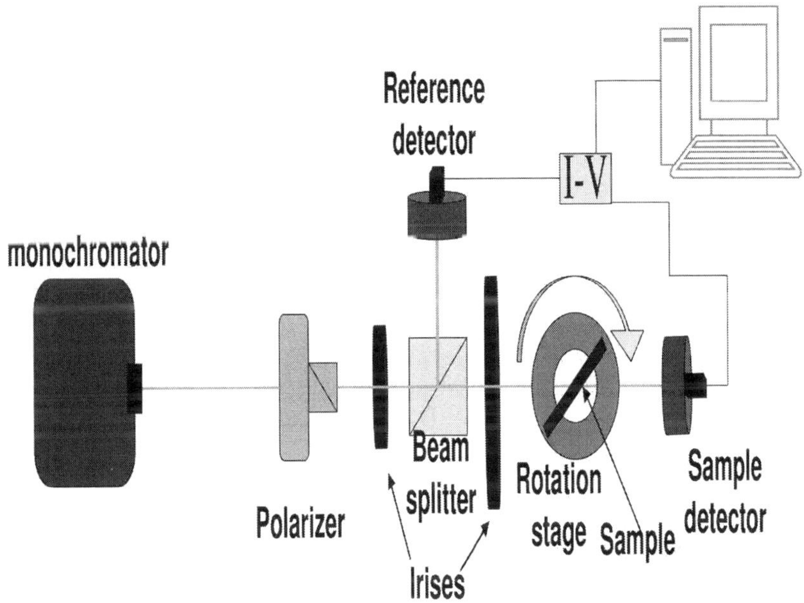

Figure 3. A schematic diagram of the RVAPAS apparatus is shown.

Table II. $<P_2>$ Determined by the RVAPAS Technique for YLD124 in PMMA as a Function of Chromophore Number Density N

N (molecules/cc) x 10^{20}	*$<P_2>$ (RVAPAS)*
0.8	0.043 ± 0.002
1.2	0.036 ± 0.007
1.6	0.025 ± 0.003

and interactions involving liquid crystal mesogens (*30*). In all cases, a significant improvement in poling efficiency (and hence maximum obtainable electro-optic activity) is observed for binary chromophore organic glasses. The magnitude of the effect on chromophore order varies from sample to sample consistent with the specific interactions in play. Nuclear repulsive (steric) interactions clearly play a role in influencing order and the fact that dopant and host chromophores have different shapes is important both from enthalpic and entropic considerations. Steric interactions and covalent bond potentials limit the number of ways in which chromophores can organize and thus the situation is quite different from that for similar high chromophore concentration single chromophore composite materials. Quantitative understanding of each material requires detailed theoretical (Monte Carlo) simulations and discussion of all of the examples reported in the literature to date is beyond the scope of this chapter. Clearly, thought must be given to designing the structures of the components of binary chromophore materials; mixing "like" chromophores which have unhindered three-dimensional freedom will not lead to an improvement in poling-induced acentric order.

Table III. The Variation of Calculated $\langle P_2 \rangle$ Order Parameter Values with Chromophore Number Density

N (molecules/cc) x 10^{20}	*$\langle P_2 \rangle$ (Theory)*
0.8	0.038
1.2	0.034
1.6	0.028

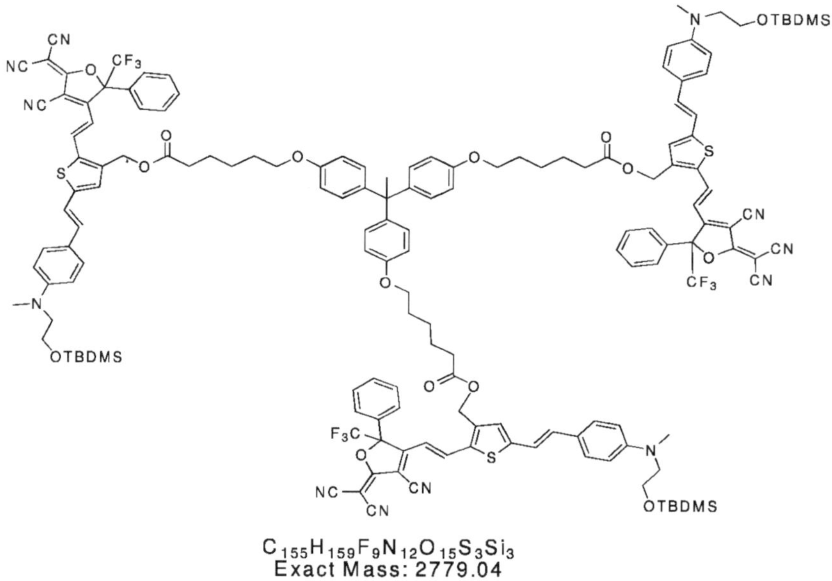

Figure 4. The structure of the PAS-33 multi-chromophore-containing dendrimer is shown.

Laser-Assisted Electric Field Poling

Doping the YLD-124 chromophore into DR1-co-PMMA yields a poling efficiency (at 1.3 microns wavelength) of r_{33}/E_{pol} = 1.65 $(nm/V)^2$ while irradiating the DR1 chromophore with polarized 532 nm light leads to an increase in poling efficiency of r_{33}/E_{pol} = 4.0 $(nm/V)^2$. The YLD-124/DR1-co-PMMA binary chromophore organic glass affords another advantage. The charge-transfer (HOMO-LUMO) transitions of the two chromophores do not overlap significantly. No line broadening or solvatochromic shifts are observed for these glasses as a function of varying the concentration of the YLD-124 chromophore (*13*). This experiment provides rather straightforward proof of the interaction between quest and host chromophores in this material. RVAPAS measurements permit the order of guest and host chromophores to be obtained independently and thus permits assessment of the intermolecular interactions between these species. A more detailed discussion of these measurements will be presented elsewhere.

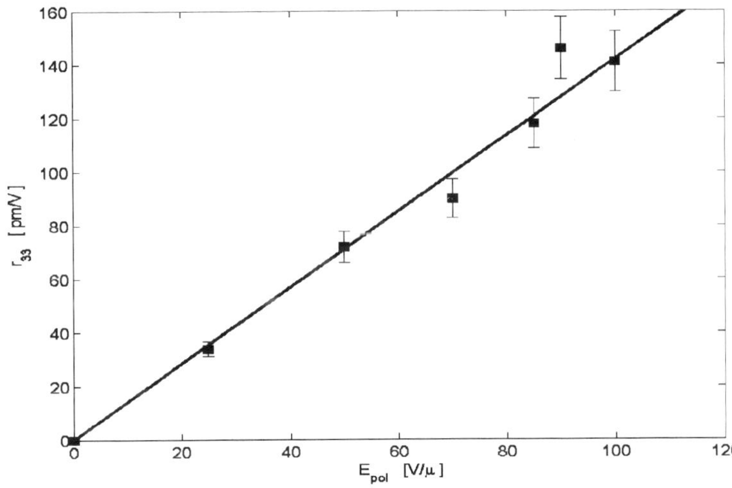

Figure 5. The variation of electro-optic activity with poling field (E_p or E_{pol}) is shown. The slope (poling efficiency) is $r_{33}/E_{pol} = 1.35\ (nm/V)^2$. Data are for a wavelength of 1.3 microns.

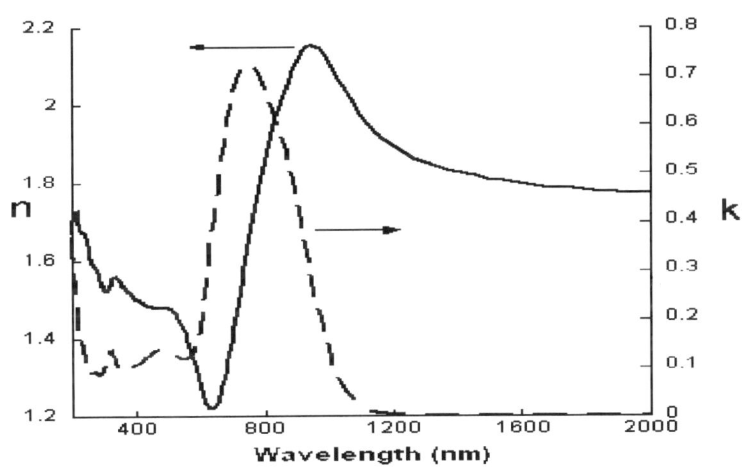

Figure 6. Real and imaginary components of the index of refraction are shown.

Binary chromophore organic glasses provide a variety of routes to enhancing poling efficiency and electro-optic activity through the systematic exploitation of spatially-anisotropic interactions. At the same time, binary chromophore organic glasses permit minimization of optical loss through control of spectral line broadening and solvatochromic (bathochromic) shifts. It is not our intention to suggest that these are the ultimate electro-optic material. In fact, the most obvious lesson from the above studies is to keep an open mind. For some applications (e.g., incorporation of OEO materials into silicon photonic structures with nanoscale critical dimensions), sequential synthesis, rather than electric field poling, may be the desired processing approach. Whatever the approach, theoretical guidance can be important for reducing unproductive synthetic effort.

Device Requirements

A discussion of efforts to optimize electro-optic activity and auxiliary properties must be set in the context of device requirements. The electro-optic activity of organic materials has already exceeded that of lithium niobate and the insertion loss of devices based on organic materials can be less than that of III-V semiconductor electro-absorptive modulators; nevertheless, these inorganic materials continue to be the materials of choice for most device applications. The question must be asked as to what material parameters are critical for organic electro-optic materials to achieve wide scale utilization.

An undisputed advantage of organic electro-optic (second order nonlinear optical) materials is the intrinsic ultrafast (tens of femtosecond) response time to applied electric field variations. The intrinsic response is the phase relaxation time of the π-electron system. Difference frequency (terahertz) generation establish the potential intrinsic bandwidth of organic second order nonlinear optical materials to be tens of terahertz (*31*). This intrinsic bandwidth has also been measured by femtosecond pulse techniques (*32*) and in all-optical stripline modulator devices (*33*). Since bandwidth of EO devices is typically limited by the electrical resistivity of metal (e.g., gold) electrodes, the intrinsic bandwidth of organic electro-optic materials can only be exploited for very short stripline devices. For example, realization of a 100 GHz bandwidth using a Mach Zehnder interferometer device configuration requires a device length on the order of 0.5 cm. Of course, device length, L, influences device drive voltage (e.g., the voltage, V_π, required to effect electrical-to-optical signal transduction using a Mach Zehnder device) as can be seen from the equation: $V_\pi = \lambda d/(2n^3 r_{33} L\Gamma)$ where λ is the optical wavelength, n is the index of refraction, d is the electrode gap, and Γ is the electrical/optical overlap integral. For telecommunication wavelengths and normal electrode spacings (e.g., 8 microns), a V_π of 0.75 V would require an electro-optic activity of 300 pm/V. If the material optical loss is 2 dB/cm and the fiber coupling loss is 0.8 dB per facet then an insertion loss of 2.6 dB would be observed which is below the maximum acceptable value of about 6 dB. A frequently-stated desired objective of RF photonics is to achieve gain in the electrical-to-optical-to-electrical signal transduction process. For this to be achieved, a V_π of approximately 0.2 V is required. For the device dimensions just considered, an electro-optic coefficient of nearly 1200 pm/V would be required to achieve gain.

To the present, the best V_π achieved with a Mach Zehnder device employing organic electro-optic materials is 0.25 V. This has required device lengths of 2-3 centimeters (with insertion loss above 6 dB). From this brief discussion, it is clear that further improvement of the electro-optic activity of organic materials is an important objective for simultaneously achieving high bandwidth and low drive voltage operation. Smaller devices are also important for chipscale integration where size is a premium.

In addition to being used to fabricate stripline device structures such as Mach Zehnder (amplitude) modulators, phase modulators, and directional couplers (*34*), organic electro-optic materials have been used to produce ring microresonator (*35–42*) and cascaded prism modulators (*43*, *44*). Ring microresonators can be used for electrical-to-optical signal transduction, active optical signal routing, and active wavelength division multiplexing (WDM). Cascaded prism devices are used for beam steering and spatial light modulation applications. Resonant devices such as etalons are also used for applications such as optical switching; the discussion of such devices has many points in common with the discussion of ring microresonators and will not be duplicated here.

Although ring microresonators permit reduction in the "footprint" (size) of devices, the effective pathlength of the device is comparable to that of stripline devices. The quality (Q) factor determines the lifetime (or distance traveled) of the photons in the device and thus the operational bandwidth of the device. Since light transits the device many times, the drive voltage requirements are correspondingly reduced. Drive voltages as low as 0.25 volts (at 1.55 microns wavelength) have been observed for slotted silicon ring microresonators filled with organic electro-optic materials (*40*). In addition, to the material absorption and scattering losses (associated with intrinsic material losses and processing-induced losses) of stripline devices, bending losses come into play for ring microresonators. Acceptable levels of bending loss define the size of ring microresonators, which is also defined by the index of refraction contrast between the core waveguide and its surroundings (cladding). If a large number of ring microresonator structures are being integrated on a single chip, insertion loss can become a dominant issue. Again, the realization of high electro-optic activity can greatly simplify design requirements; adequate drive voltage can be achieved with a lower Q factor thus permitting a higher bandwidth to be achieved for each resonator. Higher electro-optic activity can permit more attention to be paid to reducing the optical loss of a material (e.g., use of hypsochromic-shifted chromophores exhibiting smaller EO activity). For beam steering and spatial light modulation applications, organic electro-optic materials are competing against liquid crystalline materials, which have much larger effective electro-optic activity but have very slow response times. Unless the electro-optic activity of solid state organic materials (purely electronic activity) is improved above 1000 pm/V, such materials will be used only for applications requiring ultra high speed spatial light modulation.

A major issue, which has impeded the use of solution-processed (electrically-poled) organic electro-optic materials, is that of stability, both thermal and photochemical. Thermal stability is defined by final material glass transition temperature and glass transition temperatures on the order of 200°C are now routinely obtained exploiting cycloaddition chemistries (*18–21*). Such thermal stability is adequate for most applications involving optical powers of 50 mW or less. Very high optical power applications currently represent a challenge for all classes of materials, although heating (or high temperature applications) is not a problem for lithium niobate.

Photochemical stability of organic electro-optic materials is primarily a problem of singlet oxygen chemistry. Photochemical stability for a given chromophore π-electron structure can vary by up to 5 orders of magnitude

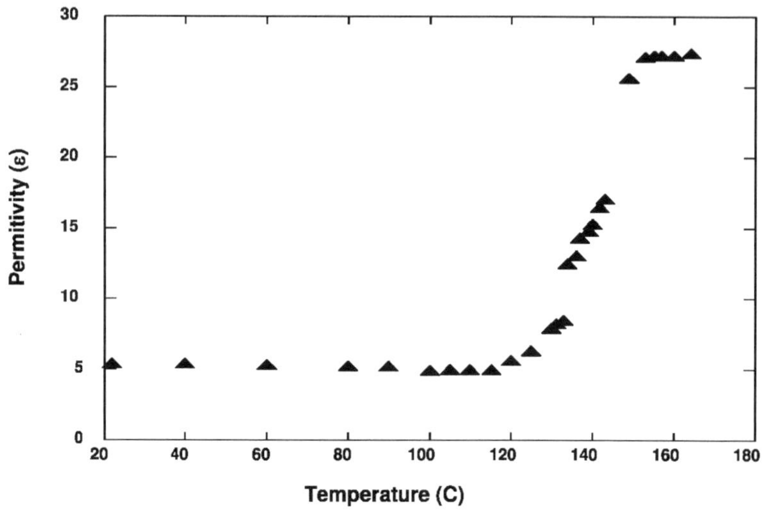

Figure 7. Measurement of dielectric permittivity of PAS-33 from ambient temperature to somewhat above the glass transition temperature (~125°C) is shown. The dwell time at each temperature (before measurement) is greater than 20 minutes so values should represent equilibrium values.

depending on structural modifications and packaging. For example, incorporating chromophores into dense dendrimer structures can lead to an order of magnitude improvement in photostability. Even minor structural modifications can have a large effect, e.g., replacement of the hydrogen at the 2 isophorone ring position of YLD-124 with an OR group (see Scheme 1) leads to a 5-fold improvement in photochemical stability. Because of dense covalently-coupled chromophore lattices (inhibiting the in-diffusion of oxygen), single crystal OEO materials and OEO materials prepared by sequential synthesis exhibit superior stability relative to electrically-poled materials.

Photochemical stability is typically measured by monitoring chromophore decomposition using input cw powers of up to 1 watt (accelerated testing) (*45*). With reasonable packaging, photochemical stability for greater than 10 years can be expected for current chromophores and telecommunication power levels of 10-20 mW (at wavelengths in 1.55 micron telecommunication band). Stability decreases as the interband electronic transition is approached. The present use of spatially-anisotropic intermolecular electrostatic interactions to systematically improve acentric order in solution-processed materials can be viewed as a step toward the systematic engineering of materials toward crystalline-level of order. Stronger intermolecular interactions will yield higher order but care must be exercise to avoid elevating material glass transition temperature above the intrinsic chromophore decomposition temperature or melt processability (electric field poling) will be lost. The absence of melt processability makes annealing of microdomain structures impossible. Crystal growth is not necessarily an answer to high performance electro-optic materials unless growth anisotropy can also be controlled since devices typically use thin films of materials.

Only recently, have organic electro-optic materials been demonstrated to meet Telcordia standards. Thus, for applications requiring bandwidths of 20 GHz or less and operating voltages of the order of 2-5 volts, long (4-8 cm) devices fabricated from lithium niobate remain attractive since the reliability of these devices is well-established. An aspect of organic materials not easily realized with lithium niobate and III-V semiconductor materials is amenability to various processing options (solution, vapor phase, nanoimprint and soft lithograph (*36*)) and ease of integration with diverse materials (e.g., silicon photonics (*33*, *38*, *40–42*)). This feature of organic materials make them particularly attractive for the chipscale integration of photonics and electronics. Organic electro-optic materials are also amenable to the fabrication (by lift-off techniques) of conformal and flexible devices (*46*, *47*).

Conclusions

Correlated quantum (TD-DFT) and statistical (pseudo-atomistic Monte Carlo) mechanical calculations, supported by HRS, ATR, and RVAPAS measurements, have been employed to guide the improvement of organic electro-optic materials and have permitted "first-principles" rationalization of observed electro-optic coefficients (although the large experimental uncertainty of HRS measurements leave something to be desired in the comparison of experimental and theoretical molecular first hyperpolarizability data). Poling efficiencies, r_{33}/E_p, have been improved for a number of chromophore materials to values on the order of 4.0 $(nm/V)^2$. Although it has been impossible, to the present time, to obtain the same degree of poling efficiency in device structures (due to issues associated with cladding and electrode materials), improved electro-optic activities have permitted the realization of drive (V_π) voltages on the order of 0.25 volts (at 1.55 microns wavelength) for Mach Zehnder stripline and ring microresonator devices.

RVAPAS measurements, together with ATR and HRS measurements, establish that acentric order parameters are still low (typically less than 0.2) and can be further improved by the use of additional spatially-anisotropic intermolecular electrostatic interactions. It is also likely that chromophore molecular first hyperpolarizability will be further improved by the synthesis of new chromophore structures (*48–51*) and by exploiting the tuning of $\beta(\omega,\varepsilon)$ that can be achieved through control of dielectric permittivity. Such further improvement in electro-optic activity is likely necessary for wide application of organic electro-optic materials. The development of organic electro-optic materials can still be viewed as immature although paradigms for the realization of acceptable thermal and photochemical stability now appear more well-defined. We do not mean to suggest that the materials discussed in this chapter should be materials of choice for device applications; indeed, we do not view the improvements noted here as sufficient. Electro-optic activity is likely to ultimately be improved beyond 1000 pm/V but the specific chromophore and material structures that will yield such values remain to be defined. With achievement of high electro-optic activity, greater attention will likely be paid to reducing optical loss but it is unlikely that optical loss in cost-effective materials will be reduced below 1 dB/cm (as such reduction would require isotopic modification of chromophores to reduce hydrogen vibrational overtone absorption at telecommunication wavelengths). With the use of short optical pathlength devices, such low optical loss may become much less important but for now optical loss is an important component of the figure-of-merit of an EO material.

Significant device demonstrations have been achieved for both all-organic devices and for organic electro-optic materials integrated into silicon photonic structures. While many issues remain to be addressed, particularly those related to improving poling efficiency in device structures, organic electro-optic materials appear to afford the possibility of millivolts drive voltage, high bandwidth devices. The processability and compatibility of organic electro-optic materials with disparate materials is also a distinct advantage of these materials.

Electro-optic device technology is critical to photonic/electronic integration with chipscale integration presenting special challenges. The options for electro-optic technology include all-silicon modulators (*52–57*), crystalline inorganic materials such as lithium niobate, semiconductor electro-absorptive modulators, modulated semiconductor lasers, and organic electro-optic materials. Liquid crystalline materials and thermo-optic modulators are now generally viewed as too slow for many high speed applications in computing, telecommunications, and sensing. Each technology option has its advantages and disadvantages. Organic electro-optic materials may be unique in still affording the possibility of dramatic (orders of magnitude) improvement in performance.

Single crystalline OEO materials typically arise when a small number of spatially-anisotropic strong (e.g., ionic) interactions overwhelm chromophore dipolar interactions. These strong interactions typically elevate material glass transition temperatures above material decomposition temperatures so that melt processability is lost. Growth anisotropic is a general problem with single crystalline OEO materials and has, together with modest electro-optic activity reflecting the modest molecular first hyperpolarizability of constituent chromophores, inhibited the commercial utilization of such materials. The approach discussed in this chapter has focused on the use of intermediate strength spatially-anisotropic interactions to achieve melt processable materials with some of the enhanced order feature of single crystalline materials (although the order is substantially less than single crystalline materials). A critical question to be defined in future studies is "How much can order be improved without losing processability?" It may also be useful to direct theory to optimization of materials prepared by sequential synthesis/self assembly methods, which may be particularly relevant to silicon photonic device applications. This approach affords some of the advantages of single crystalline materials but is more amenable to the fabrication of thin film EO materials required for device applications.

Acknowledgments

The authors gratefully acknowledge financial support provided by the National Science Foundation (DMR-0551020 and DMR-0120967) and by the Air

Force Office of Scientific Research (F49620-1-0110-P000). Financial support is also acknowledged from the DARPA MORPH program.

References

1. Halter, M.; Liao, Y.; Plocinik, R. M.; Coffey, D. C.; Bhattacharjee, S.; Mazur, U.; Simpson, G. J.; Robinson, B. H.; Keller, S. L. *Chem. Mater.* **2008**, *20*, 1778.
2. Zhu, P.; van der Boom, M. E.; Kang, H.; Evmenenko, G.; Dutta, P.; Marks, T. J. *Chem. Mater.* **2002**, *14*, 4982.
3. Ishorn, C. M.; Leclercq, A.; Vila, F. D.; Dalton, L. R.; Bredas, J. L.; Eichinger, B. E.; Robinson, B. H. *J. Phys. Chem. A* **2007**, *111*, 1319.
4. Marques, M. A. L.; Gross, E. K. U. *Ann. Rev. Phys. Chem.* **2004**, *55*, 427.
5. Takimoto, Y.; Villa, F. D.; Rehr, J. J. *J. Chem. Phys.* **2007**, *127*, 154114.
6. Cannuccia, E.; Pulci, O.; Palummo, M.; Garbuio, V.; Del Sole, R. *Phys. Status Solidi* **2008**, *5*, 2543.
7. Takimoto, Y.; Isborn, C. M.; Eichinger, B. E.; Rehr, J. J.; Robinson, B. H. *J. Phys. Chem.* **2008**, *112*, 8016.
8. Kim, W. K.; Hayden, L. M. *J. Chem. Phys.* **1999**, *111*, 5212.
9. Leahy-Hoppa, M. R.; Cunningham, P. D.; French, J. A.; Hayden, L. M. *J. Phys. Chem. A* **2006**, *110*, 5792.
10. Robinson, B. H.; Dalton, L. R. *J. Phys. Chem. A* **2000**, *104*, 4785.
11. Rommel, H. L.; Robinson, B. H. *J. Phys. Chem. C* **2007**, *111*, 18765.
12. Sullivan, P. A.; Rommel, H.; Liao, Y.; Olbricht, B. C.; Akelaitis, A. J. P.; Firestone, K. A.; Kang, J. W.; Luo, J.; Choi, D. H.; Eichinger, B. E.; Reid, P. J.; Chen, A.; Jen, A. K. Y.; Robinson, B. H.; Dalton, L. R. *J. Am. Chem. Soc.* **2007**, *129*, 7523.
13. Olbricht, B. C.; Sullivan, P. A.; Wen, G. A.; Mistry, A.; Davies, J. A.; Ewy, T. R.; Eichinger, B. E.; Robinson, B. H.; Reid, P. J.; Dalton, L. R. *J. Phys. Chem. C* **2008**, *112*, 7983.
14. Kim, T. D.; Kang, J. W.; Luo, J.; Jang, S. H.; Ka, J. W.; Tucker, N.; Benedict, J. B.; Dalton, L. R.; Gray, T.; Overney, R. M.; Park, D. H.; Herman, W. N.; Jen, A. K. Y. *J. Am. Chem. Soc.* **2007**, *129*, 488.
15. Dalton, L. R.; Sullivan, P. A.; Bale, D. H.; Olbricht, B. C. *Solild-State Electron.* **2007**, *51*, 1263.
16. Pereverzev, Y. V.; Gunnerson, K. N.; Prexhdo, O. V.; Sullivan, P. A.; Liao, Y.; Olbricht, B. C.; Akelaitis, A. J. P.; Jen, A. K. Y.; Dalton, L. R. *J. Phys. Chem. C* **2008**, *112*, 4355.
17. Kim, T. D.; Lao, J.; Cheng, Y. J.; Shi, Z.; Hau, S.; Jang, S. H.; Zhou, X. H.; Tian, Y.; Polishak, B.; Huang, S.; Ma, H.; Dalton, L. R.; Jen, A. K. Y. *J. Phys. Chem. C* **2008**, *112*, 8091.
18. Iacono, S. T.; Budy, S. M.; Jin, J.; Smith, D. W., Jr. *J. Polym. Sci., Part A-1: Polym. Chem.* **2007**, *45*, 5706.
19. Luo, J.; Haller, M.; Ma, H.; Liu, S.; Kim, T. D.; Tian, Y.; Chen, B.; Jang, S. H.; Dalton, L. R.; Jen, A. K. Y. *J. Phys. Chem. B* **2004**, *108*, 8523.

20. Shi, Z.; Hau, S.; Luo, J.; Kim, T. D.; Tucker, N. M.; Ka, J. W.; Sun, H.; Pyayt, A.; Dalton, L. R.; Chen, A.; Jen, A. K. Y. *Adv. Funct. Mater.* **2007**, *17*, 2557.
21. Sullivan, P. A.; Olbricht, B. C.; Akelaitis, A. J. P.; Mistry, A. A.; Liao, Y.; Dalton, L. R. *J. Mater. Chem.* **2007**, *17*, 2899.
22. Firestone, K. A.; Lao, D. B.; Casmier, D. M.; Clot, O.; Dalton, L. R.; Reid, P. J.Bale, D. *Proc. SPIE-Int. Soc. Opt. Eng.* **2005**, *5935*, 59350P1, to be published.
23. Teng, C. C. *Appl. Opt.* **1993**, *32*, 1051.
24. Park, D. H.; Kang, J. W.; Luo, J D.; Lim, T. D.; Jen, A. K. Y.; Lee, C. H.; Herman, W. N. *Proc. SPIE-Int. Soc. Opt. Eng.* **2005**, *5935*, 593506.
25. Park, D. H.; Lee, C. H.; Herman, W. N. *Opt. Express* **2006**, *14*, 8737.
26. Chen, A.; Chuyanov, V.; Garner, S.; Steier, W.; Dalton, L. *Tech. Dig. Ser. - Opt. Soc. Am.* **2001**, *14*, 158.
27. Davies, J. A.; Elangovan, A.; Sullivan, P. A.; Olbricht, B. C.; Bale, D. H.; Ewy, T. R.; Isborn, C. M.; Eichinger, B. E.; Robinson, B. H.; Reid, P. J.; Li, X.; Dalton, L. R. *J. Am. Chem. Soc.* **2008**, *130*, 10565.
28. Graf, H.. M.; Zobel, O.; East, A. J.; Haarer, D. *J. Appl. Phys.* **1994**, *75*, 3335.
29. Mansuripur, M. *J. Appl. Phys.* **1990**, *67*, 6466.
30. Benight, S. Unpublished results.
31. McLaughlin, C. V.; Hayden, L. M.; Polishak, Huang, S.; Luo, J.; Kim, T. D.; Jen, A. K. Y. *Appl. Phys. Lett.* **2008**, *92*, 151107.
32. Drenser, K. A.; Larsen, R. J.; Strohkendl, F. P.; Dalton, L. R. *J. Phys. Chem.* **1999**, *103*, 2290.
33. Hochberg, M.; Baehr-Jones, T.; Wang, G.; Shearm, M.; Harvard, K.; Liu, J.; Chen, B.; Shi, Z.; Lawson, R.; Sullivan, P. A.; Jen, A. K. Y.; Dalton, L. R.; Scherer, A. *Nat. Mater.* **2006**, *5*, 703.
34. Dalton, L. R.; Harper, A. W.; Wu, B.; Ghosn, R.; Laquindanum, J. G.; Hubbel, A.; Xu, C. *Adv. Mater.* **1995**, *7*, 519.
35. Rabiei, P.; Steier, W. H.; Zhang, C.; Dalton, L. R. *J. Lightwave Technol.* **2002**, *20*, 1968.
36. Huang, Y.; Paloczi, G. T.; Yariv, A.; Zhang, C.; Dalton, L. R. *J. Phys. Chem. B* **2004**, *108*, 8606.
37. Chen, A.; Dalton, L. R.; Sherwood, T. J.; Jen, A. K. Y.; Rabiei, P.; Steier, W. H.; Huang, Y.; Paloczi, G. T.; Poon, J. K.; Scherer, A.; Yariv, A. *Proc. SPIE-Int. Soc. Opt. Eng.* **2005**, *5708*, 187.
38. Baehr-Jones, T.; Hochberg, M.; Wang, G.; Lawson, R.; Sullivan, P. A.; Dalton, L. R.; Jen, A. K. Y.; Scherer, A. *Opt. Express* **2005**, *13*, 5216.
39. Pyayt, A.; Dalton, L. R.; Chen, A. *Proc. SPIE-Int. Soc. Opt. Eng.* **2005**, *5935*, 59350W1.
40. Baehr-Jones, T.; Penkov, B.; Huang, J.; Sullivan, P. A.; Davies, J.; Takayesu, J.; Luo, J.; Kim, T. D.; Dalton, L. R.; Jen, A. K. Y.; Hochberg, M.; Scherer, A. *Appl. Phys. Lett.* **2008**, *92*, 1633031.
41. Takayesu, J.; Hochberg, M.; Baehr-Jones, T.; Chan, E.; Wang, G.; Sullivan, P. A.; Liao, Y.; Davies, J.; Dalton, L. R.; Scherer, A.; Krug, W. *J. Lightwave Technol.* **2008**, *26*, 440.

42. Dalton, L. R.; Lao, D.; Olbricht, B. C.; Benight, S.; Bale, D. H.; Davies, J. A.; Ewy, T.; Hammond, S. R.; Sullivan, P. A. *Opt. Mater.* 2008, in press, doi:10.1016/j.optmat.2009.02.002, available on-line.
43. Sun, L.; Kim, J. H.; Jang, C. H.; An, D.; Lu, X.; Zhou, Q.; Taboada, J. M.; Chen, R. T.; Maki, J. J.; Tang, S.; Zhang, H.; Steier, W. H.; Zhang, C.; Dalton, L. R. *Opt. Eng.* **2001**, *40*, 1217.
44. Kim, J. H.; Sun, L.; Jang, C. H.; An, D.; Taboada, J. M.; Zhou, Q.; Lu, X.; Chen, R. T.; Li, B.; Han, X.; Tang, S.; Zhang, H.; Steier, W. H.; Ren, A. S.; Dalton, L. R. *Proc. SPIE-Int. Soc. Opt. Eng.* **2001**, *4279*, 37.
45. Rezzonico, D.; Jazbinsek, M.; Gunter, P.; Bosshard, C.; Bale, D. H.; Liao, Y.; Dalton, L. R.; Reid, P. J.Bale, D. H. *J. Opt. Soc. Am. B* **2007**, *24*, 2199unpublished results.
46. Steier, W. H.; Chen, A.; Lee, S.-S.; Garner, S. M.; Zhang, H.; Chuyanov, V.; Dalton, L. R.; Wang, F.; Ren, A. S.; Zhang, C.; Todorova, G.; Harper, A. W.; Fetterman, H. R.; Chen, D.; Udupa, A. H.; Bhattacharya, D.; Tsap, B. *Chem. Phys.* **1999**, *245*, 487.
47. Song, H.; Oh, M.; Ahn, S.; Steier, W. H. *Appl. Phys. Lett.* **2003**, *82*, 4432.
48. Kang, H.; Facchetti, A.; Zhu, P.; Jiang, H.; Yang, Y.; Cariati, E.; Righetto, S.; Ugo, R.; Liu, Z.; Ho, S.; Marks, T. J. *Angew. Chem., Int. Ed. Engl.* **2005**, *44*, 7922.
49. Kang, H.; Facchetti, A.; Jiang, H.; Cariati, E.; Righetto, S.; Ugo, R.; Zuccaccia, A.; Macchioni, A.; Stern, C. L.; Liu, Z.; Ho, S. T.; Brown, E. C.; Ratner, M. A.; Marks, T. J. *J. Am. Chem. Soc.* **2007**, *129*, 3267.
50. Brown, E. C.; Marks, T. J.; Ratner, M. A. *J. Phys. Chem. B* **2008**, *112*, 44.
51. Wang, Y.; Frattarelli, D.; Facchetti, A.; Cariati, E.; Tordin, E.; Ugo, R.; Zuccaccia, C.; Macchioni, A.; Wegener, S. L.; Stern, C. L.; Ratner, M. A.; Marks, T. J. *J. Phys. Chem. C* **2008**, *112*, 8005.
52. Barrios, C. A.; Almeida, V. R.; Lipson, M. *J. Lightwave Technol.* **2003**, *21*, 1089.
53. Barrios, C. A.; Almeida, V. R.; Panepucci, R.; Lipson, M. *J. Lightwave Technol.* **2003**, *21*, 2332.
54. Liu, A.; Jones, R.; Liao, L.; Samara-Rubio, D.; Cohen, R. O.; Nicolaescu, R.; Paniccia, M. *Nature* **2004**, *407*, 615.
55. Xu, Q.; Schmidt, B.; Pradhan, S.; Lipson, M. *Nature* **2005**, *435*, 325.
56. Zhou, L.; Poon, A. W. *Opt. Express* **2006**, *14*, 6851.
57. Chuang, R. W.; Liao, Z. L.; Hsu, M. T.; Liao, J. C.; Cheng, C. C. *Jpn. J. Appl. Phys.* **2008**, *47*, 2945.

Chapter 3

Substituent Effects on Interior Locations of Large Dyes: Changes in First Hyperpolarizability As Predicted by DFT

G. A. Lindsay* **and A. P. Chafin**

Research Department, Chemistry Branch, U.S. Navy, NAVAIR, NAWCWD, China Lake, CA 93555
***geoffrey.lindsay@navy.mil**

Quantum chemistry computations using Density Functional Theory (DFT) on organic asymmetric chromophores (hereafter called dyes) containing 33 to 53 atoms recommend certain internal locations for placing electron-donating and electron-withdrawing substituents to increase the first hyperpolarizability (β). Computations were performed on mono-substituted dyes for two π-scaffolds containing three thiophenylene groups, and for one π-scaffold containing a *trans*-tetraene. Dipole moments and λ_{max} were also calculated for each dye. Trade-off scenarios for improving the optical loss are recommended. DFT results are compared to predictions from Dewar's rules.

Introduction

Recently reported large asymmetric dyes (*1–3*) are very promising candidates for use in electro-optic modulators (*4–6*). The success of these new dyes is largely due to their extraordinarily large first hyperpolarizability (β) and newly developed molecular superstructures that enable increased concentration of dye and improved thin-film poling efficiency (*7*). Most of the previous work on developing improved dyes has focused on increasing the strength of the electron donating (push) and electron-withdrawing (pull) groups at the ends of the dye, and on increasing the length of the π-conjugated scaffold (which is the intra-dye charge-transfer conduit). Until recently there have only been a few investigations on the effect of placing push and pull substituents on interior locations of the π-scaffold of large dyes, which is the subject of the present study.

Over twenty years ago Nicoud and Twieg (*8*) investigated the small dimethyl-aminostilbene π-scaffold in which they substituted azomethine (*9*) for various internal carbon methines (including SP^2 benzenoid carbons). Their PPP-SCF-CI calculations predicted changes in β that were generally in accord with Dewar's rules, which were originally proposed for predicting shifts in the electronic absorption spectra of cyanine dyes (*10*). Tsunekawa *et al.* (*11*) validated the work by Nicoud and Tweig on benzylidene-anilines, and later, using coupled-perturbed Hartree-Fock (CPHF) and full singly-excited configuration interactions (SCI) calculations, Tsunekawa *et al.* predicted that internal azomethine substituents on a push-pull *trans*-hexaene scaffold should give the same trend in β as predicted by Dewar's rules (*12*). A recent DFT study also reported distinct internal patterns of electron-withdrawing and electron-donating substituents for increasing (and decreasing) the β of large dyes based on a π-scaffold containing phenylene-hexaene (*13*).

Dewar's rules were based on Perturbational Molecular Orbital theory, and were first used to describe qualitative changes in the HOMO – LUMO bandgap when the internal methine carbons of dyes are variously substituted (*10*). A general feature of Dewar's rules is that adjacent methine carbons behave oppositely (an odd-even effect). Dewar described one set of methines as "active" and the other set of methines as "inactive." The active positions were those on which a formal charge can reside in a principal resonance structure. The active and inactive positions were designated "unstarred" and "starred," respectively.

In comparing dyes that have various push-pull end groups, an increase in β usually results when there is a decrease in the HOMO-LUMO band gap. This relationship is an important element of the Two-State Model (Equation 1), first proposed by Oudar and Chemla (*14*):

$$\beta \propto \frac{\mu_{01}^2 \times (\mu_{11} - \mu_{00})}{E_{01}^2} \qquad (1)$$

where E_{01} is the transition energy between the ground and first excited state, μ_{01} is the transition dipole moment, μ_{00} and μ_{11} are the ground-state and excited-state dipole moments, respectively. The Two-State Model predicts that β should be proportional to λ^2_{max}. However, Equation 1 also reminds that substitutions on the dye will change the dipole moments, which can cause deviations from Dewar's rules as applied to predicting changes in β. Equation 1 has generally been found to be successful in approximating the β of small dyes in which a single excited state dominates the response to an electric field (*15*). An exception to that was the Gregory and Thorp (*16*) study on substitutions at the 2 position of 4-diethylamino-4'-nitroazobenzene, which found for the most part that shifts in λ_{max} were in the opposite direction than Dewar's rules predict. It should be noted for their dye that fluorine was one of the more electron withdrawing substituents producing a strong hypsochromic shift, as did cyano. However, in other dyes, fluorine is predicted to be an electron-donating group, as will be discussed below.

The well-known bond-length alternation (BLA) treatment of Seth Marder *et al.*, (*15*, *17*) shows that changes in β as a function of changes in the strength of the push-pull end groups on the dye (which was simulated with a change in the static electric field applied along the long axis of the polyene dye) is attended by changes in the bond lengths [Table 1 in (*17*)], and also by changes in the Mulliken ground-state charge density on the internal methines in the polyene [Figure 3 in (*17*)]. On alternating methines, first there is an increase in charge density (and then a decrease) as the polyene is transformed towards the cyanine state (and on to the zwitterionic state). Hence, it should not be surprising that internal substitutions would have an effect on β.

Isborn *et al.* (*18*) have recently tested the reliability of density functional theory (DFT) for calculating dipole moments, polarizabilities, and first-order hyper-polarizabilities (βs) for a number of large dyes and found that the relative properties of the dyes are consistently predictable with DFT. DFT was the quantum chemistry computational method used in the present study.

The following sections present a range of probable changes in β and λ_{max} for mono-substitutions on three different π-scaffolds as computed by DFT. Several new dyes are recommended for further consideration in the quest for better EO modulator materials.

Background and DFT Results for a Polyenic π-Scaffold

A new target for real-world applications requires adequate thermal and photo-chemical stability. To achieve better stability, at least every other ene in the π-scaffold should be ring-locked, as for example, with the isophorone unit (*19*). Ring locking is typified by the so-called CLD class of dye shown in Figure 1, developed principally by the Dalton and Jen groups (*20*).

Figure 1 shows CLD with the dimethyl-tricyanofuran electron-withdrawing end group. Replacing one of the methyl groups on the tricyanofuran (TCF) with a trifluoromethyl group increases electro-optic activity and red-shifts the absorption spectrum (*21*, *22*). The authors have not found a report comparing the experimental value of β for CLD based on (CF_3,CH_3)-TCF with that of CLD based on dimethyl-TCF; however, DFT computations predicted that the trifluoromethyl group in that position could increase the first hyperpolarizability by 9% (*13*). An even larger increase in β was measured by hyper-Rayleigh scattering on a large bisthiophene class of dye containing the (CF_3,phenyl)-TCF end group compared to the same π-scaffold with the dimethyl-TCF end group (*23*). However, for the present study dimethyl-TCF was deemed appropriate since its blue-shifted absorption spectrum would somewhat compensate for the red-shifted spectra caused by internal substitutions described below that should increase β. Furthermore, the dimethyl-TCF end group is reported to be more thermally stable (*23*).

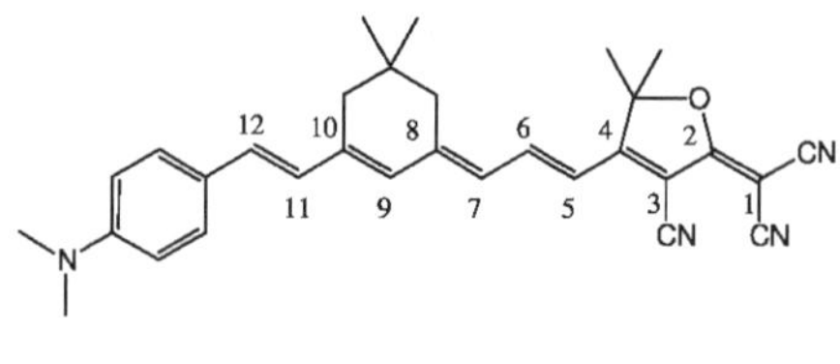

Figure 1. The CLD dye (20) containing the isophorone and furan ring-locking units.

Before looking at specific examples, it should be pointed out that only static βs were calculated, and only for the geometry-optimized (minimal-energy) conformations of the dyes. See the Modeling Procedure section below for details on the functionals and basis set used. Both the total magnitude of β and β_μ were calculated. The β_μ is the portion of β that projects on the ground-state dipole moment vector. β_μ is appropriate to consider when the nonlinear optical film containing the dye is to be processed by electric-field poling.

Using a numbering scheme for each position in the π-scaffold of a large dye is more useful than using the starred notation because one can refer to each location on the dye. For the present study the terminal methine containing the electron withdrawing group(s) is given the #1 position. Using this numbering system, the odd-numbered methines correspond to Dewar's inactive (or starred) methines and the even-numbered methines correspond to the active (or unstarred) methines.

Interior Substitutions on a π-Scaffold Containing a Bare Tetraene

The substituent pattern for which Dewar's rules would predict an increase β is the {eW/oD} pattern, where e and o stand for even- and odd-numbered methines, and W and D stand for withdrawing and donating substituents. However, when referring to the "substituent pattern," the methines at the two extreme ends of the dye are not considered part of the pattern because, of course, they are oppositely substituted compared to the interior methines. In summary, placing an electron donating group on an interior odd-numbered methine would increase β, placing an electron withdrawing group on an interior even-numbered methine would increase β, and the opposite substitution pattern, {oW/eD}, would decrease β.

Earlier DFT computations on a large polyenic dye predicted that placing fluorine on methine carbons should shift β in the same direction and nearly the same magnitude as the electron-donating methoxy substituent (*13*); therefore, for a *trans*-polyene dye DFT predicts fluorine would be resonantly donating as it is in aromatic electrophilic substitutions. In the following examples of substituent effects, fluorine was used as the electron donating group and cyano and azomethine were used as the electron withdrawing groups.

For the purpose of demonstrating the effect of substituents on a *trans*-polyene, the present study examined a dye with a bare tetraene between 4-(dimethylamino)phenyl and dimethyl-TCF (Figure 2) with the TCF group in the low-energy $Z^{4,5}$ conformation (*13*). For mono-substituted dyes, Figure 2 shows the effect of the beneficial {eW/oD} and the detrimental {oW/eD} patterns on

β. The changes in β obey Dewar's rules perfectly (β is called "β magnitude" in Figure 2). Recommendations for new target dyes based on this π-scaffold, after examining Figures 345, are discussed below.

First Hyperpolarizability Projected along the Dipole Moment, β_μ

Figure 3 shows an example of how the vectors of β and μ can diverge. Figure 4 shows the values for β_μ, which is the portion of the total magnitude of β that projects along the dipole moment vector. For cyano at position 12, DFT predicts an anomalously low value of β_μ. This can be explained by comparing changes in the angle between the β and the dipole moment vectors. The calculations show that the electronegative nitrogen of the cyano group perturbs the dipole vector enough to increase the angle between β and the dipole vector, thereby decreasing the projection of the hyperpolarizability tensor on the dipole moment vector. This is the same explanation given by Marder *et al* for the drop in β_μ on going from 4-nitro-4'-dimethylaminostilbene to 2,4-dinitro-4'-dimethylaminostilbene.

A 28% increase in β_μ is predicted when substituting a cyano group at the 6 position; a 20% increase when fluorine is at the 11 position; and an 18% increase when fluorine is at the 9 position – all three would be desirable targets if isophorone were used to ring-lock positions 8-10 (*13*). Jen *et al* have shown that the chemical stability is improved with alkoxy at the 9 position for the same π-scaffold containing the (CF_3,phenyl)-TCF end group (*1*).

Also apparent in Figures 2 and 4 is a gradient in first hyperpolarizability predicted by DFT. The gradient effect is superimposed on Dewar's odd-even effect; namely, the magnitudes of the change in β_μ (and β) are predicted to be larger near the ends of the dye (*13*).

Absorption Maximum, λ_{max}

This class of dye is well suited for electro-optic modulators operating in the 1550nm telecommunication window. Optimizing performance in an optical modulator requires a trade-off between maximizing the electro-optic coefficient and minimizing optical absorption loss (minimizing the red tail at 1550nm). To reduce the absorption loss, one may blue-shift the electronic absorption by choosing substituents that decrease λ_{max} and/or steepen the low-energy band edge of the absorption spectrum. Jen *et al.* (*2*) showed that placing a slightly electron-withdrawing thiol group at position 9 may afford an adequate trade off (using the Figure-1 π-scaffold except with the (CF_3,phenyl)-TCF end group). The CF_3- group increases λ_{max} by about 60nm compared to dimethyl-TCF (*21*, *23*). The present DFT study recommends substituent strategies promising small increases in λ_{max} (< 30nm) and a large increase in β. Figures 4 and 5 recommend placing a strong electron-donating group at positions 9 and 11, and a strong electron-withdrawing group at position 6; however, in both cases an isophorone should be used to ring-lock positions 8-10 for adequate stability – the isophorone unit would slightly lower the β values given in Figure 4 (*13*).

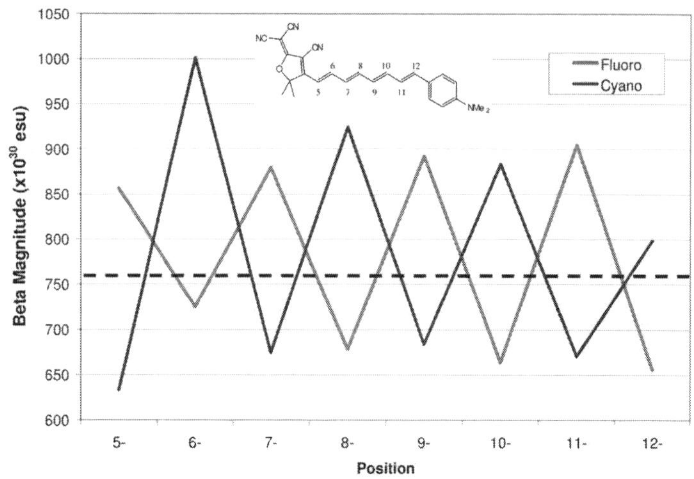

Figure 2. Calculated total magnitude of β versus position for the trans-polyene dye. Black dashed line is beta for the unsubstituted dye, 760 × 10^{-30} esu. The IUPAC name of this dye scaffold is 2-{3-cyano-4-[8-(4-(dimethylamino)phenyl)-octa-13,5,7-tetraenyl]-5,5-dimethyl-5H-furan-2-ylidene}-malononitrile. (see color insert)

It should be noted that the DFT-computed λ_{max} values in Figure 5 are blue-shifted compared to experimental results (about 20% lower), i.e., the λ_{max} of CLD actually lies between 640 and 670nm (*24*) depending on the solvent environment, compared to about 510nm predicted by this DFT calculation (using Time-Dependent DFT and the BH&B functional) that assumes the dye is in vacuum. However, the relative changes in predicted values should still provide valid comparisons.

Dipole Moment

Figure 6 shows that the shifts in ground-state dipole moment (μ_{00}) due to fluorine substitution are in the same direction found for cyano substitutions, which means fluorine is predicted to have an inductive withdrawing effect on the dipole moment, in contrast to its resonance donating effect on β and λ_{max}. Therefore, to improve dye alignment in highly loaded films by lowering the dipole moment, the recommendation is to place fluorine (or other electron-donating substituents) at odd-numbered positions (as this should also give a higher β).

The ground-state dipole moments calculated by DFT in the present study are probably larger than what would be experimentally observed (*25*). However, trends in the DFT calculations for the various substituents should still provide relatively valid comparisons.

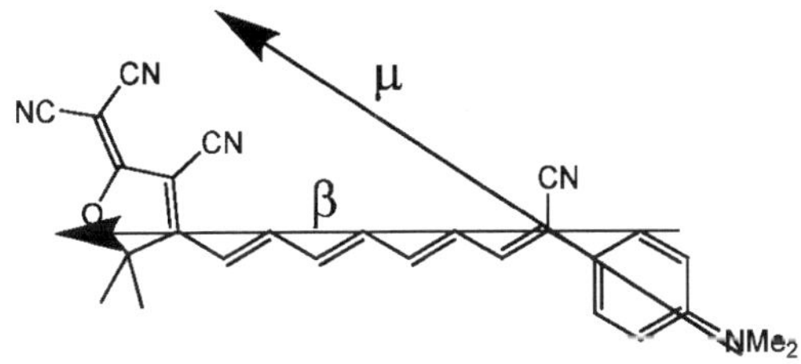

Figure 3. An example of diverging vectors of β and μ (a 33° angle in this case was calculated by DFT).

Background and Results for a Few Thiophenic Dyes

In general, thiophenylene in large π-scaffolds contributes good chemical stability at high temperatures (*26*, *27*) and also good photo-chemical stability (*28*, *29*). The enhanced stability is likely due to the degree of aromatic stabilization in the thiophenylene ring, which may also reduce the ground-state polarization compared to a *trans*-polyene. Enhanced stability was the motivation to further investigate new thiophenic dyes.

Ring-locked polyenes have been shown to have enhanced stability compared to bare polyenes (*30*). It could be said that thiophene has two ring-locked enes held in a *Z* conformation by the sulfur bonds. Thus, for internally-substituted thiophenic π-scaffolds, the question arose: would DFT computations reveal the same odd-even pattern (Dewar's rules) predicted for the polyenic π-scaffold? Breitung *et al.* (*31*) predicted an affirmative answer to this question using ZINDO computations in their study on orientational effects of thiazolene in push-pull dyes containing two-rings; they revealed that one thiazolene orientation was beneficial to β ("matched") and the other was detrimental ("mismatched") – essentially in accord with the beneficial Dewar's-rule {eW/oD} pattern. Ten years before that, Wong *et al.* (*32*) showed for several two-ring push-pull dyes that replacing the phenylene ring with a thiophene ring enhanced the β. These authors also found a gradient effect; namely, thiophene has a stronger positive effect on β when attached to the dialkylamino end of the dye. Also pertinent to the gradient effect, Jen *et al.* (*33*, *34*) showed that placing electron-withdrawing thiazolene next to the electron-withdrawing end of the dye has a strong enhancing effect on β.

The present DFT study analyzed the effects of various push-pull substituents on dyes containing three thiophenes within the π-scaffolds.

Interior Substitutions on Dyes Containing Thiophenylenes and Vinlyenes

The first thiophenic dye of the present study used a vinylene between each thiophenylene ring (Figure 7). Here again, Dewar's "starred" positions are the odd-numbered positions. The average β_μ calculated for this class of dye was about 20% smaller than that of the polyenic dye discussed above even though this thiophenylene-vinylene π-scaffold has only 11% fewer π electrons. Evidently, the

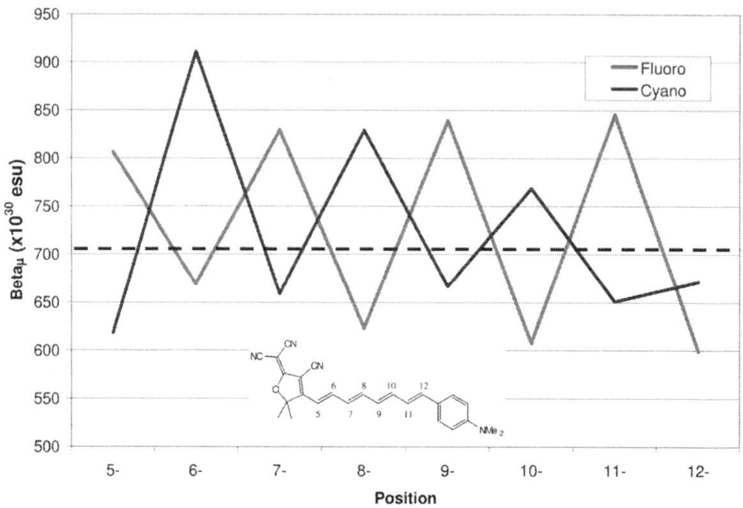

Figure 4. Calculated β_μ versus methine position for fluorine and cyano mono-substitutions. The black dashed line is β_μ for the unsubstituted dye. Taken from reference (13). (see color insert)

aromatic stabilization of the thiophenylenes reduced the polarizability. [Note: the results for the total magnitude of β are in the Appendix, Table 2.]

Overall, electron-donating substituents at the odd-numbered positions 5, 9 and 11 are predicted to increase β_μ, as are electron-withdrawing substituents at the even-numbered positions 2 and 6 (in accord with Dewar's {eW/oD} pattern). Looking from one end of the dye to the other, DFT predicts a strong gradient effect making it unprofitable to replace hydrogen with either of these substituents in the #12 vinylene position and the #14 and #15 thiophenylene positions.

As for the vinylenes, fluorine on the 5 and 11 positions is predicted to give the largest increase in β_μ (22% and 17% increases, respectively). As for the rings, a central thiazolene oriented with its azomethine in the 8 position and a central thiophenylene with fluorine on the 9 position should give significant increases in β_μ (which is in accord with Dewar's rules). Substituting fluorine for hydrogen in position 3 would also give a significant increase in β_μ.

Absorption Maximum, λ_{max}

Figure 8 shows that for fluorine on vinylene position 11, DFT predicts a zero increase in λ_{max} (in opposition to Dewar's rules). Hence, fluorine (and presumably other donors) at position 11 would have the most attractive combination of λ_{max} and β_μ. DFT also predicts that azomethine at position 8 should give an attractive combination of λ_{max} and β_μ.

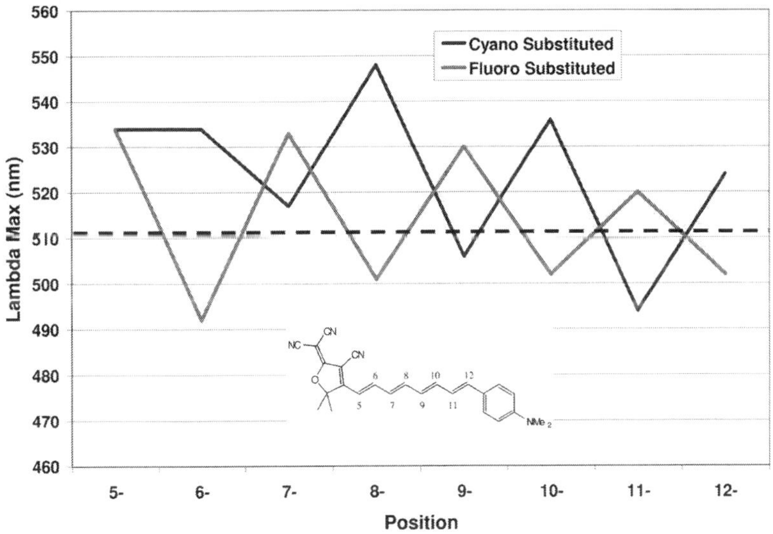

Figure 5. Calculated λ_{max} versus position for the trans-polyene dye. The black dashed line is the calculated λ_{max} for the unsubstituted dye (512 nm). The BH&B functional was used for the λ_{max} calculation. (see color insert)

Interior Substitutions on the *tris*thiophenylene Dye

The next dye scaffold considered was 12-dimethyamino-1-nitro-*tris*thiophene (Figure 9). The *tris*thiophene π-scaffold should be even more resistant to photo-thermal-chemical degradation, but the average value of β_μ for this class of dye was predicted to be half that of the previous class of dye (even though all three rings were calculated to be coplanar). [Note: the results for the total magnitude of β are in the Appendix, Table 2.]

Figure 9 shows that fluorine is highly recommended for the 3 and 7 positions. Fluorine benefited those same {eW/oD} ring positions for the thiophenylene-vinylene dye in Figure 7, but the percent increase in β_μ is predicted to be even larger for the *tris*thiophene π-scaffold. For fluorine in position 11, the reduced β_μ is not due to modification of the dipole vector, but perhaps due to an inductive withdrawing effect of fluorine. Cyano in the 2 position gives the largest increase in β_μ; and, as expected the λ_{max} is red-shifted (Figure 10). For electron withdrawing substituents at the other even-numbered positions, the gradient effect should completely overpower Dewar's {eW/oD} pattern and are not recommended.

The same strong gradient effect in λ_{max} is predicted for the electron-withdrawing substituents as was seen for β_μ. The electronic absorption tail at 1550nm for this class of dye is expected to be negligible. Therefore, even though fluorine at positions 3 or 7 would give a considerable red-shift, the large increases in β_μ would recommend these positions for electron-donating substituents (and cyano at position 2 is recommended).

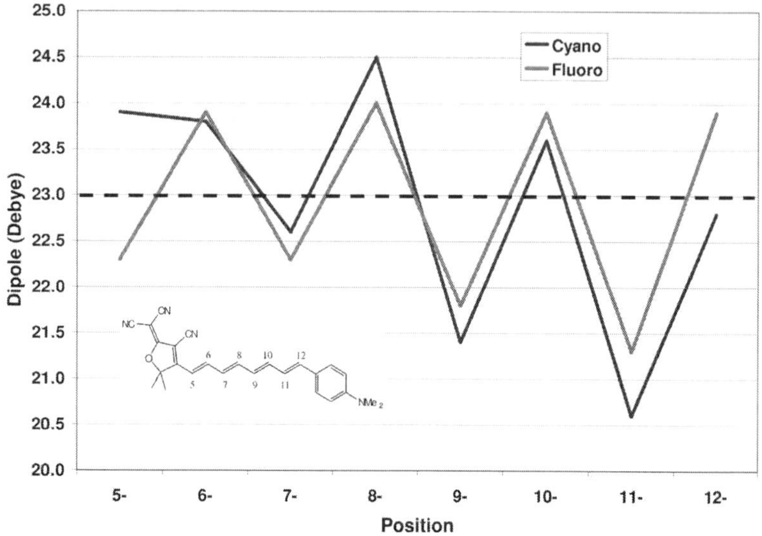

Figure 6. Calculated ground state dipole moment versus position for the trans-polyene (black dashed line is the dipole for the unsubstituted dye). (see color insert)

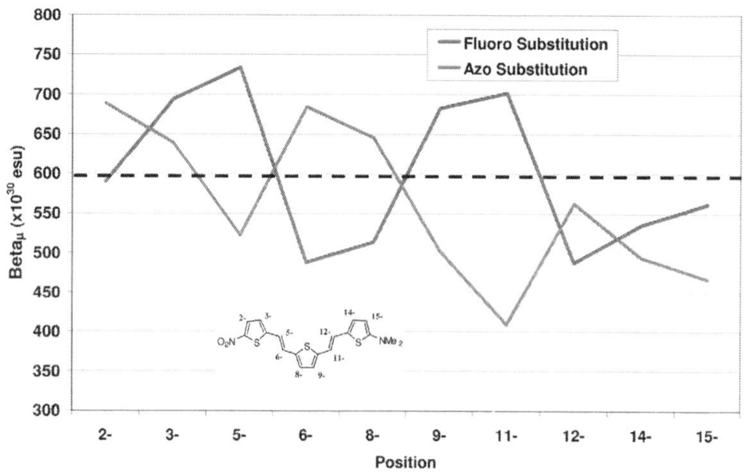

Figure 7. The calculated β_μ versus position for fluorine and azo mono-substituted dimethyl-[5-(2-{5-[2-(5-nitro-thiophen-2-yl)-vinyl]-thiophen-2-yl}-vinyl)-thiophen-2-yl]-amine. The black dashed line is β_μ for the bare dye (600×10^{-30} esu). (see color insert)

Modeling Procedure

All β and dipole calculations were performed using the DFT functional PBE1PBE with the basis set 6-31G(d,p) in the program Gaussian03 (*35*). Gaussian03 is based on the Taylors series convention. All calculations are for

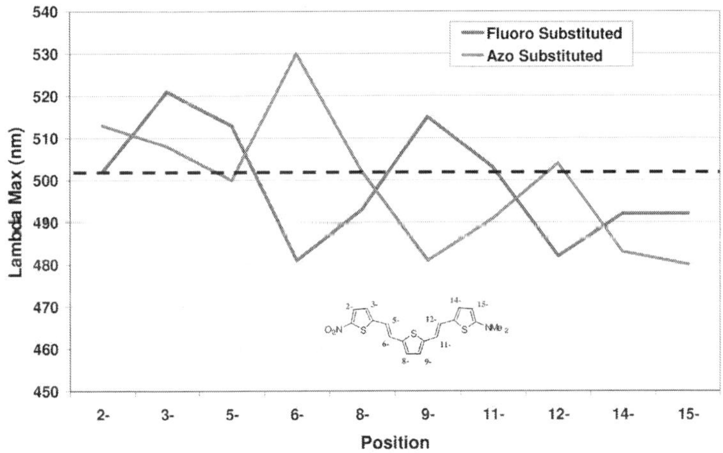

Figure 8. Calculated λ_{max} versus position for fluoro and azo mono-substituted dimethyl-[5-(2-{5-[2-(5-nitro-thiophen-2-yl)-vinyl]-thiophen-2-yl}-vinyl)-thiophen-2-yl]-anime. The black dashed line is λ_{max} for the bare dye (502 nm). The BH&B functional was used for the λ_{max} calculation. (see color insert)

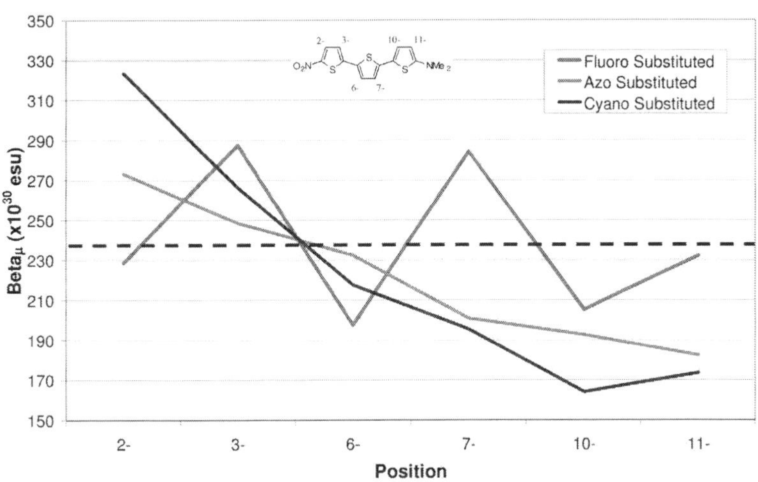

Figure 9. DFT-calculated β versus position for fluoro, azo and cyano mono-substitutions on 12-dimethylamino-1-nitrotristhiophene. The black dashed line is β_μ for the bare dye (237×10^{-30} esu). (see color insert)

the gas phase (no solvation interactions were used). The geometry for all dyes was fully optimized before the static first hyperpolarizability (β) was calculated using the "polar=cubic" keyword. The irreducible part of the β tensor, which is the β vector, $\vec{\beta}$, was calculated according to equation 2 and its total magnitude according to equation 3:

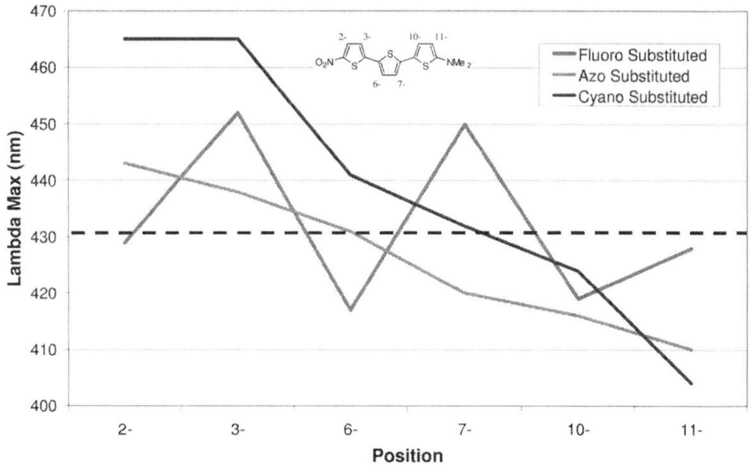

Figure 10. Calculated λ_{max} versus position for fluoro, azo and cyano substituted 12-dimethylamino-1-nitrotristhiophene. The black dashed line is λ_{max} for the bare dye (431 nm). The BH&B functional was used for the λ_{max} calculation. (see color insert)

$$\beta_i = \tfrac{1}{3}\sum_{j=1}^{3}\left(\beta_{ijj} + \beta_{jij} + \beta_{jji}\right) \tag{2}$$

$$\beta \equiv \left|\vec{\beta}\right| = \sqrt{\sum_{i=1}^{3}\beta_i^2} \tag{3}$$

The results of these calculations were found to be comparable to other DFT functionals, such as B3LYP. The value of β along the molecular dipole direction, β_μ, was calculated by equation 4, where μ_i is a component of the ground-state dipole moment vector of the dye ($\vec{\mu}$):

$$\beta_\mu = \frac{\sum_{i=1}^{3}(\beta_i \cdot \mu_i)}{\sqrt{\sum_{i=1}^{3}\mu_i^2}} \tag{4}$$

The λ_{max} calculations were performed by Time-Dependent DFT using the the 6-31G(d,p) basis set and BH&H functional, which tended to give less blue-shifted values than the PBE1PBE DFT functional.

The calculation of static β using various DFT methods has recently been reviewed (*18*). The DFT calculation usually gives larger values for β_μ and μ and smaller values of λ_{max} than has been found experimentally, however it provides an accurate relative ranking of similar dyes. The calculations can be performed overnight on desk-top computers.

Conclusions

For the dyes investigated, when placing electron-donating (D) and electron-withdrawing (W) substituents along the internal portion of the π-electron scaffolds, DFT predicted two distinct property-changing patterns: the {eW/oD} and {oW/eD} patterns, where e and o stand for even- and odd-numbered methine positions substituted with W and D groups. These two patterns usually changed the first-hyperpolarizability (β_μ) in opposite directions, which is in accord with Dewar's rules. Most often, for mono-substitutions (relative to hydrogen), DFT predicted the {eW/oD} pattern would increase β_μ and give a bathochromic (red) shift to the absorption spectra. The fluorine substituent acted as an electron-donating group (resonantly) in most cases. For the tetraene π-scaffold in a CLD-type dye, about a 19% increase in β_μ was predicted by DFT when adding the electron-donating fluorine in place of hydrogen at the #9 or #11 methine positions, and a 28% increase in β_μ was predicted when adding the electron-withdrawing cyano at the #6 methine position – all these substitutions were calculated to give relatively small increases in λ_{max}. Dewar's rules were also observed for substituents on the dimethylamino-(thiophenylene-vinylene)$_2$-thiophenylene-nitro π-scaffold, for which recommended substitutions were fluorine at position 11 (predicted to have essentially no increase in λ_{max} and a 17% increase in β_μ), and a central thiazolene ring with its azomethine in the 8 position was predicted to give similar results. A gradient effect was predicted for both of these π-scaffolds wherein DFT showed the Dewar patterns would have a stronger effect on β_μ near the ends of the π-scaffold. For the third class of dye investigated, which contained the *tris*thiophene π-scaffold, Dewar's-rule changes in β_μ were generally observed; however, the only substitutions recommended were fluorine for the #3 and #7 methine positions, and cyano for the #2 position.

Acknowledgments

The authors would like to thank the Laboratory for Physical Sciences and DARPA-MTO for partial support of this study.

Appendix

Table 1. DFT-calculated total magnitude of β (x 10^{30} esu) for fluorine and azomethine mono-substituted dimethyl-[5-(2-{5-[2-(5-nitro-thiophen-2-yl)-vinyl]-thiophen-2-yl}-vinyl)-thiophen-2-yl]-anime (structure shown in Figure 7)

Position	*Azo*	*Fluoro*
Unsubstituted ->	*(600.4)*	*(600.4)*
2-	690.3	590.6
3-	640.8	693.8
5-	523.2	735.0
6-	697.8	491.5
8-	666.4	520.2
9-	519.0	696.3
11-	416.1	712.4
12-	562.9	488.4
14-	494.9	535.7
15-	467.1	562.5

Table 2. DFT calculated magnitude of total β (x 10^{30} esu) for 12-dimethylamino-1-nitro*tris*thiophene (structure shown in Figure 9)

Position	*Azo*	*Cyano*	*Fluoro*
Unsubstituted ->	*(238.5)*	*(238.5)*	*(238.5)*
2-	273.8	325.6	228.9
3-	249.0	271.6	288.0
6-	239.7	230.8	200.0
7-	207.1	222.2	290.7
10-	192.8	169.3	205.3
11-	182.4	193.1	232.7

References

1. Luo, J.; Huang, S.; Cheng, Y.-J.; Kim, T.-D.; Shi, Z.; Zhou, X.-H.; Jen, A. K.-Y. *Org. Lett.* **2007**, *9*, 4471–4474.

2. Cheng, Y.-J.; Luo, J.; Huang, S.; Zhou, X.; Shi, Z.; Kim, T.-D.; Bale, D. H.; Takahashi, S.; Yick, A.; Polishak, B. M.; Jang, S.-H.; Dalton, L. R.; Reid, P. J.; Steier, W. H.; Jen, A. K.-Y. *Chem. Mater.* **2008**, *20*, 5047–5054.
3. Schmidt, K.; Barlow, S.; Leclercq, A.; Zojer, E.; Jang, S.-H.; Marder, S. R.; Jen, A. K.-Y.; Brédas, J.-L. *J. Mater. Chem.* **2007**, *17*, 2944–2949.
4. (a) Bosshard, Ch.; Bösch, M.; Laikatas, I.; Jäger, M.; Günter, P. In *Nonlinear Optical Effects and Materials*; Günter, P., Ed.; Springer-Verlag: Berlin, Germany, 2000; pp 163–248. (b) Kwon, S.-J.; Kwon, O-P.; Seo, J.-I.; Jazbinsek, M.; Mutter, L.; Gramlich, V.; Lee, Y.-S.; Yun, H.; Günter, P. *J. Phys. Chem. C* **2008**, *112*, 7846–7852.
5. Ashley, P. R.; Temmen, M. G.; Diffey, W. M.; Sanghadasa, M.; Bramson, M. D.; Lindsay, G. A.; Guenthner, A. J. *Proc. of SPIE* **2006**, *6314*, 63140J-1–11.
6. Baehr-Jones, T. W.; Hochberg, M. J. *J. Phys. Chem. C* **2008**, *112*, 8085–8090.
7. Kim, T. D.; Luo, J.; Cheng, Y.-J.; Shi, Z; Hau, S.; Jang, S.-H.; Xing-Zhou, H.; Tian, Y.; Polishak, B.; Huang, S.; Ma, H.; Dalton, L. R.; Jen, A. K.-Y. *J. Phys. Chem. C* **2008**, *112*, 8091–8098.
8. Nicoud; J. F.; Twieg, R. J. Nonlinear Optical Properties of Organic Molecules and Crystals. In *Quantum Electronics—Principles and Application*; Chemla, D. S., Zyss, J., Eds.; Academic: New York, 1987; Vol. 1, pp 227–296.
9. Azomethine is an SP^2 nitrogen (–N=), and it is electronegative (usually withdrawing) relative to a carbon methine (–CH=).
10. (a) Dewar, M. J. S. *J. Chem. Soc.* **1950**, 2329–2334. (b) Dewar, M. J. S. *Chem. Soc.* **1956**, Special Publication No. 4, 64–83.
11. Tsunekawa, T.; Gotoh, T.; Iwamoto, M. *Chem. Phys. Lett.* **1990**, *166*, 353–357.
12. Tsunekawa, T.; Yamaguchi, K. *J. Phys. Chem* **1992**, *96*, 10268–10275.
13. Chafin, A. P.; Lindsay, G. A. *J. Phys. Chem. C* **2008**, *112*, 7829–7835.
14. (a) Oudar, J. L.; Chemla, D. S. *J. Chem. Phys.* **1977**, *66*, 2664–2668. (b) Oudar, J. L. *J. Chem. Phys.* **1977**, *67*, 446–457.
15. (a) Tiemann, B. G.; Marder, S. R.; Perry, J. W.; Cheng, L.-T. *Chem. Mater.* **1990**, *2*, 690–695. (b) Marder, S. R.; Beratan, D. N.; Cheng, L.-T. *Science* **1991**, *252*, 103–106.
16. Gregory, P.; Thorp, D. *J. Chem. Soc., Perkin Trans. 1* **1979**, 1990–2000.
17. Meyers, F.; Marder, S. R.; Pierce, B. M.; Brédas, J. L. *J. Am. Chem. Soc.* **1994**, *116*, 10703–10714.
18. Isborn, C. M.; Leclercq, A.; Vila, F. D.; Dalton, L. R.; Brédas, J. L.; Eichinger, B. E.; Robinson, B. H. *J. Phys. Chem. A* **2007**, *111*, 1319–1327.
19. Shu, C.-F.; Tsai, W. J.; Chen, J.-Y.; Jen, K.-Y.; Zhang, Y.; Chen, T.-A. *Chem. Commun.* **1996**, 2279–2280.
20. (a) Robinson, B. H.; Dalton, L. R.; Harper, A. W.; Ren, A.; Wang, F.; Zhang, C.; Todorova, G.; Lee, M.; Aniszfeld, R.; Garner, S.; Chen, A.; Steier, W. H.; Houbrecht, S.; Persoons, A.; Ledoux, I.; Zyss, J.; Jen, A. K.-Y. *Chem. Phys.* **1999**, *245*, 35–50. (b) Lee, S.-S.; Garner, S. M.; Chuyanov, V.; Zhang, H.; Steier, W. H.; Wang, F.; Dalton, L. R.; Udupa, A. H.; Fetterman, H. R. *IEEE J. Quantum Electron.* **2000**, *36*, 527–532. (c) Zhang, C.; Dalton, L. R.; Oh, M.-C.; Zhang, H.; Steier, W. H. *Chem. Mater.* **2001**, *13*, 3043–3050.

21. Liu, S.; Haller, M. A.; Ma, H.; Dalton, L. R.; Jang, S.-H.; Jen, A. K.-Y. *Adv. Mater.* **2003**, *15*, 603–607.
22. He, M.; Leslie, T. M.; Sinicropi, J. A. *Chem. Mater.* **2002**, *14*, 2393–2400.
23. Hammond, S. R.; Clot, O.; Firestone, K. A.; Bale, D. H.; Lao, D.; Haller, M.; Phelan, G. D.; Carlson, B.; Jen, A. K.-Y.; Reid, P. J.; Dalton, L. R. *Chem. Mater.* **2008**, *20*, 3425–3434.
24. Guenthner, A. J.; Wright, M. E.; Fallis, S.; Lindsay, G. A.; Petteys, B. J.; Yandek, G. R.; Zang, D.-Y.; Sanghadasa, M.; Ashley, P. R. *Proc. SPIE* **2006**, *6331*, 63310M-1–11.
25. Champagne, B.; Perpète, E. A.; Jacquemin, D.; van Gisbergen, S. J. A.; Baerends, E.-J.; Soubra-Ghaoui, C.; Robins, K. A.; Kirtman, B. *J. Phys. Chem. A* **2000**, *104*, 4755–4763.
26. Wang, F. Ph.D. Thesis, University of Southern California, Los Angeles, CA, 1998.
27. Lindsay, G. A.; Guenthner, A. J.; Wright, M. E.; Sanghadasa, M.; Ashley, P. R. *Polymer* **2007**, *48*, 6605–6616.
28. DeRosa, M. E.; He, M.; Cites, J. S.; Garner, S. M.; Tang, Y. R. *J. Phys. Chem. B* **2004**, *108*, 8725–8730.
29. Rezzonico, D.; Jazbinsek, M.; Günter, P.; Bosshard, C.; Bale, D.H.; Liao, Y.; Dalton, L.R.; Reid, P.J. *J. Opt. Soc. Am. B* **2007**, *24*, 2199–2207.
30. (a) Shu, C.-F.; Shu, Y.-C.; Gong, Z.-H.; Peng, S.-M.; Lee, G.-H.; Jen, A. K.-Y. *Chem. Mater.* **1998**, *10*, 3284−3286. (b) Shu, Y.-C.; Gong, Z.-H.; Shu, C.-F.; Breitung, E. M.; McMahon, R. J.; Lee, G.-H.; Jen, A. K.-Y. *Chem. Mater.* **1999**, *11*, 1628−3632.
31. Breitung, E. M.; Shu, C.-F.; McMahon, R. J. *J. Am. Chem. Soc.* **2000**, *122*, 1154–1160.
32. Wong, K. Y.; Jen, A. K-Y.; Rao, V. P.; Drost, K. J. *J. Chem. Phys.* **1994**, *100*, 6818–6825.
33. Varanasi, P. R.; Jen, A. K.-Y.; Chandrasekhar, J.; Namboothiri, N. N.; Rathna, A. *J. Am. Chem. Soc.* **1996**, *118*, 12443–12448.
34. Leclercq, A.; Zojer, E.; Jang, S.-H.; Barlow, S.; Geskin, V.; Jen, A. K.-Y.; Marder, S. R.; Brédas, J. L. *J. Chem. Phys.* **2006**, *124*, 044510/1–044510/8.
35. Frisch, M. J. et al. Gaussian 03, revision C.02; Gaussian, Inc.: Wallingford, CT, 2004.

Chapter 4

Molecular Design and Supramolecular Organization of Highly Efficient Nonlinear Optical Chromophores for Exceptional Electro-Optic Properties

Xing-Hua Zhou,[1] Jingdong Luo,[1] Tae-Dong Kim,[1] Sei-Hum Jang,[1] René M. Overney,[2] Larry R. Dalton,[3] and Alex K.-Y. Jen[*,1,3]

[1]Department of Materials Science and Engineering, University of Washington, Seattle, WA 98195
[2]Department of Chemical Engineering, University of Washington, Seattle, WA 98195
[3]Department of Chemistry, University of Washington, Seattle, WA 98195
[*]ajen@u.washington.edu

Arene-perfluoroarene stacking interactions are explored in a new series of dendronized nonlinear optical (NLO) chromophores to create extended network structures and overcome intermolecular electrostatic interactions for improving their poling efficiency and alignment stability. These supramolecular self-assembled molecular glasses showed ultrahigh electro-optic (EO) activity (r_{33} > 300 pm/V) and excellent thermal stability (at 85 °C). These unprecedented EO activities could inspire the vigorous development of new integrated optical devices.

Organic and polymeric electro-optic (EO) materials have been intensively studied for several decades due to their potential applications for the high speed/broad bandwidth information technologies (*1–3*), terahertz radiation generation and detection (*4*, *5*), integrated circuitry, and multifunctional micro- and nanodevices (*6–8*). Their development is motivated not only by the large nonlinear optical (NLO) susceptibilities, but also by the versatility, ease of processability, and possibility of tailoring the physicochemical properties through molecular engineering. Recently, significant advances (*9*, *10*) in the design and synthesis of new-generation organic EO materials have led to the successful demonstration of ultralarge EO coefficients (r_{33} values) of greater than 300 pm/V. This breakthrough result allowed organic NLO materials to significantly surpass the EO properties of lithium niobate, the dominant inorganic material currently used in modern EO modulator technology. These materials with such high EO activities and unprecedented performance in conventional device formats provide a powerful platform for building innovative devices such as hybrid systems using EO polymers integrated with silicon photonics (*11*, *12*).

Introduction

It is well-known that second-order NLO properties originate from the non-centosymmetric alignment of NLO chromophores, conventionally either doped as a guest or covalently bound as side chains in poled polymers. For the fabrication of practical EO devices, several material requirements need to be optimized simultaneously: (1) the design and synthesis of chromophores with high-molecular nonlinearity along with efficient translation of microscopic to macroscopic optical nonlinearity, (2) long-term stability of polar order, (3) minimal optical loss at operation wavelengths, and (4) good mechanical strength and processability for multilayer fabrication.

In the past decade, the search for organic EO materials with very large r_{33} values has led to extensive efforts (*13–16*) in exploring "push-pull" type chormophores with high dipole moment (μ) and molecular hyperpolarizability(β). However, the most challenging problem in the optimization of organic EO materials lies in the translation of exceptional molecular nonlinearity into macroscopic EO activity by poling-induced acentric ordering of the dipolar chromophores. Experimental and theoretical studies (*17–19*) have shown strong intermolecular electrostatic interaction among high-μ chromophores particularly at high concentrations to be the major factor defining the EO activity of electric field-poled organic materials. These studies also suggest that the maximum realizable EO activity could be enhanced by controlling the chromophore shape to prevent stacking.

In our continuous pursuit of efficient organic EO materials, our major efforts have focused on developing shape-engineered chromophores with high-$\mu\beta$ and increasing polar order through careful control of the nanoscale architecture of macromolecules (*20*, *21*). Owning to the unique characteristics of dendrimers, such as globular shape, core encapsulation, structural uniformity, and high end-group functionality, dendritic chromophore shape modification represents a powerful approach for molecular and supramolecular engineering to efficiently improve poling efficiency for realization of large EO activity (*22–26*). Dendritic EO materials are also able to introduce other auxiliary properties, including thermal and chemical stability, optical transparency, and compatibility in composite materials. For example, in comparison with conventional guest-host EO polymers, improved poling efficiency and excellent temperal alignment stability were demonstrated with several dendritic systems. In particular, multichromophore dendrimers and dendronized chromophores showed improved poling efficiency and excellent temperal alignment stability through a novel lattice-hardening approach using a sequential poling and cross-linking process (*22*, *27–29*).

Molecular engineering of chromophores and polymers can be used to modulate the strong electrostatic interactions between chromophores and improve poling efficiency of organic E-O materials. However, such modification of molecular structures can add significant passive weight to the materials and effectively lower the E-O activities of organic E-O materials. Additionally, some unique processes during electric field poling, so-called in situ crosslinking and postfunctionalization, add a significant complication to both molecule design and the poling processing for these highly efficient dendritic EO materials (*27–30*). Both of these factors limit the orientational flexibility of chromophores to some extent.

In this contribution, we present our recent progress in a new generation of dendronized EO chromophores as supramolecular self-assembled molecular glasses (*9*, *31*). Ultralarge EO activity and enhanced thermal stability are achieved simultaneously through the careful manipulation and organization of intricate supramolecular interactive forces for self-assembly. Reversible and fine-tuned supramolecular interactions are introduced to create extended network structure, overcome intermolecular electrostatic interaction and improve their poling efficiency and stability. The success of these material developments could inspire the exploration of new device concepts trying to take full advantage of the organic EO materials with ultrahigh r_{33} values.

Supramolecular Self-Assembly of Dendronized NLO Chromophores Directed by Arene-Perfluoroarene Interactions

The ability to employ supramolecular interactive forces in order to manipulate self-organized nanostructures and create desired functional properties is very important for practical applications. Among various non-covalent intermolecular interactions, arene-perfluoroarene (Ar^H-Ar^F) interaction (*32*, *33*), known as a strong face-to-face interaction between perfluorinated and nonfluorinated aromatic rings. The approach was recently widely utilized to control supramolecular organization in reversible media (*34–36*) and also to create tightly packed molecular assemblies for electronic and optoelectronic applications (*37*, *38*). Although some other non-covalent interactions were well investigated for the realization of supramolecular self-assembled structures for use in NLO applications (*39–42*), there are few reports concerning the use of the Ar^H-Ar^F interaction in EO materials to both assist dipole alignment during electric-filed poling and provide physical cross-linking for lattice hardening.

Recently, a series of systematically designed NLO chromophores were developed for the self-assembly (*9*) by our group to explore the Ar^H-Ar^F interaction. Our aim was to construct a reversible supramolecular self-assembly of NLO chromophores with improved material properties. The molecular structures of the chromophores in the study are shown in Figure 1. Both phenyl and pentafluorophenyl rings are incorporated diagonally as peripheral dendrons on the π-bridge and the donor-end of the chromophores **1**, **2**, and **3** to facilitate the formation of intermolecular Ar^H-Ar^F interactions.

The thermal properties of chromophores **1–3** were evaluated by differential scanning calorimetry (DSC) and shear-modulation force microscopy (SM-FM), a nanorheological characterization method (*43*) for the detection of structural transitions and thermally induced relaxations in constrained thin NLO films not obtainable by conventional means such as DSC. All dendronized chromophores showed a thermodynamically stable amorphous state and did not undergo crystallization or any other phase transition above T_g up to the decomposition temperature. A glass transition temperatures (T_g) of chromophores **1–3** was noticed in a DSC thermogram at 57 °C, 75 °C, and 76 °C, respectively. SM-FM analyses revealed that each of these glassy chromophores exhibited two thermal transitions, of which only the higher-temperature transition (T_2), corresponding to the main glass transition temperature of the chromophores could be observed by DSC (Table 1). The low-temperature relaxation (T_1) is ascribed to the cooperative local reorganizations in the systems (*44*). As expected from the Ar^H-Ar^F interactions, chromophore **3** shows a significantly increased T_1 and T_2 compared to those obtained from **1**. It is worthwhile to also note that these thermal transitions are also consistent with those of composite **5**, which were generated by blending chromophore **1** and **2** (1:1) to maximize the Ar^H-Ar^F interactions. Figure 2 shows the SM-FM curve of chromophore **3** along with its DSC thermogram for elucidation.

Structural insight of the Ar^H-Ar^F interactions in these chromophore glasses has been provided by a single crystal x-ray structure of dimeric co-crystal of the model compounds, phenyl group-terminated and pentafluorophenyl group-terminated dendron carboxylic acids (1:1). They form co-crystals in the triclinic *P-1* space group (No. 2). The crystal parameters are a = 8.054, b = 13.011, c = 18.507 Å, α = 77.022°, β = 79.146°, and γ = 92.596°, with two residues of each component in the asymmetric unit by alternating heterodimer stacks with an interplanar distance of 3.5 to 3.8 Å. A cylindrical bond representation viewed to reveal the Ar^H-Ar^F interaction of a dimer pair and a space filling structure of with head-to-tail alternating hetero-dimer stacks are shown in Figure 3. This structure represents a clear solid-state evidence of Ar^H-Ar^F interactions between peripheral dendron groups to form extended structures when the chromophores are substituted with a chemo-specific combination of dendrons.

The EO properties of these molecular glassy chromophores in thin films were evaluated after electric filed poling using a device configuration of ITO/Chromophore glasses/Au. EO activity was measured using the Teng-Man simple reflection technique (*45*, *46*) at a wavelength of 1.31 μm. Chromophore **3** gave the highest r_{33} (108 pm/V) among all monolithic molecular glasses. This value is more than 2 times those obtained from chromophores **1** or **2** that lack Ar^H-Ar^F interactions. Moreover, the composite **5** (1:1 blend of **1** and **2**) showed an r_{33} of 130 pm/V. The enhanced r_{33} in this composite indicates that it is also possible to form supramolecular self-assembly between complementary

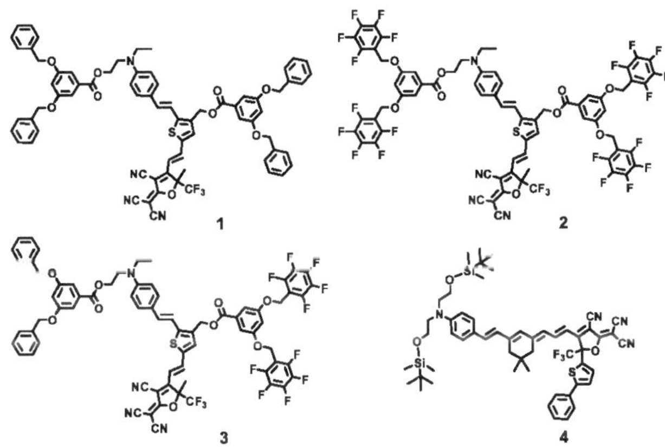

*Figure 1. Chemical structures of the studied chromophores **1–4** for supramolecular self-assembly.*

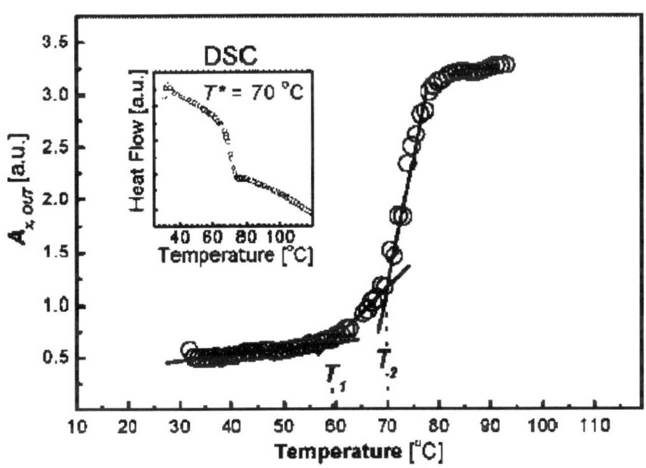

*Figure 2. SM-FM profile of chromophore **3** film on ITO exhibiting two transitions at T_1 and T_2. Inset: DSC thermogram of **3**.*

but different chromophores with extended networks. The application of high electric fields (greater than 100 V/μm) to pole chromophore **3** and composite **5** to their resultant high r_{33} values, also suggests improved dielectric strength in the materials. In contrast, the highest reported r_{33} value for a undendronized chromophore **AJL8**, with the same core stucture as that of **1**–**3**, doped (20 wt%) in a amorphous polycarbonate (APC) showed only 94 pm/V using a high poling field of 100~125 V/μm (*47*). These poled films of chromophore **3** and composite **5** could retain over 90 % of their original r_{33} values at room temperature for more than 2 years. On the contrary, the temporal stability of the glasses without the Ar^H-Ar^F interactions deteriorated dramatically within 1 month (Table 1). A graphical illustration of the poling process of the supramolecular self-assembly of chromophore **3** is shown in Figure 4.

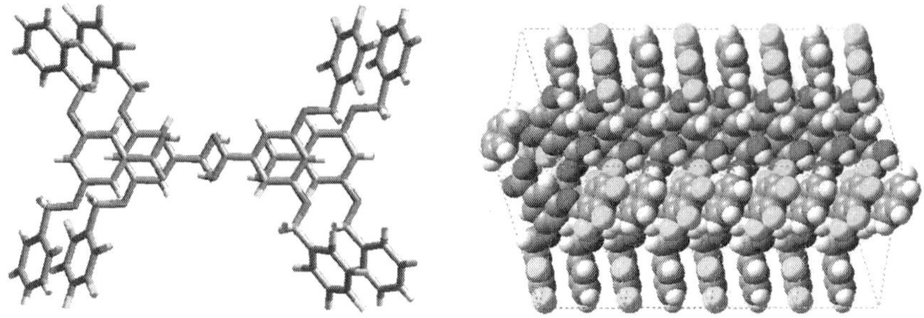

Figure 3. The structure of 1:1 cocrystal of phenyl dendron acid and pentafluorophenyl dendron acid: (a) a cylindrical bond representation viewed to reveal the Ar-Ar$_F$ interaction of a dimer pair (b) a space filling structure of with head-to-tail alternating hetero-dimer stacks.

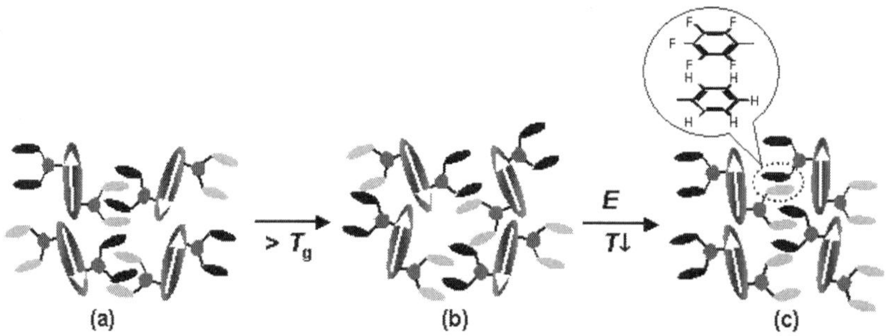

*Figure 4. Graphical illustration of the alignment formation of self-assembled chromophore **3** by Ar^H-Ar^F interactions: (a) locked random dipoles (shown as arrows) before poling (b) unlocked random dipoles before poling (c) locked acentric dipoles after poling followed by cooling.*

Although the chromophore number density (N = 3.8–5.0 × 10^{20} cm^{-3}) in these molecular glasses is already very high (around twice the maximum loading density in typical guest-host polymers), further attempts to enhance the r_{33} values of these materials were conducted by doping a second more efficient glassy NLO chromophore **4** in these glasses. By this approach, higher hyperpolarizability and loading level could be achieved simultaneously in the same materials. The resultant binary glass composite **6** (**3**/**4** = 3:1) exhibited an ultralarge r_{33} value of 275 pm/V. The applied electric filed as high as 90 V/μm suggested that the supramolecular self-assembly of host glass matrix **3** based on the Ar^H-Ar^F interactions is robust enough and was not disturbed by the addition of 25 wt% of guest chromophores. This results in little change of the desired dielectric properties of these self-assembled glasses. Even higher r_{33} values could be obtained from the higher-loading binary composite (**3**/**4** = 1:1), however the poling results were not easily repeatable due to their extremely high intrinsic conductivity.

Table 1. Physical and optimal EO properties of molecular glasses and polymers based on chromophore 1–4

Materials	M_n	*SM-FM*[a]		T_g[b]	λ_{max}[c]	*Number Density*[d]	*Poling Voltage*	r_{33}[e]	*Temp. Stab.*[f]
		T_1 *[°C]*	T_2 *[°C]*	*[°C]*	*[nm]*	*[×10*[20]*/cm*[3]*]*	*[V/μm]*	*[pm/V]*	*(%)*
1	1215	36	57	57	719	5.0	75	52	0
2	1575	47	69	75	689	3.8	75	51	85
3	1395	59	70	76	703	4.3	100	108	92
4[g]	963	–	101	100	803	1.6	100	150	86
5[g]	–	50	69	72	704	4.3	120	130	93
6[g]	–	62	76	75	727	4.8	90	275	86

[a] Thermal transition temperature measured by SM-FM under N_2 at the heating rate of 5 °C/min. [b] Glass transition temperature measured by DSC under N_2 at the heating rate of 10 °C/min. [c] Absorption maxima of thin films by UV/vis spectroscopy. [d] Core chromophoric moiety (formular of **1**–**3**: $C_{29}H_{25}F_3N_4OS$, molecular weight 534.6; formular of **4**: $C_{36}H_{30}F_3N_4O$, molecular weight 623.7) counted by total loading weight. [e] EO coefficient measured at 1310 nm by simple reflection technique. [f] Temporal alignment stability at RT after 3 months. [g] **4**: 25 wt% of **4** in a PMMA; **5**: composite of 1:1 mixture of **1** and **2**, **6**: composite of 3:1 mixture of **3** and **4**.

It is noteworthy that the E-O activities of these poled composites do not show the saturation maxima even with such high chromophore loading densities and can retain ~90% of their initial value for more than 500 hours at room temperature (Table 1). Interestingly, it was also found (*9*) that the r_{33} values (150 pm/V) of a guest-host polymer based on 25 wt% chromophore **4** in poly(methyl methacrylate) (PMMA, with much higher T_g of 100 °C) showed a 20 % decrease in temporal alignment stability (~60% of initial r_{33} value) at 50 °C in 25 hours compared to the binary composite **6** (~80% of initial r_{33} value). Such behavior is remarkably different from the commonly observed fast relaxation of poling induced polar order in a low T_g guest-host polymer.

The normalized analysis of EO activities showed that the binary glass composite **6** exhibited significantly enhanced r_{33} values far beyond the simple summation of the individual components of guest and host chromophores. The origin of this enhancement is still under investigation (*48*, *49*). To further understand these results, we applied the measurement of the order parameters based on linear optical dichroism of intramolecular charge transfer bands. These results indicate that the improved dipole alignment and polar order might play a role in the enhanced r_{33} values and improved poling efficiencies of the binary chromophore glass composites. The self-assembled chromophore matrices may act as pseudo-Ising lattices to help organize doped chromophores and improve their polar order (*50*). The physical and optical properties of molecular glasses and polymers based on chromophore **1**–**4** are summarized in Table 1.

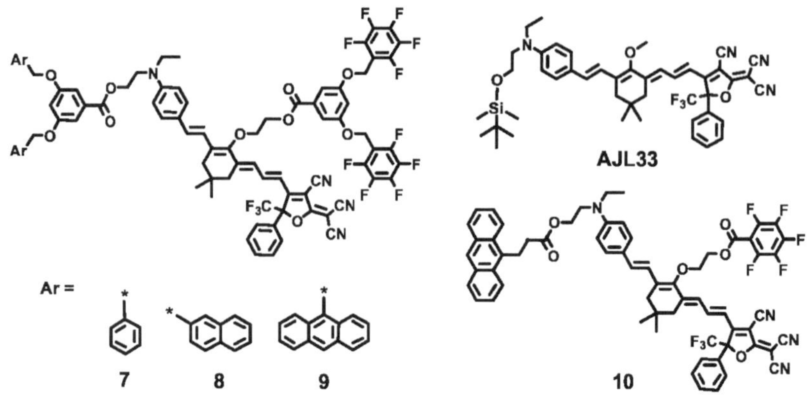

*Figure 5. Chemical structures the dendronized polyene chromophores **7–9** as well as other studied chromophores.*

Although ultrahigh EO activity and improved thermal stability were achieved using these supramolecular self-assembled dendronized chromophores glasses and their binary composites, relatively low glass transition properties of these materials render the materials inadequate for device application. As a result, further studies are needed to improve the thermal properties. Among the various approaches are increasing the strength of Ar^H-Ar^F interactions and the molecular weight of chromophores with more rigid structures. Another approach would entail the application of novel chemical crosslinking reactions to constrain the aligned chromophore glasses after poling. The incorporation of more efficient polyene chromophores, such as **4**, into the supermolecular self-assembled molecular glasses is another approach being pursued. Some of these approaches, molecular designs, and studies are discussed in the following sections.

Fine-Tuning of Arene-Perfluoroarene Interactions between Dendronized Polyene Chromophores for Ultralarge EO Activity and Enhanced Thermal Stability

To achieve ultrahigh EO activity, organic EO materials generally include a conjugated (4-dialkylamino)phenyltetraene bridge with terminal CF_3-TCF acceptors. This core design serves a primary class of high-$\mu\beta$ chromophores with reasonably high synthetic efficacy (*51*, *52*). Here, we develop a new series of supramolecular self-assembled dendronized NLO chromophores **7–9** derived from a highly efficient and chemically stable polyene chromophore (*52*) as shown in Figure 5. By the similar strategy of molecular design decribed before, complementary aryl/perfluoroaryl dendrons are introduced diagonally on the periphery allowing the desired intermolecular Ar^H-Ar^F interactions. Three different aromatic rings, including benzene, naphthalene, and anthracene, were previously used to demonstrate gradually increased Ar^H-Ar^F interactions with hexafluorobenzene (*53*). These ring systems were selected and applied to construct new aryl dendrons. Such structural variation provides the possibility of fine-tuning the Ar^H-Ar^F interaction involved in the supramolecular self-assembly to systematically investigate the effect of such interactions on EO activity and thermal stability.

As shown in Figure 6, in comparison with the highly crystalline undendronized chromophore, **AJL33**, all three dendronized chromophores of **7**–**9** also showed thermodynamically stable amorphous states and no other phase transition above T_g up to the decomposition temperature around 200 °C. This decomposition temperature was roughly 20 °C higher than that of **AJL33** suggesting improved thermal stability by encapsulation of the active chromophore moiety in aryl/perfluoroaryl dendrons (*54*). The glass transition temperatures (T_g) as measured by DSC of **7**–**9** were observed at 80 °C, 92 °C, and 124 °C, respectively. The increase of T_g was ascribed to their gradually increased structural rigidity and much stronger Ar^H-Ar^F interaction with the aromatic rings in the aryl-dendrons increasing from benzene to naphthalene to anthracene, with the highest T_g observed for chromophore **9** with the anthracene-containing dendron.

SM-FM revealed that each of these glassy chromophores also exhibited two thermal transitions, of which the higher-temperature transition (T_2) corresponded to the main glass transition temperature of the chromophores observed by DSC. The low-temperature transitions (T_1) were found to take place at temperatures of 59 °C, 75 °C, and 118 °C for **7**–**9**, respectively. By comparison of the T_1 between chromophore **7** and **1** with the same types of aryl/perfluoroaryl dendrons, it is interesting to note that the structure change of the core chromophore moiety hardly affected T_1, which indicates that the low-temperature transition of these dendronized chromophores was primarily associated with the relaxation of the interactive aryl/perfluoroaryl dendron pairs of the chromophore. The significant increase of T_1 from **7** to **9** was attributed to the steadily increasing cohesive energy of the Ar^H-Ar^F interactions and decreasing flexibility of aryl-dendrons due to the more rigid and bulky naphthalene and anthracene units, which could potentially lead to improvement of the temporal stability, especially for **9**.

The normalized UV-vis-NIR absorption spectra of chromophores **7**–**9** in thin films are shown in Figure 7. In comparison with the absorption in solutions, thin films of **7**–**9** showed significant broadening of the characteristic π–π^* charge-transfer band with a red-shifted peak at 790, 795, and 804 nm, respectively. A weak shoulder located in the range of 900–1000 nm was observed for all three chromophores. Such a dramatic difference in the absorption intensity found in the long-wavelength spectral region when changing from solution to solid has been widely observed in NLO chromophore-doped guest-host (G-H) materials (*55, 56*). This electronic absorption behavior has been attributed to the inhomogeneous broadening and solvatochromism of the electronic absorption characteristics associated with dye-dye and dye-polymer interactions. By comparing the absorption spectra of **7**–**9** with various aromatic rings from benzene, naphthalene, and anthracene incorporated in the peripheral aryl-dendron, a red shift of the absorption peak and spectral broadening, especially in the long-wavelength region, were also observed. This was most obvious for **9** even with the lowest content of the active chromophore moiety. Such absorption spectral difference may result from the more polarizable microenvironment around the central dipolar chromophore moiety provided by the more electron-rich anthracene-containing dendron (*28*).

To investigate the effect of the molecular structure on the electronic absorption in solid state, a comparative study of the absorption spectra was conducted between the dendritic chomophore **7** and a composite of a non-dendronized chromophore **AJL33** (50 wt%) blended with PMMA having a similar concentration of the active chromophore moiety. As shown in Figure 7, the monolithic glass of **7** exhibited a much narrower absorption spectrum with a λmax similar to the **AJL33**-PMMA blend system but a significantly blue-shifted (by about 37 nm) long-wavelength tail. This blue-shifted absorption tailing may contribute to a lower optical loss at the operation wavelengths for telecommunication (1.31 and 1.55 μm).

Poling conditions and measured r_{33} values of chromophores **7**–**9** from the optimal poling processes are summarized in Table 2. The ITO substrates have been carefully selected with low reflectivity and good transparency in order to minimize the measurement errors (*46*). High poling field strength (greater than 100 V/μm) could be applied to the films of the monolithic glasses of **7**–**9** despite the high number density of high-$\mu\beta$ chromophores. This is attributed to the good dielectric strength associated with the formation of the extended supramolecular network structure by the Ar^H-Ar^F interactions. An ultrahigh r_{33} value of 318 pm/V was achieved for chromophore **7**, which was almost 2 times higher than that of **3** with a different chromophore core. This unprecedented r_{33} value also represents a 50% improvement over those of binary chromophore composites based on **3** under the same processing and poling conditions. Chromophores **8** and **9** showed slightly lower r_{33} values of 280 and 249 pm/V, respectively. This series of dendronized chromophores showed dramatically increased EO activities compared to the maximum r_{33} value observed for the **AJL33**-doped PMMA composites (about 168 pm/V with a chromophore loading of 25 wt%). This result illustrated again that these self-assembled dendritic chromophores had significantly improved poling efficiencies at high chromophore number density far beyond the saturating chromophore loading of the guest-host polymers.

Furthermore, the temporal stability study of **7**–**9** reveals that chromophore **9** showed excellent alignment stability with ~90% of the original EO activity maintained even at 85 °C for over 300 hours after an initial fast decay. This excellent temporal stability is attributed to the significantly higher T_1 of **9** (118 °C). Thus, ultralarge EO activity and good temporal stability can be achieved simultaneously with this class of dendritic chromophores that are physically crosslinked through supramolecuar self-assembly by strong Ar^H-Ar^F interactions.

We also designed and synthesized an amorphous undendronized chromophore **10** with one anthracene ring on the donor-end and one pentafluorobenzene ring on the π-bridge as the complementary aryl/perfluoroaryl substituent. Although it has a high chromophore number density (5.3×10^{20} cm^{-3}), the poled films of this material only yielded an r_{33} of 207 pm/V. This relatively low r_{33} value is due to a reduced poling field at only 50 V/μm, dictated by the low dielectric strength of the material. Lower dielectric strength arises from dramatically reduced Ar^H-Ar^F interactions and higher intrinsic conductivity in comparison with the dendronized chromophores **7**–**9**.

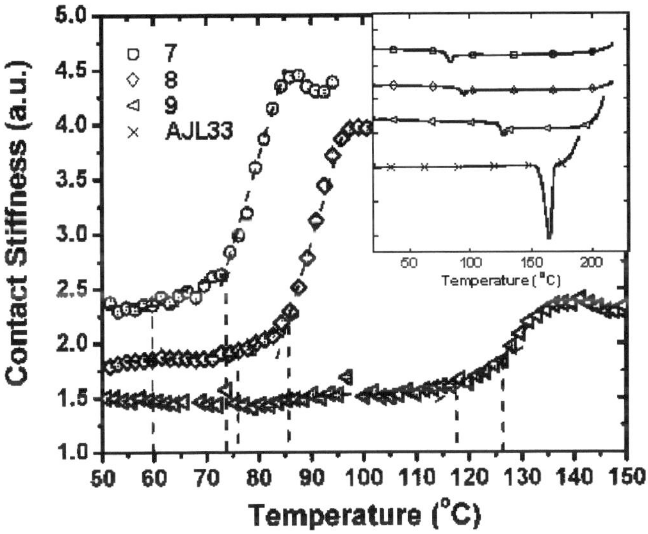

*Figure 6. SM-FM curves for chromophores **7–9** measured at a heating rate of 1°C/min in nitrogen. Inset: DSC curves of chromophores **7–9** and **AJL33** with a heating rate of 10 °C/min in nitrogen.*

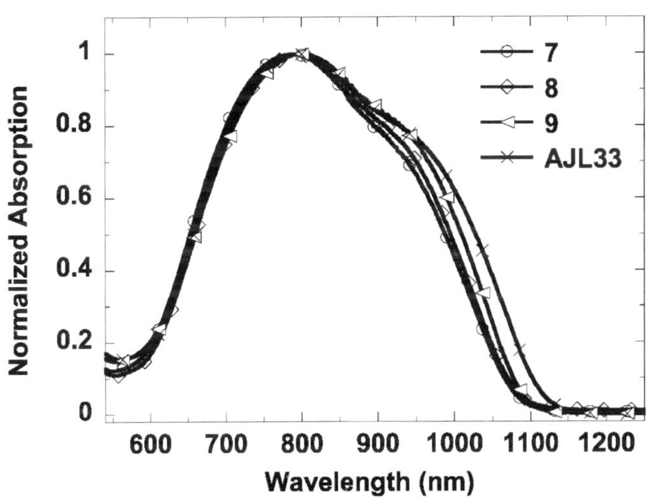

*Figure 7. Normalized absorption spectra of chromophores **7–9** in thin films. The absorption spectra of 50 wt% of an undendronized chromophore **AJL33** blended with PMMA is also shown for comparison.*

By way of summarizing the poling results of all the systems studied above, the variation of the EO activity normalized to the poling field strength, r_{33}^{max}/V_p, i.e. the poling efficiency, is shown in Figure 8 as a function of the chromophore number density (N). The increase in poling efficiency displayed by **7** ($N = 4.0 \times 10^{20}$ cm^{-3})

and **10** ($N = 5.3 \times 10^{20}$ cm^{-3}) samples was found to be approximately proportional to the chromophore number density even in such high regions. This is relatively consistent with medium-loading **AJL33**-doped PMMA composites ($N \sim 2.0 \times 10^{20}$ cm^{-3}). A similar linear relationship between chromophore number density and EO activity at high chromophore densities was also recently reported from both experimental results and theoretical analysis using Monte Carlo modeling in the study of multichromophore dendrimers (*26*). In contrast, the poling efficiency of **8** and **9** dropped much faster than a linear decrease in chromophore number density would suggest. The modelling suggests that the increased cohesive energy of the Ar^H-Ar^F interactions and decreased flexibility of aryl-dendrons when varying from benzene to anthracene units may result in the attenuation of poling efficiency of the chromophore. Further enhancement of EO activity could be expected upon doping a next-generation guest chromophore, which is needed to be much more efficient than polyene ones, into this new series of self-assembled polyene chromophore glasses.

Table 2. Physical and optimal EO properties of chromophores 7–10 in comparison with AJL33

Materials	M_n	*SM-FM*[a]		T_g[b] *[°C]*	λ_{max}[c] *[nm]*	*Number Density*[d] *[×10²⁰/cm³]*	*Poling Voltage [V/μm]*	r_{33}[e] *[pm/V]*
		T_1 *[°C]*	T_2 *[°C]*					
7	1509	59±3	75±2	80	790	4.0	107	318
8	1609	75±2	85±2	92	795	3.7	110	280
9	1710	118±2	125±2	124	804	3.5	110	249
10	1123	–	–	90	792	5.3	50	207
AJL33	781	–	–	–	796[f]	1.9[g]	100[g]	168[g]

[a] Thermal transition temperature measured by SM-FM under N_2 at a heating rate of 1 °C/min. [b] Glass transition temperature measured by DSC under N_2 at the heating rate of 10 °C min^{-1}. [c] Absorption maxima of thin films by UV-vis spectroscopy. [d] Core chromophore moiety (formular of **7–10** and **AJL33**: $C_{38}H_{33}F_3N_4O$, molecular weight 618.7) [e] EO coefficient measured at 1310 nm by simple reflection technique. [f] Values from composite of 1:1 mixture of **AJL33** and PMMA. [g] Values for 25 wt % of **AJL33** doped in PMMA.

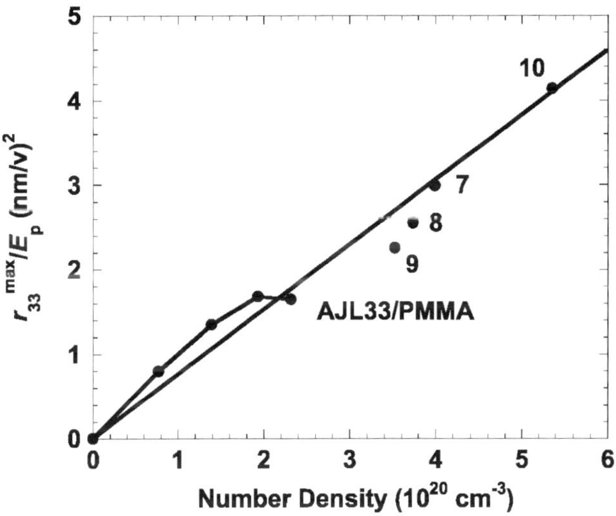

Figure 8. Variation of the maximum electro-optic activity of normalized to the poling field strength as a function of chromophore number density.

Conclusion and Perspectives

Ultralarge EO activity (r_{33} = 249–318 pm/V) and good temporal stability (at 85 °C) were achieved simultaneously through rational molecular engineering of a new generation of supramolecular self-assembled dendronized chromophores using Ar^H-Ar^F interactions. However, the long-term temporal stability of the dipolar alignment is still a major concern for these supramolecular self-assembled chromophore glasses due to relatively low glass transiton temperature and physically reversible lattice formation. The most effective method to overcome such problems is to crosslink the poled materials without sacrifice of poling efficiency at the same time somehow. It is highly desired to develop some novel crosslinking reactions induced by themal and photochemical activations as well as other mechanisms which cound be ideally separated from the poling process under mild conditions. In addition, it is very interesting to investigate other suitable non-covalent interactions, instead of the Ar^H-Ar^F interaction, for developing new types of exceptional supramolecular EO materials. These studies can also provide a good platform to deeply understand the intermolecular interactions of complex systems.

References

1. Kajzar, F.; Lee, K.-S.; Jen, A. K.-Y. *Adv. Polym. Sci.* **2003**, *161*, 1.

2. Enami, Y.; DeRose, C. T.; Mathine, D.; Loychik, C.; Greenlee, C.; Norwood, R. A.; Kim, T. D.; Luo, J.; Tian, Y.; Jen, A. K.-Y.; Peyghambarian, N. *Nat. Photonics* **2007**, *1*, 183.
3. Lee, M.; Katz, H. E.; Erben, C. D.; Gill, M.; Gopalan, P.; Heber, J. D.; McGee, D. J. *Science* **2002**, *298*, 1401.
4. Zheng, X.; Sinyukov, A.; Hayden, L. M. *Appl. Phys. Lett.* **2005**, *87*, 081115.
5. McLaughlin, C. V.; Hayden, L. M.; Polishak, B.; Huang, S.; Luo, J.; Kim, T.-D.; Jen, A. K.-Y. *Appl. Phys. Lett.* **2008**, *92*, 151107.
6. Baehr-Jones, T.; Hochberg, M.; Wang, G.; Lawson, R.; Liao, Y.; Sullivan, P. A.; Dalton, L.; Jen, A. K.-Y.; Scherer, A. *Opt. Express* **2005**, *13*, 5216.
7. Kim, S.-K.; Hung, Y.-C.; Seo, B.-J.; Geary, K.; Yuan, W.; Bortnik, B.; Fetterman, H. R.; Wang, C.; Steier, W. H.; Zhang, C. *Appl. Phys. Lett.* **2005**, *87*, 061112.
8. Huang, Y.; Paloczi, G. T.; Yariv, A.; Zhang, C.; Dalton, L. R. *J. Phys. Chem* **2004**, *108*, 8006.
9. Kim, T.-D.; Kang, J.-W.; Luo, J.; Jang, S.-H.; Ka, J.-W.; Tucker, N.; Benedict, J. B.; Dalton, L. R.; Gray, T.; Overney, R. M.; Park, D. H.; Herman, W. N.; Jen, A. K.-Y. *J. Am. Chem. Soc.* **2007**, *129*, 488.
10. Kang, H.; Facchetti, A.; Jiang, H.; Cariati, E.; Righetto, S.; Ugo, R.; Zuccaccia, C.; Macchioni, A.; Stern, C. L.; Liu, Z.; Ho, S.-T.; Brown, E. C.; Ratner, M. A.; Marks, T. J. *J. Am. Chem. Soc.* **2007**, *129*, 3267.
11. Baehr-Jones, T. W.; Hochberg, M. J. *J. Phys. Chem* **2004**, *108*, 8006.
12. Hochberg, M.; Baehr-Jones, T.; Wang, G.; Shearn, M.; Harvard, K.; Luo, J.; Chen, B.; Shi, Z.; Lawson, R.; Sullivan, P.; Jen, A. K. Y.; Dalton, L.; Scherer, A. *Nat. Mater.* **2006**, *5*, 703.
13. He, M. Q.; Leslie, T. M.; Sinicropi, J. A. *Chem. Mater.* **2002**, *14*, 4662.
14. Liu, S.; Haller, M. A.; Ma, H.; Dalton, L. R.; Jang, S. H.; Jen, A. K.-Y. *Adv. Mater.* **2003**, *15*, 603.
15. Jang, S.-H.; Luo, J.; Tucker, N. M.; Leclercq, A.; Zojer, E.; Haller, M. A.; Kim, T.-D.; Kang, J.-W.; Firestone, K.; Bale, D.; Lao, J.; Benedict, B.; Cohen, D.; Kaminsky, W.; Kahr, B.; Brédas, J. L.; Reid, P.; Dalton, L. R.; Jen, A. K.-Y. *Chem. Mater.* **2006**, *18*, 2982.
16. Davies, J. A.; Elangovan, A.; Sullivan, P. A.; Olbricht, B. C.; Bale, D. H.; Ewy, T. R.; Isborn, C. M.; Eichinger, B. E.; Robinson, B. H.; Reid, P. J.; Li, X.; Dalton, L. R. *J. Am. Chem. Soc.* **2008**, *130*, 10565.
17. Dalton, L. R. *Pure Appl. Chem.* **2004**, *76*, 1421.
18. Shi, Y.; Zhang, C.; Zhang, H.; Bechtel, J. H.; Dalton, L. R.; Robinson, B. H.; Steier, W. H. *Science* **2000**, *288*, 119.
19. Dalton, L. R.; Harper, A. W.; Robinson, B. H. *Proc. Natl. Acad. Sci. U.S.A.* **1997**, *94*, 4842.
20. Luo, J.; Haller, M.; Ma, H.; Liu, S.; Kim, T.-D.; Tian, Y.; Chen, B.; Jang, S.-H.; Dalton, L. R.; Jen, A. K.-Y. *J. Phys. Chem. B* **2004**, *108*, 8523.
21. Ma, H.; Liu, S.; Luo, J.; Suresh, S.; Liu, L.; Kang, S. H.; Haller, M.; Sassa, T.; Dalton, L. R.; Jen, A. K.-Y. *Adv. Funct. Mater.* **2002**, *12*, 565.
22. Ma, H.; Chen, B.; Sassa, T.; Dalton, L. R.; Jen, A. K-Y. *J. Am. Chem. Soc.* **2001**, *123*, 986.

23. Luo, J.; Ma, H.; Haller, M.; Jen, A. K.-Y.; Barto, R. R. *Chem. Commun.* **2002**, 888.
24. Luo, J.; Liu, S.; Haller, M.; Liu, L.; Ma, H.; Jen, A. K.-Y. *Adv. Mater.* **2002**, *14*, 1763.
25. Pereverzev, Y. V.; Prezhdo, O. V.; Dalton, L. R. *Chem. Phys. Lett.* **2003**, *373*, 207.
26. Sullivan, P. A.; Rommel, H.; Liao, Y.; Olbricht, B. C.; Akelaitis, A. J. P.; Firestone, K. A.; Kang, J.-W.; Luo, J.; Davies, J. A.; Choi, D. H.; Eichinger, B. E.; Reid, P. J.; Chen, A.; Jen, A. K-Y.; Robinson, B. H.; Dalton, L. R. *J. Am. Chem. Soc.* **2007**, *129*, 7523.
27. Shi, Z.; Luo, J.; Huang, S.; Zhou, X.-H.; Kim, T.-D.; Cheng, Y.-J.; Polishak, B.; Younkin, T. R.; Block, B. A.; Jen, A. K.-Y. *Chem. Mater.* **2008**, *20*, 6372.
28. Shi, Z.; Hau, S.; Luo, J.; Kim, T.-D.; Tucker, N. M.; Ka, J.-W.; Sun, H.; Pyajt, A.; Dalton, L. R.; Chen, A.; Jen, A. K.-Y. *Adv. Funct. Mater.* **2007**, *17*, 2557.
29. Sullivan, P. A.; Olbricht, B. C.; Akelaitis, A. J. P.; Mistry, A. A.; Liao, Y.; Dalton, L. R. *J. Mater. Chem.* **2007**, *17*, 2899.
30. Kim, T.-D.; Luo, J.; Cheng, Y.-J.; Shi, Z.; Hau, S.; Jang, S.-H.; Zhou, X.-H.; Tian, Y.; Polishak, B.; Huang, S.; Ma, H.; Dalton, L. R.; Jen, A. K.-Y. *J. Phys. Chem. C* **2008**, *122*, 8091.
31. Zhou, X.-H.; Luo, J.; Huang, S.; Kim, T.-D.; Shi, Z.; Cheng, Y.-J.; Jang, S.-H.; Knorr, D. B.; Overney, R. M.; Jen, A. K.-Y. *Adv. Mater.*, in press.
32. Patrick, C. R.; Prosser, G. S. *Nature* **1960**, *187*, 1021.
33. Williams, J. H. *Acc. Chem. Res.* **1993**, *26*, 593.
34. Weck, M.; Dunn, A. L.; Matsumoto, K.; Coates, G.W.; Lobkovsky, E. B.; Grubbs, R. H. *Angew. Chem., Int. Ed.* **1999**, *38*, 2741.
35. Kilbinger, A. F. M.; Grubbs, R. H. *Angew. Chem., Int. Ed.* **2002**, *41*, 1563.
36. Coates, G. W.; Dunn, A. R.; Henling, L. M.; Ziller, J. W.; Lobkovsky, E. B.; Grubbs, R. H. *J. Am. Chem. Soc.* **1998**, *120*, 3641.
37. Watt, S. W.; Dai, C.; Scott, A. J.; Burke, J. M.; Thomas, R. L.; Collings, J. C.; Viney, C.; Clegg, W.; Marder, T. B. *Angew. Chem., Int. Ed.* **2004**, *43*, 3061.
38. Yoon, M.-H.; Facchetti, A.; Stern, C. E.; Marks, T. J. *J. Am. Chem. Soc.* **2006**, *128*, 5792.
39. Khan, R. U. A.; Kwon, O.-P.; Tapponnier, A.; Rashid, A. N.; Günter, P. *Adv. Funct. Mater.* **2006**, *16*, 180.
40. Datta, A.; Pati, S. K. *Chem. Soc. Rev.* **2006**, *35*, 1305.
41. Facchetti, A.; Annoni, E.; Beverina, L.; Morone, M.; Zhu, P.; Marks, T. J.; Pagani, G. A. *Nat. Mater.* **2004**, *3*, 910.
42. Verbiest, T.; Van Elshocht, S.; Kauranen, M.; Hellemans, L.; Snauwaert, J.; Nuckolls, C.; Katz, T. J.; Persoons, A. *Science* **1998**, *282*, 913.
43. Ge, S.; Pu, Y.; Zhang, W.; Rafailovich, M.; Sokolov, J.; Buenviaje, C.; Buckmaster, R.; Overney, R. M. *Phys. Rev. Lett.* **2000**, *85*, 2340.
44. Gray, T.; Kim, T.-D.; Knorr, D. B.; Luo, J.; Jen, A. K.-Y.; Overney, R. M. *Nano Lett.* **2008**, *8*, 754.
45. Teng, C. C.; Man, H. T. *Appl. Phys. Lett.* **1990**, *56*, 1734.

46. Park, D. H.; Lee, C. H.; Herman, W. N. *Opt. Express* **2006**, *14*, 8866.
47. Luo, J.; Liu, S.; Haller, M. A.; Kang, J.-W.; Kim, T.-D.; Jang, S.-H.; Chen, B.; Tucker, N.; Li, H.; Tang, H.-Z.; Dalton, L. R.; Liao, Y.; Robinson, B. H.; Jen, A. K.-Y. *Proc. SPIE* **2004**, *5351*, 36.
48. Kim, T.-D.; Luo, J.; Ka, J.-W.; Hau, S.; Tian, Y.; Shi, Z.; Tucker, N. M.; Jang, S.-H.; Kang, J.-W.; Jen, A. K.-Y. *Adv. Mater.* **2006**, *18*, 3038.
49. Pereverzev, Y. V.; Gunnerson, K. N.; Prezhdo, O. V.; Sullivan, P. A.; Liao, Y.; Olbricht, B. C.; Akelaitis, A. J. P.; Jen, A. K.-Y.; Dalton, L. R. *J. Phys. Chem. C* **2008**, *112*, 4355.
50. Pereverzev, Y. V.; Prezhdo, O. V.; Dalton, L. R. *ChemPhysChem* **2004**, *5*, 1821.
51. Jen, A.; Luo, J.; Kim, T.-D.; Chen, B.; Jang, S.-H.; Kang, J.-W.; Tucker, N. M.; Hau, S.; Tian, Y.; Ka, J.-W.; Haller, M.; Liao, Y.; Robinson, B. H.; Dalton, L.; Herman, W. *Proc. SPIE* **2005**, *5935*, 593506.
52. Luo, J.; Huang, S.; Cheng, Y.-J.; Kim, T.-D.; Shi, Z.; Zhou, X.-H.; Jen, A. K.-Y. *Org. Lett.* **2007**, *9*, 4471.
53. Collings, J. C.; Roscoe, K. P.; Robins, E. G.; Batsanov, A. S.; Stimson, L. M.; Howard, J. A. K.; Clark, S. J.; Marder, T. B. *New. J. Chem.* **2002**, *26*, 1740.
54. Faccini, M.; Balakrishnan, M.; Diemeer, M. B. J.; Hu, Z.-P.; Clays, K.; Asselberghs, I.; Leinse, A.; Driessen, A.; Reinhoudt, D. N.; Verboom, W. *J. Mater. Chem.* **2008**, *18*, 2141.
55. Barto, R. R.; Frank, C. W.; Bedworth, P. V.; Ermer, S.; Taylor, R. E. *J. Chem. Phys.* **2005**, *122*, 234907/1.
56. Le Duff, A.-C.; Ricci, V.; Pliska, T.; Canva, M.; Stegeman, G. I.; Chan, K. P.; Twieg, R. *Appl. Opt.* **2000**, *39*, 947.

Chapter 5

Surface Plasmon Polariton Waveguides in Nonlinear Optical Polymer

Jin Hyoung Lee,* Jiuzhi Xue, Wounjhang Park, and Alan Mickelson

Department of Electrical and Computer Engineering, University of Colorado, Boulder, CO 80309
***Jin.H.Lee@Colorado.edu**

Nonlinear optical materials are used as substrates for electrically addressable surface plasmon polariton waveguides. Materials preparation including the mixing of chromophores and polymers and their coating and poling is described. The film quality is studied by various techniques including atomic force microscopy. Techniques for fabricating surface plasmonic guides in the nonlinear polymer films are detailed. Measurements of surface plasmon polariton propagation in the resulting structures are presented. The measured field structures of the plasmonic guides are displayed and compared with fields calculated with theoretical models.

Introduction

Plasmonics (as recently reviewed in (*1–4*) is a field in which practitioners exploit the electromagnetic properties of metal dielectric interfaces. The surface plasmon polariton (SPP) is the simplest example of a plasmonic excitation. The SPP is a transversely (to the plate surface) propagating disturbance that couples an oscillating charge distribution that exists within a skin depth of the surface of a metal plate (*5*, *6*) with an exponentially decaying normal (to the plate surface) electromagnetic field both interior and exterior to the metal surface (*7–9*). When the plate is sufficiently thin, the SPP couples such polaritonic excitations at both the top and bottom of the plate with either even or odd symmetry about the mean depth of the plate (*8*, *9*). The even configuration, the one in which the normal (transverse magnetic) field has no null, exhibits smaller field gradient than the odd mode. A consequence is the low (Ohmic) loss of this even mode often called the long range (LR) SPP (*10*) mode. When the thin metal plate is of finite width (of at most a few wavelengths), the spot size of the resulting LR-SPP can be controlled by the stripe thickness (*10–13*). Matching the spot size to an optical fiber can result in LR-SPP excitation by a 1550 nm single mode fiber with less than 0.5 dB loss. The stripes can be lithographically patterned into all of the usual integrated optic device structures (*14, 15*). The resulting propagation loss remains low (6-8 dB/cm in silicon (*16*) and correspondingly less in lower index materials such as silica or polymer) so long as the bending radius is kept to less than wavelength divided by the transverse spot size in radians.

In the present work, we take up a problem of electro-optic (EO) addressing of plasmonic structures. The background literature cited in the last paragraph discussed surface plasmon polaritons that exist as a collective excitation of electronic oscillation in a metal with an associated propagating electromagnetic disturbance that exists in dielectric materials above and below the nanometrically thin metallic stripe. Electrooptic materials, materials in which the application of a low frequency electrical field along certain symmetry axes can affect the relation between the optical frequency electric field and displacement current is a type of dielectric material. Although various SPP devices have been demonstrated, none have been demonstrated in an electro optic material. In this work, we discuss the fabrication of SPP structures in a number of different electro optic materials. Our choice of electro optic material in this work is that of electro optic polymer. In all of the electro optic polymers to be discussed here, the polymer is a composite of a long chain polymer with some density of dissolved small molecule dopants. These dopants consist of chromophores, that is, dye molecules, molecules whose absorption gives rise to wavelength dispersion and, therefore, color. As far as electrooptic activity is concerned, though, the polar nature of the dye molecules is more important than the dispersive absorption. A polar molecule evidently will have different electrical response to a field aligned along its axis than to one anti-aligned to the axis. An individual polar molecule will, therefore, exhibit a nonlinear electrical response when incident harmonic waves are polarized along the polar axis of the molecule. In a conjugated structure, the bonds are perfectly singly unsaturated. That is, each of the bonding sites possess a σ bonded electron as well as π bonded electron. The latter electrons are significantly more delocalized than the former electrons that are the primary ones that bond the structure together. When the dye is such a conjugated organic dye, the electrical response (linear and nonlinear) can be greatly enhanced over the response of a dye with saturated bonds as a result of the delocalization that allows for a greatly enhanced charge mobility over that exhibited by structures with saturated bonds. As we will see in the first subsection of the next section, this molecular response is characterized by a third order tensor that we refer to as the second order hyperpolarizability tensor β. When the dye molecules are randomly oriented, the local order is complex, but macroscopic nonlinear effect of the dye molecules is minimized. When the dye molecules are aligned, however, there will be a macroscopic nonlinear response. In a system of conjugated dye molecules, this response may be significantly larger (orders of magnitude) than the response obtained from a saturated or inorganic system. As we will soon see, this response is characterized by a third order tensor that we refer to as χ_2. This tensor is related to the second order hyperpolarizability tensor β and the degree of macroscopic ordering that can be achieved by poling the medium. When the response is specialized to the case of a low frequency (10^{11} Hz or less) modulation field and a high frequency 2×10^{14} Hz or higher) optical carrier field, then the third order tensor characterizing the effect is referred to as the electro optic tensor r and its elements referred to as electro optic coefficients.

Because the electro optic moieties that we dissolve into the electro optic polymer are conjugated dye molecules, the second order nonlinear polarizability β (which after poling gives rise to the the χ_2 macroscopic nonlinearity) that gives rise to the electrooptic effect is electronic and, therefore, fast (femtosecond response time). There are advantages to fast modulation and reconfiguration of devices from the systems perspective. There are also numerous advantages to using an SPP structure for a polymeric electro optic device. Electro optic devices generally are interferometric, that is, device operation is based on a tunable phase delay of one optical path relative to another optical path. The amount of phase delay that is produced is dependent on the tunable change of effective length along the optical path. The change in index of refraction of an electro optic material is proportional to the applied low frequency field. When the optical wave is confined to the highest field region within the electro optic material, we say the overlap (of low and high frequency fields in electro optic material) is large. This is seldom the case in conventional core – cladding – electrode electrooptic devices although may well be in plasmonic electro optic devices. Conventional integrated optical (IO) devices have employed index guidance. That is, the electromagnetic (optical) wave is confined between cladding layers of index of refraction lower than that of the core. In order that the mode approximates the optical fiber mode to which it will be coupled, the guidance is weak. In order that the optical field not overlap the electrode significantly (to minimize Ohmic loss), the cladding must be thick in terms of wavelengths. Much energy is then guided within the cladding. The core contains the electrooptic material. The overlap of the optical mode field with the location of the electrooptic material is often small as the cladding where the much of the optical energy is located must be undoped and therefore non electrooptic material. Further, the core must be poled, that is, subjected to an ordering field at elevated temperature in order to align the electrooptic moieties in order to have a high electrooptic coefficient. A high degree of order requires efficient poling. But the electrostatic voltage drop is proportional to the inverse of the dielectric constant. Most of the (high magnitude) poling voltage is dropped across the cladding. The efficiency of poling then would be much improved in the SPP structure where there need to be no cladding but only electrooptic material as would the overlap of the modal field and modulating field. This work serves as a first step in realizing highly efficient electrooptic addressed plasmonic devices.

Material Properties

Electrooptic Materials

The propagation of electromagnetic waves is described by Maxwell's equations.

$$\nabla \times \mathbf{E}(\mathbf{r},t) = -\frac{\partial \mathbf{B}(\mathbf{r},t)}{\partial t}$$

$$\nabla \times \mathbf{H}(\mathbf{r},t) = \mathbf{J}(\mathbf{r},t) + \frac{\partial \mathbf{D}(\mathbf{r},t)}{\partial t}$$

$$\nabla \cdot \mathbf{D}(\mathbf{r},t) = \boldsymbol{\rho}(\mathbf{r},t)$$

$$\nabla \cdot \mathbf{B}(\mathbf{r},t) = 0$$

which relate the electric field $\mathbf{E}(\mathbf{r},t)$, magnetic field $\mathbf{H}(\mathbf{r},t)$, displacement current $\mathbf{D}(\mathbf{r},t)$ and magnetic induction $\mathbf{B}(\mathbf{r},t)$ with the current density $\mathbf{J}(\mathbf{r},t)$ and charge density $\boldsymbol{\rho}(\mathbf{r},t)$ that generate them. At optical frequencies, naturally occurring materials are non magnetic such that $\mathbf{B}(\mathbf{r},t)=\mu_0\mathbf{H}(\mathbf{r},t)$ where μ_0 is the permeability of free space. Generally, we can write $\mathbf{D}(\mathbf{r},t)=,\varepsilon_0\mathbf{E}(\mathbf{r},t)+\mathbf{P}(\mathbf{r},t)$ where ε_0 is the permittivity of free space and $\mathbf{P}(\mathbf{r},t)$ is the macroscopic material polarization. In a source free ($\mathbf{J}(\mathbf{r},t)=0$ and $\boldsymbol{\rho}(\mathbf{r},t)=0$) region then wave propagation can be described by the partial differential equation

$$\nabla \times \nabla \times \mathbf{E}(\mathbf{r},t) = \frac{1}{c^2}\frac{\partial^2 \mathbf{E}(\mathbf{r},t)}{\partial^2 t} + \mu_0 \frac{\partial^2 \mathbf{P}(\mathbf{r},t)}{\partial^2 t}$$

where the other relations serve to define the polarization (relation between $\mathbf{E}(\mathbf{r},t)$ and $\mathbf{H}(\mathbf{r},t)$) of any electromagnetic disturbance existing in the medium defined by the response $\mathbf{P}(\mathbf{r},t)$ to the field $\mathbf{E}(\mathbf{r},t)$. In order to make the propagation problem more tractable, we generally expand the time harmonic response at frequency, $\mathbf{P}(\mathbf{r},\omega)$ defined by $\mathbf{P}(\mathbf{r},t)=\mathrm{Re}\{\mathbf{P}(\mathbf{r},\omega)\exp(-i\omega t)\}$ as a power series in the time harmonic field $\mathbf{E}(\mathbf{r},\omega)$ defined by $\mathbf{E}(\mathbf{r},t)=\mathrm{Re}\{\mathbf{E}(\mathbf{r},\omega)\exp(-i\omega t)\}$. Each Cartesian component $P_i(\mathbf{r},\omega)= \bar{e}\cdot\mathbf{P}(r,\omega)$ of the time harmonic polarization can be expressed as

$$\begin{aligned} P_i(\mathbf{r},\omega) = {} & P_i^{(0)}(\mathbf{r},\omega) \\ & + \sum_j \chi_{ij}^{(1)} E_j(\mathbf{r},\omega) \\ & + \sum_{jk} \chi_{ijk}^{(2)} E_j(\mathbf{r},\omega_1) E_k(\mathbf{r},\omega_2) \\ & + \sum_{jkl} \chi_{ijk}^{(3)} E_j(\mathbf{r},\omega_1) E_k(\mathbf{r},\omega_2) E_l(\mathbf{r},\omega_3) \\ & + \cdots \end{aligned}$$

in terms of the Cartesian components $E_i(r) = \bar{e} \cdot \mathbf{E}(\mathbf{r}, \omega)$ of fields at other frequencies where the constituent frequencies, ω_i combine to produce the frequency ω. The $\chi^{(1)}$ is known as the electric susceptibility and each $\chi^{(i)}$'s is know as the ith order nonlinear susceptibility. The electro-optic coefficient r_{ijk} is defined by the relation

$$D_i(\mathbf{r}, \omega_0) = \sum_{jk} r_{ijk} E_j(\mathbf{r}, \omega_m) E_k(\mathbf{r}, \omega_0 - \omega_m)$$

where $D_i(\mathrm{r}) = \bar{e} \cdot \mathbf{D}(\mathbf{r}, \omega)$ is the ith component of the displacement current. Evidently, comparison of the definitions of the electro optic coefficient and the higher order susceptibility are linearly related. The macroscopic susceptibilities may be expressed in terms of molecular polarizabilities by making the identification

$$\mathbf{P}(\mathbf{r}, t) = \sum_{molecules_i} n_i \langle \mathbf{p}(\mathbf{r}, t) \rangle$$

Where the n_i is the density of dipoles in the neighborhood of the ith molecular dipole, the $\mathbf{p}(\mathbf{r},t)$ is the molecular polarizability of ith molecule. The angular brackets indicate a macroscopic average over a volume surrounding the ith molecule. The local field correction is associated with this averaging. Evidently, the alignment is crucial as the molecular polarizability is a vector quantity. Without alignment, the sum will be zero as the effect of any molecule polarized in a given direction will be cancelled by a molecule polarized in the opposite direction. A time harmonic component of the microscopic polarizability can be expressed as

$$\begin{aligned} p_i(\mathbf{r}, \omega) = p_i^{(0)}(\mathbf{r}, \omega) \\ &+ \sum_j \alpha_{ij} \mathrm{E}_j(\mathbf{r}, \omega) \\ &+ \sum_{jk} \beta_{ijk} \mathrm{E}_j(\mathbf{r}, \omega_1) \mathrm{E}_k(\mathbf{r}, \omega_2) \\ &+ \sum_{jkl} \gamma_{ijk} \mathrm{E}_j(\mathbf{r}, \omega_1) \mathrm{E}_k(\mathbf{r}, \omega_2) \mathrm{E}_l(\mathbf{r}, \omega_3) \\ &+ \cdots \end{aligned}$$

in terms of the Cartesian components the microscopic local field $\mathrm{E}_i(\mathrm{r}) = \bar{e} \cdot \mathbf{E}(\mathbf{r}, \omega)$ of microscopic local fields at other frequencies where the constituent frequencies, ω_i combine to produce the frequency ω, where the α is known as the molecular polarizability and the β, γ, ... are known as hyperpolarizabilities with β being the second order hyperpolarizability, γ being the third order hyperpolarizability, etc.

Polymers for SPP WG

Several different types of polymers have been exploited for fabricating SPP WG device. Polyimide (PI-2555, HDMicroSystems) was used for the passive waveguide structure. Polyimide has been used for optical waveguide with low loss (0.3 dB/cm (*17*)) and it has a strong chemical resistance so that the structure is mechanically stable during the photolithography and the lift-off process. Moreover, polyimide is compatible with CMOS technique enabling the use of SPPs in the integrated optical device. For the nonlinear optical polymer, we used side chained poly (methyl methacrylate)-disperse red 1 (PMMA-DR1). We have guest hosted nonlinear chromophores such as CLD (*18*) to obtain composites, for example, PMMA-DR1/CLD42A, which are known to exhibit high electro-optic coefficients (~120 pm/V). PMMA-DR1 (22.5 wt%, IBM Almaden Research Center) was used as received. To prepare the PMMA-DR1/CLD42A solution, we first dissolved CLD42A (from NAVAIR) in cyclopentanone and stirred it for overnight with heating. After complete dissolution, PMMA-DR1 was added so that the CLD42A solid fraction is about 30%. A single layer of PMMA-DR1/CLD42A (CLD42A:29%) film was fabricated by spin-coating the polymer solution on a glass substrate followed by a curing process. We measured the absorption spectrum to characterize resulting material. The absorption spectrum of a doped material will consist of the original absorption spectrum and that of the dopant. The spectra of the constituents will be ever more distorted at higher concentrations. However, the most important features of the spectra are the absorption peaks. The most noticeable peak of the bridged dyes that we are using is the cis trans isomerization of the molecules involved. That is, when a dye is comprised of a donor on one end, a bridge segment attaching benzene molecules to which the donor and acceptors are attached and an acceptor at the far end from the donor, there will be two configurations of the molecule, one with donor and acceptor maximally separated and one where the bridge turns to allow the donor and acceptor to be more closer to each other. The transition between these configuration leads to a dominant absorption feature that in absence of broadening may appear Lorentzian. The polymer host generally greatly broadens absorption features. In a uniform dispersion one would then assume that the broadening would approach a Gaussian in shape in wavelength space. The absorption spectrum of PMMA-DR1/CLD42A (Figure 1) clearly shows two distinct absorption peaks which correspond to the absorption from DR1 (at 480 nm) and CLD42A (at 660 nm). The absorption peak location and intensity indicates CLD42A was well-dispersed and homogeneously doped in PMMA-DR1 host. A second optical polymer was synthesized by doping 4-dicyanomethylene-2-methyl-6-P-dimethylaminostyryl-4H-pyran (DCM) into polyimide (PI-2555). DCM was doped into polyimide solution (N-Methyl-2-Pyrrolidone (NMP)) and stirred for overnight with heating. DCM solid fraction was 15%.

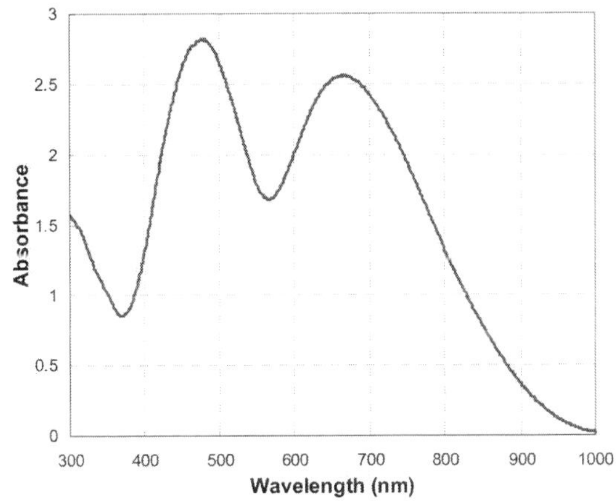

Figure 1. Absorption Spectrum of 1.9 um thick PMMA-DR1/CLD42A layer on glass substrate. The absorption spectrum shows the characteristic absorption peak of DR1 centered around 480nm, and then a second peak centered around 660nm that is contributed to CLD42A. The height of the absorption peaks are directly related to the chromophore concentration in the film, and in this case the solid concentration of CLD42A is 29%. (see color insert)

Poling

Poling is a technique used to align the molecules or moieties of the material in order to maximize the electrooptic properties of nonlinear optical polymers. The development of new chromophores with high molecular polarizabilities has the potential to greatly improve the second-order nonlinearity of material systems containing the new chromophores. To use the large molecular polarizability of the chromophores in a material system, they must be poled to achieve a quasi-permanent alignment of the molecules. The degree to which poling can be carried out can be as important as the intrinsic nonlinearity of the microscopic material in determining the eventual electro optic coefficient. As we previously discussed, conventional waveguides use cladding. Plasmonic waveguides do not use cladding. In the case of static field poling, the alignment is achieved by heating the polymer near its glass transition temperature and exposing the material to a strong static electric field. The static electric field preferentially aligns the molecules, and then the polymer is cooled well below its glass transition temperature to lock in the order. The existence of cladding can greatly reduce the field applied. This is not the case in plasmonic guide in a homogeneous electro optic material. Parallel-plate poling or contact poling is performed by placing the polymer film between two electrodes. Typically the polymer is placed between electrodes by spin coating the polymer onto a substrate with the first electrode and evaporating the second electrode onto the film itself. The polymer film is then raised near its glass transition temperature, and an electric field is applied to the electrodes. A small current is also established in the case of parallel-plate poling owing to the conductivities of the polymers. Such effects are described by Song et al (*19*). When the electric field is high enough, the chromophores inside the polymer begin to align. After a set period of time, the film is cooled below its glass transition temperature, with the field applied to lock in the order. In our studies, we have adopted parallel-plate poling techniques. Our sample (nonlinear polymer film) is placed on the hotplate and heated up higher than the glass transition temperature. The electric voltage is applied to the sample by connecting the bottom ITO and top gold electrode to the high voltage supplier. To avoid bleaching of the active polymer, poling is done under a constant flow of nitrogen. The poling schedule for PMMA-DR1/CLD42A film (1.5 μm thick) is shown in Figure 2.

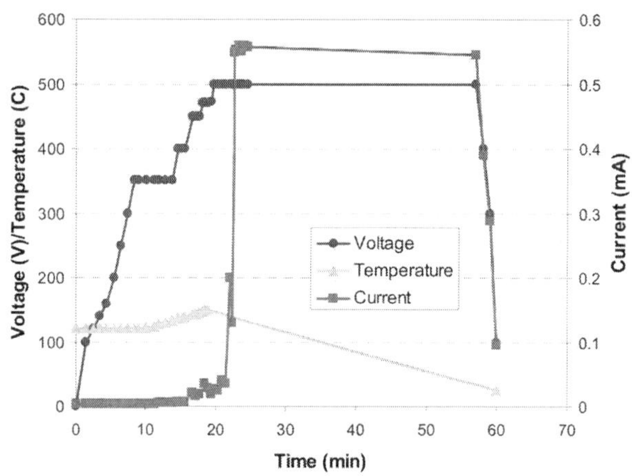

Figure 2. The poling schedule of PMMA-DR1/CLD42A film (1.5 μm). Blue curve is for the applied electric voltage, red curve is for the current flowing through the film, and green curve is for the temperature of the sample. (see color insert)

The current shows dramatic increase at 20 min when the voltage is around 500 V (~333 V/μm) which is an indicative of the molecular alignment inside the polymer film. After poling, we measured the absorption spectrum and it showed a significant change. Figure 3 compares the absorption spectrum of PMMA-DR1/CLD42A film before and after poling.

After poling, the absorption at 660 nm which is corresponding to the absorption of CLD42A is strongly suppressed. This is another indication of molecule alignment inside the polymer because the optical measurement was done by normal incident light to the sample surface. We also poled polyimide/DCM film (3.8 μm) and Figure 4 shows its poling schedule.

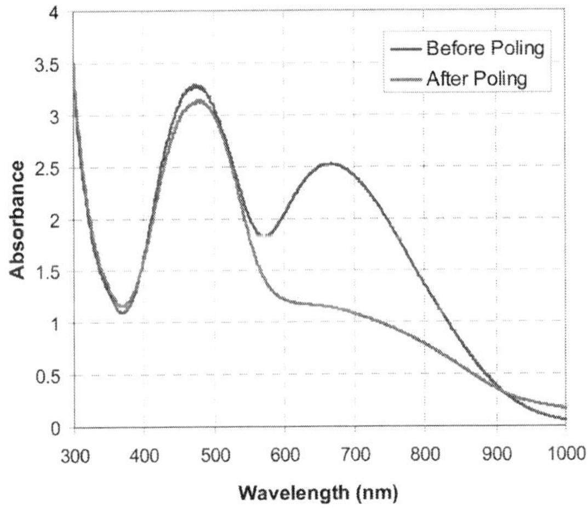

Figure 3. The absorption curve of PMMA-DR1/CLD42A film (1.5 μm) before (blue) and after (red) poling for normally incident light. As the CLD molecules are poled to be aligned normal to the film plane, light propagating along the film normal will not be absorbed by the CLD molecules, as is indicated by the disappearance of the absorption peak around 660nm. The absorption spectrum measured here also clearly indicates that the DR1 chromophores are not poled using this poling schedule, and additional poling would be required to pole the DR1 chromophores. (see color insert)

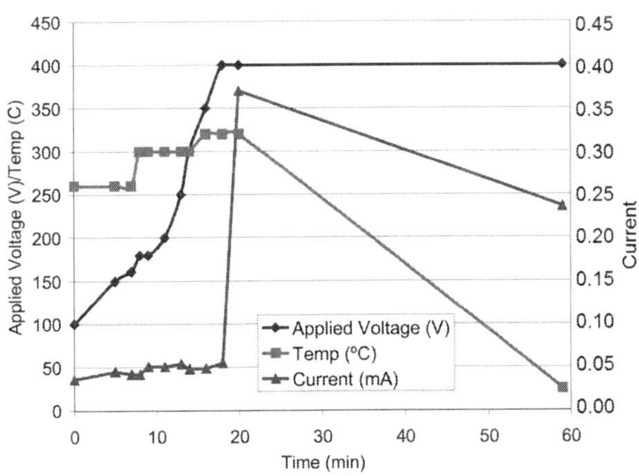

Figure 4. Poling schedule for a 15% DCM in polyimide film. The glass transition of the polyimide is at 325°C, forcing a much higher poling temperature than that for the PMMA/DR1 CLD. The polymer film is 3.8 μm thick. (see color insert)

Fabrication

The typical SPP WG structure is a thin and long metal strip embedded in dielectric host which supports the LR-SPP in the transverse direction and transmits energy into the longitudinal direction. For the passive structure, polyimide was spin-coated on a glass substrate to form the bottom cladding layer. Polyimide layer was pre-baked at 80 °C for 10 min and then increased to a temperature of 180 °C at which it was held for 1 hour in nitrogen back-filled oven for complete solvent evaporation and imidization. The thickness was 3.8 μm and the film showed a light yellow color. We made a metal (gold or silver) stripe pattern on top of polyimide layer using photolithography. Positive photoresist (S1813, Shipley) was spin-coated and exposed to UV light through prepatterned optical mask. After pattern was developed, silver was thermally evaporated on the pattern with proper thickness (20 nm) and then lift-off was done by dissolving photoresist (PR) pattern. The metal stripe thickness determines the size of mode profile and propagation loss in SPP WG (*10*). Thinner metal stripes generally provide lower loss. However, for evaporated metal surface smoothness, typically 15~20 nm metal stripe thickness has been employed (*20*, *21*). A fine metal stripe (narrow line) was formed on the polyimide layer. The upper cladding layer was made by the same process used for the bottom layer. Passive SPP WGs with three different widths (2, 5, and 10 μm) were fabricated. All devices were 20 nm thick and 3 cm long. Then we cut the sample with proper length by using the wafer dicing saw. Since the glass is easy to chip during the dicing process, resulting in severe degradation of end-face quality of WG, we carefully optimized the dicing condition with extremely slow blade feed speed (~ 5 mil/sec) and small cutting depth (5 mil/cut). Figure 5 shows the top view and cross sectional view of the end face of polyimide SPP WG. As we can see, the silver stripe was intact inside polyimide layers and the endface showed clean interface which is a crucial requirement for good optical coupling.

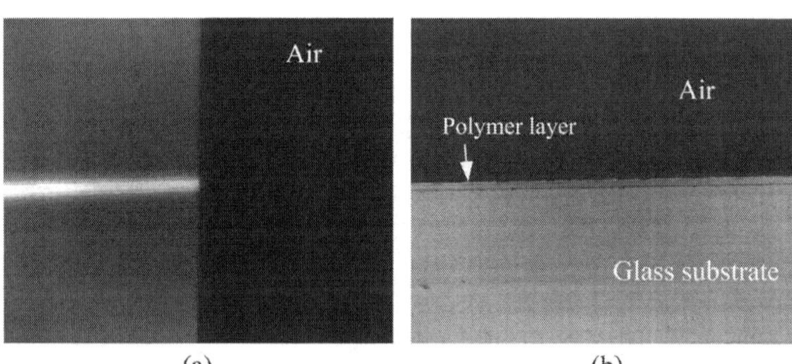

Figure 5. Microscope image of polyimide SPP WG. (a) top-view of the end face of SPP WG. The 5 μm width silver stripe is cleared seen through the polyimide layer. (b) cross-sectional view of the end face. Both pictures show that careful dicing of the polymer on top of the glass substrate enabled a clear cut and at the same time polishing of the end face of the polymer waveguide, which is crucial to the end fire coupling of the light wave into the waveguide. (see color insert)

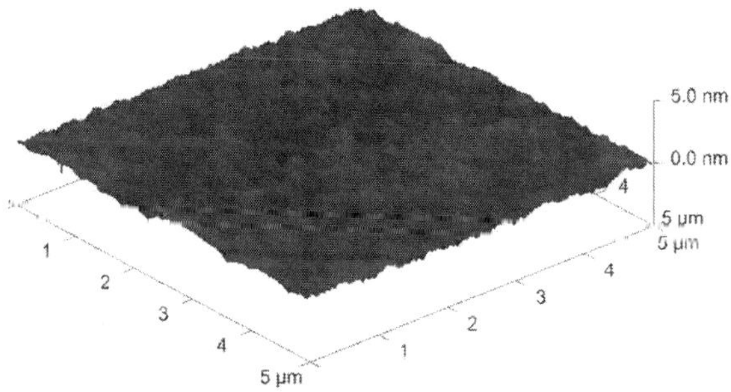

Figure 6. The AFM image of PMMA/DR1 film on ITO coated glass substrate. The image shows that the coated film is smooth to nanometer scales. (see color insert)

(a) (b)

Figure 7. Gold strip deposition results using shadow mask patterning on PMMA-DR1 (a) and on PMMA-DR1/CLD42A (b). The gold stripes are 10μm wide. (see color insert)

For an active structure, polyimide/DCM can be used in exactly the same procedure as used for the passive structure. However, many other nonlinear optical polymer including PMMA-DR1 and PMMA-DR1/CLD42A are not suitable for the photolithographic method because they are not resistant to the solvents used in standard photolithography and lift-off process. Therefore, we need to modify the procedure for metal stripe patterning on the nonlinear optical polymer film. PMMA-DR1/CLD42A (or PMMA-DR1) solution was spin-coated on ITO coated glass substrate and then baked at 90 °C for 3 hours and 120 °C for 12 hours subsequently. The surface uniformity was investigated by the AFM scanning on the polymer surface. The AFM result is shown in Figure 6 and the roughness was estimated less than 1 nm. For metal stripe patterning, we used a shadow mask which has a pattern of waveguide (10 μm wide and 3 cm long). This shadow masking technique eliminates the need for the photolithography and for the lift-off process. The shadow mask was contact-aligned on top of the nonlinear polymer film and then metal (gold) was thermally evaporated to a desired thickness (~20 nm). Figure 7 shows 10 μm wide gold stripe patterned on nonlinear optical film using shadow mask ((a) for PMMA-DR1 film, (b) for PMMA-DR1/CLD42A film).

When the upper cladding of nonlinear optical polymer layer was subsequently deposited, another complication arose because the organic based PMMA-DR1/CDL42A solution (cyclopentanone and diglyme) melts the bottom polymer layer and subsequently destroys the gold stripe. To preserve the gold SPP waveguide, we invented a new process in which a thin protective layer of poly(acrylic acid) (PAA) was spin-coated on top of the SPP waveguide and then baked at 120°C for 5 min. PAA is a water-soluble polymer exhibiting no interaction with the organic solvent and can therefore protect the bottom gold stripe and nonlinear polymer film effectively during the upper layer deposition process. In addition, we cross-linked the PAA protective layer by $CaCl_2$ solution (*22*). Since the cross-linked PAA is water-insoluble, the subsequent deposition of PMMA-DR1/CDL42A did not affect the integrity of the underlying gold waveguide and the whole structure could be preserved during wafer dicing process. The following images (Figure 8) compare the result of SPP WG fabrication with and without the PAA protective layer. Without the protective layer, the metal stripe was severely damaged (disconnected and misaligned) but when the protective layer was used the metal waveguide remained intact after upper polymer layer deposition and wafer dicing. Finally a thick gold electrode (~80 nm) was thermally evaporated for poling and biasing. The complete active SPP WG fabrication procedure is schematically shown in Figure 9. This process presents a new fabrication technique for the active SPP WG structure using electro-optic polymer which could lead a new application of SPP WG in the integrated optical device.

(a) (b)

Figure 8. Microscope image of gold strip in PMMA-DR1/CLD42A film (a) without and (b) with PAA protective layer between two NLO films. Red line is 10 μm wide gold stripe. The gold strip is broken when it is sandwiched straightforwardly between the two electrooptic polymer films. Thin layers, ~100 nm, of crosslinked PAA protective layer is deposited after the deposition of the gold strip to protect the gold strip from the electrooptic films. (see color insert)

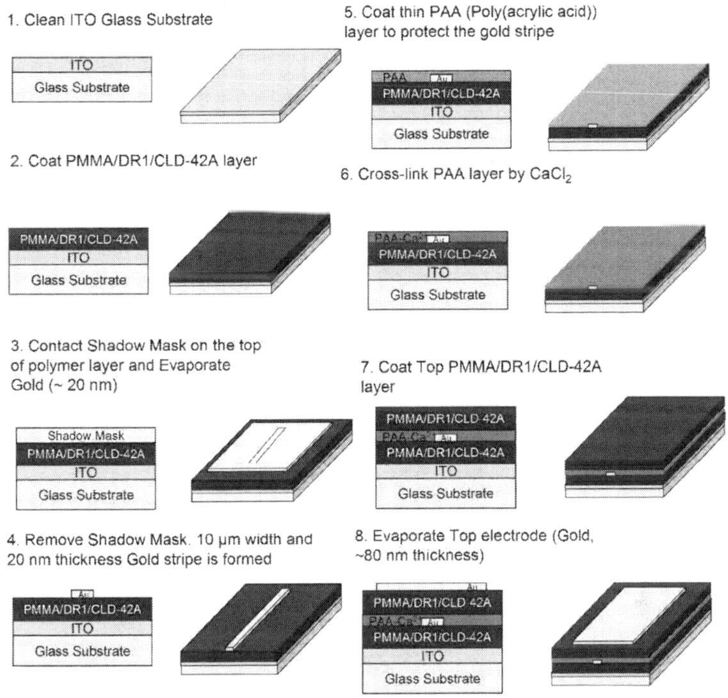

Figure 9. The illustration for the fabrication procedure of active SPP WG fabrication with NLO. (see color insert)

Measurement

For the waveguiding experiment, the laser beam is coupled into the input of SPP WG by end-fire coupling through the optical fiber. A polarization controller was placed before the coupling to SPP WG to provide perpendicular polarized (TM polarization) laser light to the sample surface. An optical fiber was carefully aligned to the input of the SPP WG with the aid of the microscope. The near-field mode at output was collected by the objective lens (×40) and imaged by CCD camera. The measurement setup is schematically shown in Figure 10. The measured mode pattern was compared with numerically simulated SPP WG mode pattern by using commercial software COMSOL. We used He-Ne laser (632 nm) and 1550 nm laser as an input source and single mode fiber for coupling to SPP WG. 632 nm laser light through single mode fiber was end-fire coupled to polyimide (passive) SPP WG which has 20 nm thick and 5 μm wide silver stripe. Since it is visible wavelength, Si CCD camera was used for the mode image capturing. Figure 11 (a) shows the mode pattern at output port of polyimide SPP WG. The mode has an elliptical shape with long axis along the metal waveguide width. In the numerical simulation, we used the same geometry as the experiment. The dielectric constant of silver at 632 nm was estimated from the experimentally measured data in the literature (*23*). The numerically simulated mode pattern (Figure 11 (b)) is well-matched with the measured mode pattern indicating the fabricated SPP WG successfully guides LR-SPP mode through the metal stripe. We also measured the near field pattern at output with 1550 nm laser. The measured near field image and numerically calculated mode field pattern are shown in Figure 12.

We also observed the light propagation through the active SPP WG. The cladding medium was PMMA-DR1/CLD42A and 20 nm thick 10 μm wide gold stripe was embedded as described in the fabrication section. Since this active polymer has high absorption in the visible and the near infrared is of interest in the optic communications, we used 1.55 μm near infrared laser as input source and InGaAs CCD infrared detector for image capture. Figure 13 (a) shows the top view of the interface between air and SPP WG where the laser light is coupled into input of active SPP WG. The bright spot between air and active SPP WG represents the strong scattering at input of SPP WG. We also observed the infrared laser light was guided through the metal stripe by seeing the scattering through the metal line in the middle of the sample (Figure 13(b)).

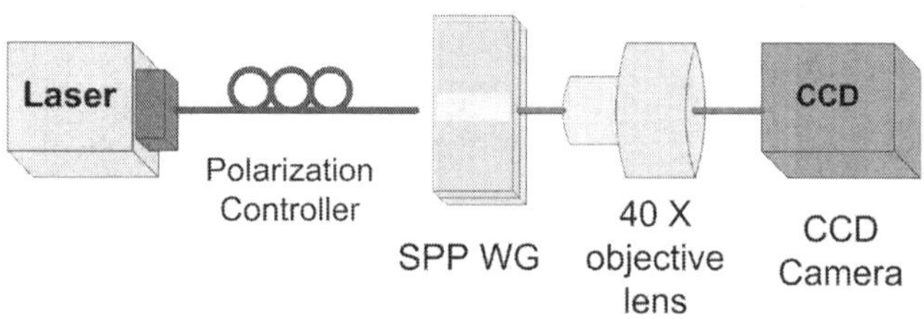

Figure 10. The near-field of SPP WG measurement setup. (see color insert)

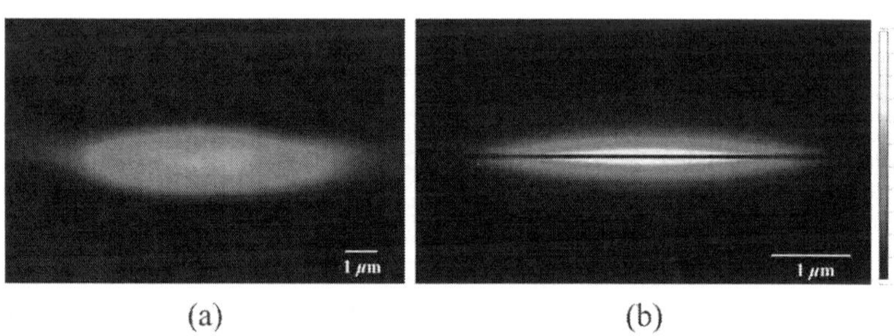

Figure 11. Experimentally measured near field image at output of SPP WG (a). The passive SPP waveguide is comprised of a silver stripe that is 5μm wide and 20 nm thick with polyimide as claddings. TEM00 single mode HeNe laser is coupled to the waveguide by endfire coupling technique and the image was taken at output end of the 2 mm SPP waveguide. (b) numerically simulated SPP WG mode pattern. (see color insert)

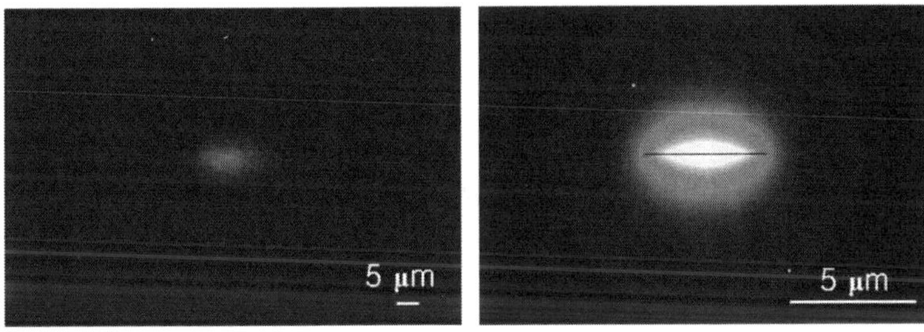

(a) (b)

Figure 12. Experimentally measured near field image at output of SPP WG with 1550 nm laser input (a). The image was taken at output end of the 2 mm SPP waveguide by the infrared camera. (b) numerically simulated SPP WG mode pattern at 1550 nm. (see color insert)

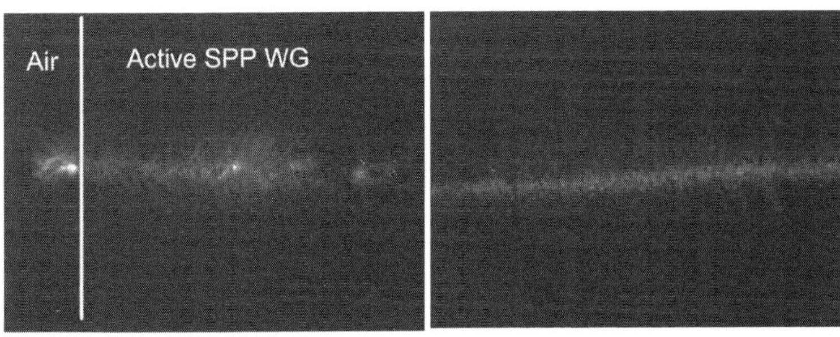

(a) (b)

Figure 13. Infrared laser light guided through the active SPP WG. (a) top view of input of WG (b) light scattering through metal stripe in the active SPP WG. (see color insert)

Conclusion

Polymer based plasmonic waveguide technology is a promising one for producing integrated optical devices. We have added the option of electro-optic addressing of plasmonic devices by incorporating nonlinear optical polymer into the waveguide themselves. In the process of incorporating nonlinear optical polymers into plasmonics, we have developed a new fabrication procedure for metal patterning in nonlinear polymer. Nano-scale metal stripes were patterned on the nonlinear polymer by using shadow mask patterning and adding the protective polymer layer between top and bottom polymer cladding layer for protecting the metal stripe during the top cladding layer deposition. We optimized the fabrication process by investigating the film surface smoothness and end-face quality of WG. We measured the surface plasmonics mode guided through nano-scale metal stripe by the optical measurement. This work can be considered a first step toward active plasmonics whereby plasmonic structures can be used to create electro-optic information processing systems.

Acknowledgments

We acknowledge the support of the Office of Naval Research under grant number N00014-08-1-0874, the National Science Foundation under grant number NSF CCF-0829950 and the graduate school of the University of Colorado. Some of the materials we have used were obtained from Geoff Lindsay of Navair China Lake.

References

1. Zia, R.; Schuller, J. A.; Chandran, A; Brongersma, M. L. Plasmonics: The next chip-scale technology. *Mater. Today* **2006**, *9*, 20–27.
2. Ozbay, E. Plasmonics: Merging photonics and electronics at nanoscale dimensions. *Science* **2006**, *311*, 189–193.
3. Maier, S. A.; Atwater, H. A. Plasmonics: Localization and guiding of electromagnetic energy in metal/dielectric structures. *J. Appl. Phys.* **2005**, *98*, 011101-1–011101-10.
4. Barnes, W. L.; Dereux, A.; Ebbesen, T. W. Surface plasmon subwavelength optics. *Nature* **2003**, *424*, 824–830.
5. Ritchie, R. H. Plasma losses by fast electrons in thin films. *Phys. Rev.* **1957**, *106*, 874–881.
6. Fano, U. Normal modes of a lattice of oscillators with many resonances and dipolar coupling. *Phys. Rev.* **1960**, *118*, 451–455.
7. Economou, E. N. Surface plasmons in thin films. *Phys. Rev.* **1969**, *182*, 539–554.
8. Sarid, D. Long-range surface-plasma waves on very thin metal films. *Phys. Rev. Lett.* **1981**, *47*, 1928–1930.
9. Burke, J. J.; Stegeman, G. I.; Tamir, T. Surface-polariton-like waves guided by thin, lossy metal films. *Phys. Rev. B* **1986**, *33*, 5186–5201.
10. Berini, P. Plasmon-polariton waves guided by thin lossy metal films of finite width: Bound modes of symmetric structures. *Phys. Rev. B* **2000**, *61*, 10484–10503.
11. Charbonneau, R.; Berini, P.; Berolo, E.; Lisicka-Shrzek, E. Experimental observation of plasmon-polariton waves supported by a thin metal film of finite width. *Opt. Lett.* **2000**, *25*, 844–846.
12. Zia, R.; Selker, M. D.; Brongersma, M. L. Leaky and bound modes of surface plasmon waveguides. *Phys. Rev. B* **2005**, *71*, 165431, 1–9.
13. Zia, R.; Chandran, A.; Brongersma, M. L. Dielectric waveguide model for guided surface polaritons. *Opt. Lett.* **2005**, *30*, 1473–1475.

14. Nikolajsen, T.; Leosson, K.; S Bozhevolnyi, I. Surface plasmon polariton based modulators and switches operating at telecom wavelengths. *Appl. Phys. Lett.* **2004**, *85*, 5833–5835.
15. Boltasseva, A.; Nikolajsen, T.; Leosson, K.; Kjaer, K.; Larsen, M. S.; Bozhevolnyi, S. I. Integrated optical components utilizing long-range surface plasmon polaritons. *J. Lightwave Technol.* **2005**, *23*, 413–422.
16. Nikolajsen, T.; Leosson, K.; Salakhutdnikov, I.; Bozhevolnyi, S. I. Polymer-based surface-plasmon-polariton stripe waveguides. *Appl. Phys. Lett.* **2003**, *82*, 668–670.
17. Sullivan, C. T.; Husain, A. Optical computing and nonlinear materials. *SPIE Proc.* **1988**, *881*, 172.
18. Shi, Y.; Zhang, C.; Zhang, H.; Bechtel, J. H.; Dalton, L. R.; Robinson, B. H. *Science* **2000**, *288*, 119–122.
19. Song, R.; Yick, A.; Steier, W. Conductivity-dependency free in-plane poling for Mach–Zehnder modulator with highly conductive electro-optic polymer. *Appl. Phys. Lett.* **2007**, *90*, 191–103.
20. Ju, J.; Park, S.; Kim, M.; Park, S.; Park, Y.; Lee, M. 40 Gbits light signal transmission in long-range surface plasmon waveguide. *Appl. Phys. Lett.* **2007**, *91*, 171117.
21. Andrew, P.; Barnes, W. L. Energy transfer across a metal film mediated by surface plasmon polaritons. *Science* **2004**, *306*, 1002–1005.
22. Winkleman, A.; Perez-Castillejos, R.; Lahav, M.; Narovlyansky, M.; Rodriguez, L. N. J.; Whitesides, G. M. Patterning micron-sized features in a cross-linked poly(acrylic acid) film by a wet etching process. *Soft Matter* **2007**, *3*, 108–116.
23. Johnson, P. B.; Christy, R. W. Optical constants of the noble metals. *Phys. Rev. B* **1972**, *6*, 4370–4379.

Chapter 6

Photorefractive Polymers for Updatable Holographic Displays

R. A. Norwood,*,[1] S. Tay,[1] P. Wang,[2] P.-A. Blanche,[1] D. Flores,[2] R. Voorakaranam,[1] W. Lin,[2] J. Thomas,[1] T. Gu,[2] P. St. Hilaire,[1] C. Christenson,[1] M. Yamamoto,[2] and N. Peyghambarian[1]

[1]College of Optical Sciences, University of Arizona, Tucson, AZ 85721
[2]Nitto Denko Technical, Oceanside, CA 92058
*rnorwood@optics.arizona.edu

Photorefractive polymers are dynamic holographic recording materials that allow for updating of images. They have been investigated over the last decade and have a wide range of applications including optical correlation, imaging through scattering media, and optical communication. Here, we review the achievement of the first updatable holographic 3D display based on photorefractive polymers and report the fabrication of a 6 × 6 inch^2 display, the largest photorefractive 3D display to date capable of recording and displaying new images every few minutes. The holograms can be viewed for several hours without the need for refreshing, and can be completely erased and updated whenever desired.

Introduction

Recently, there has been a rejuvenated effort to develop a true three dimensional (3D) display medium due to attractive applications in defense, medical, training and mapping, entertainment, 3D graphics and CAD design (*1–3*). Even though the history of 3D technology dates back several centuries, progress in this field has not been revolutionary even in the present digital technology era due to the lack of a feasible technology and recording medium. Nonetheless, some technologies have emerged like anaglyph, the polarized light method, or stereoscope goggles, all relying on special optics donned by the user, which is not as convenient as direct 3D viewing. Holographic 3D displays using photopolymers are an exception and have been well explored (*4*), demonstrating full color, high resolution holograms written over a large photopolymer area (several ft^2). The major drawback of photopolymer based 3D holograms is that the data recorded cannot be erased(i.e. not updateable) and a new static hologram has to be recorded every time the data needs to be renewed/updated.

The photorefractive (PR) effect was discovered in inorganic crystals in 1966 and remained within the crystal class of materials until the early 1990's. At that time, PR polymers were invented when a polymer composite was made with a nonlinear optical (NLO) epoxy polymer bis A-NPDA [bisphenol A-diglycidyl ether 4-nitro-1, 2-phenylenediamine] doped with 30 weight % (wt%) of the hole transporting agent DEH (diethylaminobenzaldehyde-diphenyl) (*5*). This was the first proof-of-principle that photorefraction could occur in polymeric materials by combining optical nonlinearity, charge generation, transport and trapping of charges. The second order optical nonlinearity of polymers containing NLO chromophores can be produced by electric field poling. The observation of photorefraction in polymers is very attractive because polymer systems are easier to process than inorganic crystals, flexible in formulation, readily tailored to obtain specific target properties and lower cost. Photogeneration, charge transport, optical anisotropy and mechanical properties can be tuned by mixing multiple components or by chemically modifying a single component (*6*). After the initial observations there were a series of improvements in the diffraction efficiency (*7*), response time (*8*) and two beam coupling gain coefficients (*9*).

The significant classes of organic and hybrid photorefractive materials include: (i) polymer, (ii) organic glass, (iii) hybrid organic-inorganic, and (iv) polymer dispersed liquid crystal. We now discuss each of these classes in detail.

Polymers

There are several different ways to combine all the functionalities required to obtain photorefractivity in polymers. For instance, a copolymer can be developed with the different functional groups present in the polymer backbone or attached as pendant groups. Depending on the requirement, all functionalities can be attached to the polymer or can be partially functionalized. Another widely accepted method is the so-called guest-host approach. Here the required low molecular weight molecules (guests) with the desired properties are added to the polymer (host), which is usually capable of transporting charge (usually holes). The host polymers are doped with low molecular weight electro-optic chromophores. The primary role of these molecules is to provide an index modulation in response to the electric field.

Organic Glasses

In this category of photorefractive materials, one primary component typically serves the function of both charge conductor and NLO molecule.

This approach has the advantage of minimizing the inert volume since the two functions are served by a single moiety and tight packing of the amorphous glass is possible without phase separation, which is usually a problem in polymer composites. The main disadvantage of this approach is that the ready tunability of the required properties which is possible in the polymer composite case is lost. In order to modify any of the properties, synthetic modification of the glass-forming molecule is required. Furthermore, the molecules should be designed to prevent self-crystallization so as to maintain optical clarity. Nevertheless, by engineering the molecular structure, sets of molecules with the same core moieties but with various covalently attached constituents have been developed to tune the thermal and optical properties (*10–13*).

Organic glass devices show high diffraction efficiencies and gain coefficients at low applied voltages since a high density of nonlinear functional groups is present. Some of these molecules are used for high PR performance in the near-IR wavelength region. Many of these molecules, while showing excellent steady state properties, had slow PR dynamics with response times on the order of several tens or hundreds of milliseconds (*10*, *14*). The slow dynamics of devices made with organic glasses is attributed to slow chromophore reorientation in the electric field due to intermolecular dipole-dipole interactions and low free volume. The dynamic properties can be improved by increasing the operating temperature above the T_g, but the steady state characteristics are then affected. Hence, an operating temperature range must be chosen in which steady state and dynamic performance are optimized.

Hybrid Organic-Inorganic Materials

In the process of developing high performing photorefractive materials, organic-inorganic composite materials have been investigated (*15–17*). In this technique, semiconductor quantum dots such as CdSe or CdS are introduced in small concentrations into standard hole transporting polymers such as PVK and PPV based composites. The size-dependent absorption spectra of the quantum dots invited their application as sensitizers in photorefractive polymers. In this way, properly developed devices can be sensitive over a wide range of wavelengths. It has been found that doping polymeric photoconductors with quantum dots improves their photoconductive properties as well (*18*). However, the PR performance of quantum dot doped devices is inferior to that of similar composites doped with traditional sensitizers such as C_{60} or TNFDM.

Polymer Disperse Liquid Crystals

In low T_g photorefractive polymers, the chromophores are oriented under the influence of the total electric field that is a superposition of the internal modulated field and the external applied field. Therefore, in steady state after the hologram formation, the nonlinear chromophores no longer have a uniform orientation, but have a spatially dependent orientation that varies in both magnitude and direction.

These molecular orientational effects enhance the electro-optic contribution that leads to a modulated birefringence, contributing substantially to the refractive index modulation. A drawback of this process is that a large applied voltage is required to orient the chromophore molecule. Another obvious route to induce molecular orientation is to use materials with dielectric anisotropy, such as liquid crystals (LC). In contrast to molecular dipoles, the LC molecules can be reoriented at much lower voltages. Photorefractive polymer dispersed liquid crystals (PDLC) have been developed with the goal of combining the high resolution of photorefractive polymers with the high refractive index changes obtained through field induced reorientation of nematic liquid crystals. PDLCs are prepared by dispersing liquid crystal domains of almost spherical shape in a photoconducting polymer matrix. The sensitization, charge transport and trapping properties necessary for photorefractivity are provided by the polymer and modulation of the refractive index is provided by the liquid crystal droplets.

Early studies on PDLC were done on PMMA/E49 (*19*) and PMMA/E44 (*20*) polymer/LC mixtures with C_{60} and TNFDM as sensitizers. The advantage of PDLC compared with the other polymer composites is that liquid crystal domains can be reoriented with much lower electric fields than polymer devices. The weak points of PDLC include high optical losses due to scattering, which prevent high gain and slow dynamic response in seconds or in minutes due to low charge mobility. When PMMA was replaced by photoconductive PVK, the PR response time was improved to 100 ms (for 400 mW/cm^2) at 633 nm (*21*, *22*). Even though the dynamic response was improved using photoconducting polymers like PVK, diffraction efficiency and net gain were low.

PR Polymers for 3D Display

The key advantages of photorefractive polymers for 3D display are the capability to completely erase the recorded information, the ability to re-record a new image (i.e. updatable medium), and the potential for large areas at low costs (unlike inorganic crystals). In photorefractive (PR) materials, a three-dimensional refractive index modulation is induced by non-uniform illumination. The index modulation is produced as a result of internal space charge-field creation due to selective transport of the photo-generated charges that leads to an internal field-induced index change and thereby to a phase coding of the incident light distribution. In a typical PR guest-host system, a hole-transport polymer matrix is doped with a photoreducible molecule (photo-sensitizer) that can either absorb light or form a charge-transfer complex with the hole-transport polymer. Upon excitation the photosensitizer injects a hole into the transport system and an NLO chromophore subsequently produces the field-dependent refractive index. As the highest index modulations arise from birefringence produced by the dynamic orientation of the chromophores, a glass transition temperature (T_g) close to room temperature is a desired property, as long as a constant bias field can be supplied. Small molecules are often added to the mixture to act as plasticizers thereby lowering T_g.

We have recently demonstrated updatable 3D displays in photorefractive polymers using an optimized photorefractive polymer composition and a novel technique known as voltage kick-off (*23*, *24*). Using this approach we have been able to demonstrate updatable 3D displays with high diffraction efficiency, writing times of a few minutes, erasure in less than a minute, and, most significantly, persistence of several hours. In this paper we will review these recent accomplishments with a focus on the photorefractive polymer materials that have been used to accomplish this breakthrough and also report the fabrication of a 6 × 6 in^2 display, the largest photorefractive 3D display to date, representing an increase of more than 2x in area. We note that the high fields required for the PR effect in these polymers places very demanding constraints on film quality; one defect can cause dielectric breakdown for the entire sample. These are the size of true displays, capable of delivering information that we are used to receiving in the format.

Experimental

The various components of the photorefractive composite are shown in Figure 1. A copolymer with a polyacrylic backbone was used to attach pendant groups, tetraphenyldiaminobiphenyl-type (TPD) and carbaldehyde aniline (CAAN) through an alkoxy linker in a ratio of 10:1. This is achieved by the synthetic modification of PATPD polymer (hereafter called PATPD-CAAN). Note that the composite evidenced no visible phase separation in an accelerated aging test at 60°C for 7 days. The host PATPD-CAAN copolymer provides both optical absorption and charge generation/transport at the writing wavelength (532nm). A plasticizer, 9-ethyl carbazole (ECZ) was added to the composite to reduce the glass transition temperature (T_g), which enables rapid chromophore orientation under an applied bias field. The NLO properties are achieved by adding a fluorinated dicyanostyrene (FDCST) chromophore. The composite PATPD-CAAN:FDCST:ECZ (50:30:20 wt%) was made into thin-film devices by pressing it between two indium tin oxide (ITO) coated glass electrodes at a temperature well above T_g, with the 100μm thickness set by glass spacer beads.

The PR thin-film devices were characterized by standard four-wave mixing measurements in which the diffraction of a 633nm reading beam was measured in the presence two interfering writing beams at 532nm with 1W/cm^2 total irradiance. Both static and dynamic measurements showed that this compound has an overmodulation peak of nearly 90% efficiency at 45V/μm and a short response time constant of 1.6 second. The PR nature of the grating was confirmed by two beam coupling experiments and a gain coefficient of 200 cm^{-1} was measured at 50 V/μm. The absorption spectra can be seen in Figure 2.

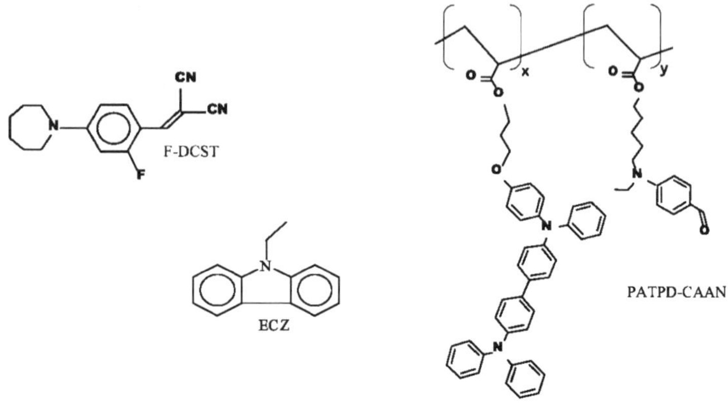

Figure 1. Molecular structures for the components in the 3D display PR composite, PATPD-CAAN copolymer, the chromophore F-DCST and the plasticizer ECZ.

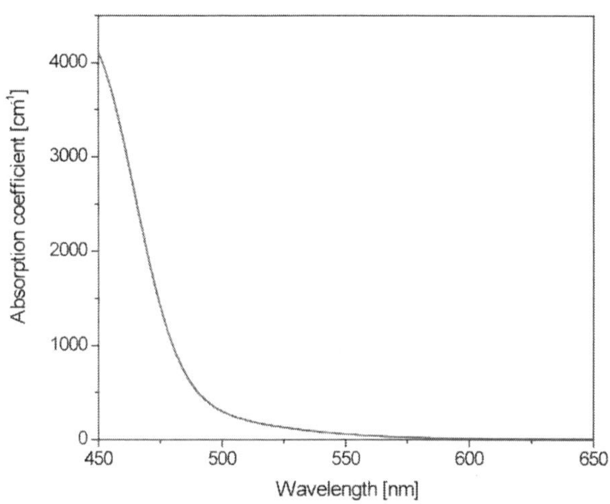

Figure 2. Absorption of the PR material in the visible region. (see color insert)

Results and Discussion

The composite was used to make larger devices, ranging from the 1-inch diameter material characterization sample up to 6×6 inch^2 active area display screen (see Figure 3). Samples did not show any evidence of degradation or dielectric breakdown for extended periods of usage (several months). The holograms recorded in the PR polymer thin-film devices can persist for up to 3 hours in the dark (without writing beams) at an applied field of 40V/μm, while continuously being probed with a red (633nm) laser beam. The 6×6 inch^2 sample is a significant milestone in this effort, representing a 2x increase in area, which is particularly challenging considering the high fields that the resulting films are subjected to; one defect can cause dielectric breakdown for the entire samples. At 6×6 inch^2 we also enter the realm of a true display, as opposed to a novelty size display suitable for advertising, for example. Furthermore, these larger samples were made using an automated lamination system, reducing the amount of touch labor involved in the fabrication of the films. The voltage kick-off technique was developed to improve the writing speed of organic PR materials based on manipulation of the applied voltage (which we call the "voltage kick-off technique"). In conventional holographic recording of PR polymers, a constant external voltage is applied across the polymer to dynamically pole the NLO chromophores. We observed that most PR polymers have a response time of the same order of magnitude as the decay time (depending on voltage and writing beam intensity). If we define a new figure of merit (FOM) as the ratio between the response and decay times, its value is generally near 1. This means that it takes about the same time to write the hologram as for it to disappear. This is not very useful for a display, where one want to write the hologram as fast as possible and then leave it for a long time. Ultimately, the FOM limits the number of elemental holograms (hogels) that can be recorded for a stereogram, and defines the maximal size of the display.

Very favorably, the photorefractive polymer composite that we have developed already has a remarkable FOM of 100, having an untypically long persistence time compared to its sensitivity at 100 mW/cm^2. We further improved the FOM thanks to the voltage kick-off approach where we apply a large field (i.e. 90V/μm) across the polymer to increase the writing speed during hologram recording, and then reduce the voltage to its optimum value of 4.5V/μm after recording is complete. The temporarily increased voltage facilitates efficient separation of electron-hole pairs, and improves the drift characteristics forcing the charges to travel faster; it also enhances the orientational order parameter and speed of the NLO chromophores. The reduction of the voltage to its optimum value after recording ensures hologram persistency. The overall benefit of the voltage kick-off is the reduction of the writing time per hogel to less than a second. By fine tuning of the applied voltage, we achieved a diffraction efficiency of 55% using a total writing time (the time during which the writing beams are turned on) of only 0.5 seconds at 1W/cm^2 irradiance with this technique, practically 40 times higher than the 1.5% efficiency achieved when writing for 0.5 seconds at 40V/μm without voltage kick-off. In this way we improved the FOM up to 1000 which, in terms of stereogram size, permits the writing of 1000 elemental hogels, each 1mm wide, for a total screen length of 1 meter.

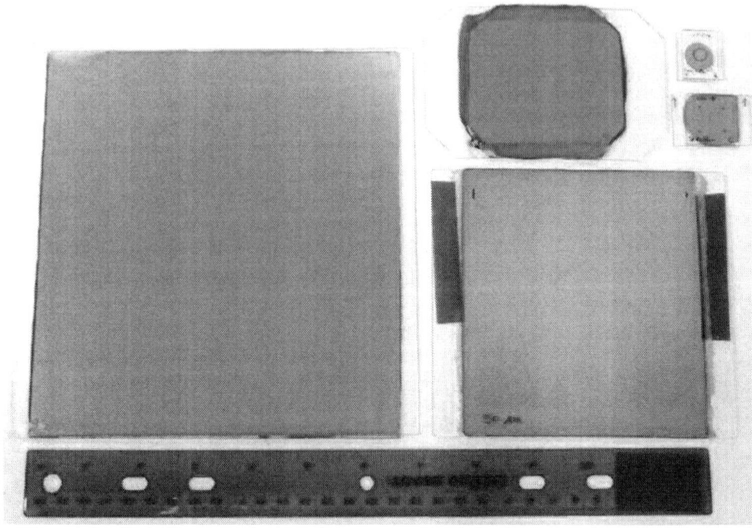

Figure 3. Sample sizes from 2 × 2 in² to large 6×6 in². (see color insert)

Based on dynamic holographic stereography, we have developed a completely automated system for 3D display. The different perspective views of the scene can be captured by a video camera or can be obtained from a 3D computer model. 2D perspective images are then subjected to a computer algorithm that creates holographic elementary pixels (known as hogels by analogy with pixels), which are then uploaded to a spatial light modulator (SLM). The SLM is illuminated with an expanded 532nm laser beam and displays the hogel data in sequence. The laser beam modulated by the SLM (object beam) illuminates a pre-defined hogel area on the polymer film where it interferes with the coherent reference beam. After one hogel is recorded, the shutter blocks the laser beams, the film is translated to the next hogel position and new image data is uploaded to the SLM. The total recording time used per hogel varies from 0.25 seconds to 2 seconds depending on laser power and required diffraction efficiency. The width of the hogels is about 0.8 mm, which gives a total number of hogels of 190 for a 6" x 6" sample. The recorded holographic display can be viewed in transmission geometry using light from an expanded low power He-Ne (633 nm) laser beam or an LED with color within the transparency region of the compound (red to green).

To demonstrate a 3D display using photorefractive polymers, we used horizontal parallax only (HPO) imaging because it significantly reduces the number of hogels required (*25*). In this way the writing time can be reduced and image depth perception is maintained as the eyes' plane is horizontal. Full parallax imaging in both horizontal and vertical directions is possible but this increases the number of hogels required by the number of HPO hogels squared.

We successfully recorded 3D images in a 6×6 inch² photorefractive polymer device within 2 minutes. The viewing angle is limited to 45 degrees and the images can be viewed with the naked eye with high resolution and brightness. Observers experience parallax and object occlusion in the horizontal plane. The pictures presented in Figure 4 were captured using a digital camera moving in the horizontal plane to render the different perspectives. Complete erasure of the recorded images was accomplished by using uniform illumination at 532 nm for 30 seconds. The record/erase process can be repeated any number of times without any loss of fidelity or material degradation. To the best of our knowledge, these are the largest photorefractive devices of any kind reported to date.

Conclusion

By combining an optimized photorefractive polymer composite with a novel voltage kick-off technique, we have developed an updatable 3D display with fast writing time, high diffraction efficiency, rapid erasure and hours of persistence. The 6×6 inch^2 display represents a major milestone for this technology, as we enter the realm of a true display of substantial area. An automated lamination process has been developed that paves the way towards large scale manufacturing of these display films. These characteristics set the stage for applications in defense, medical, and engineering industries, among other areas. We are currently working on improving our system by adding color capability, faster writing time, desensitization to ambient vibration and display sizes up to 1ft by 1ft.

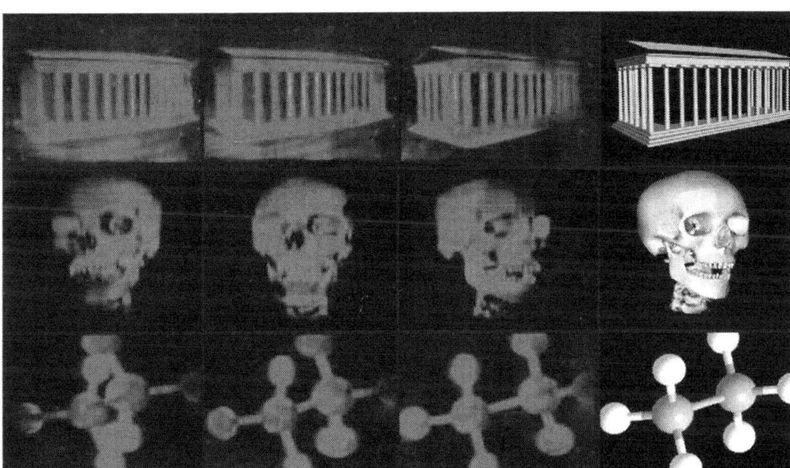

Figure 4. Holographic images of an ionic temple, human skull, and ethane molecule produced in a photorefractive polymer composite film and captured with a digital camera moving in the horizontal plane. (see color insert)

Acknowledgments

The authors of this paper would like to thank the Air Force Office for Scientific Research, the National Science Foundation CMDITR STC (Grant#-0120967) and Nitto Denko Technical for financial support of this research.

References

1. Chatterjee, M. R.; Chen, D. S. *Digital Holography and Three-Dimensional Display: Principles and Applications*; Poon, T., Ed.; Springer: New York, 2006.
2. Pastoor, S. *3D Videocommunication*; Schreer, O., Kauff, P., Sikora, T., Eds.; John Wiley & Sons, Ltd.: West Sussex, England, 2005.
3. Favalora, G. E. Volumetric 3D displays and application infrastructure. *Computer* **2005**, *38*, 37.
4. Zebra Imaging, Inc. 2005. http://www.zebraimaging.com.
5. Ducharme, S.; Scott, J. C.; Twieg, R. J.; Moerner, W. E. *Phys. Rev. Lett.* **1991**, *66*, 1846.
6. Kippelen, B.; Meerholz, K.; Peyghambarian, N. An Introduction to Photorefractive Polymers. In *Nonlinear Optics of Organic Molecules and Polymers*; Nalwa, H. S., Miyata, S., Eds.; CRC Press: Boca Raton, FL, 1997.
7. Meerholz, K.; Volodin, B. L.; Sandalphon; Kippelen, B.; Peyghambarian, N. *Nature* **1994**, *371*, 497.
8. Volodin, B. L.; Kippelen, B.; Meerholz, K.; Javidi, B.; Peyghambarian, N. *Nature* **1996**, *383*, 58.
9. Moerner, W. E.; Grunnet-Jepsen, A.; Thompson, C. L. *Annu. Rev. Mater. Sci.* **1997**, *27*, 585.
10. Wurthner, F.; Yao, S.; Schilling, J.; Wortmann, R.; Redi-Abshiro, M.; Gallengo-Gomez, F.; Meerholz, K. *J. Am. Chem. Soc.* **2001**, *123*, 2810.
11. He, M.; Tweig, R.; Ostroverkhova, O.; Gubler, U.; Wright, D.; Moerner, W. E. *Proc. SPIE* **2002**, *4802*, 9.
12. Lundquist, P. M.; Wortmann, R.; Geletneky, C.; Tweig, R. J.; Jurich, M.; Lee, V. Y.; Moylan, C. R.; Burland, D. M. *Science* **1996**, *274*, 1182.
13. He, M.; Tweig, R. J.; Gubler, U.; Wright, D.; Moerner, W. E. *Chem. Mater.* **2003**, *15*, 1156.
14. Ostroverkhova, O.; Gubler, U.; Wright, D.; Moerner, W. E.; He, M.; Tweig, R. J. *Adv. Function. Mater.* **2002**, *12*, 621.
15. Winiarz, J. G.; Zhang, L. M.; Lal, M.; Friend, C. S.; Prasad, P. N. *Chem. Phys.* **1999**, *245*, 417.
16. Winiarz, J. G.; Zhang, L. M.; Lal, M.; Friend, C. S.; Prasad, P. N. *J. Am. Chem.* **1999**, *121*, 5287.
17. Binks, D. J.; West, D. P.; Norager, S.; O'Brien, P. *J. Chem. Phys.* **2002**, *117*, 7335.

18. Choudary, K. R.; Samoc, M.; Patra, A.; Prasad, P. N. *J. Phys. Chem. B.* **2004**, *108*, 1556.
19. Golemme, A.; Volodin, B. L.; Kippelen, B.; Peyghambarian, N. *Opt. Lett.* **1997**, *22*, 1226.
20. Ono, H.; Kawatsuki, N. *Opt. Lett.* **1997**, *22*, 1144.
21. Golemme, A.; Kippelen, B.; Peyghambarian, N. *Appl. Phys. Lett.* **1998**, *73*, 2408.
22. Termine, R.; Golemme, A. *Opt. Lett.* **2001**, *26*, 1001.
23. Tay, S.; Blanche, P.-A.; Voorakaranam, R.; Lin, W.; Rokutanda, S.; Gu, T.; Flores, D.; Wang, P.; Li, G.; St. Hilaire, P.; Thomas, J.; Norwood, R. A.; Yamamoto, M.; Peyghambarian, N. *Nature* **2008**, *45*, 694.
24. Blanche, P.-A.; Tay, S.; Voorakaranam, R.; St. Hilaire, P.; Christenson, C.; Gu, T.; Lin, W.; Flores, D.; Wang, P.; Yamamoto, M.; Thomas, J.; Norwood, R. A.; Peyghambarian, N. *J. Disp. Technol.* **2008**, *4*, 424–430.
25. Benton, S. *Proc. SPIE* **1983**, *367*, 15–19.

Chapter 7

High-Voltage Poling of Bulk Guest-Host Polymers

Robert C. Hoffman,[*,1] Timothy M. Pritchett,[1] Joshua A. Orlicki,[2] Joseph M. Dougherty,[2] Robert H. Lambeth,[2] Adam M. Rawlett,[2] Warren N. Herman,[3] and Dong Hun Park[3]

[1]U.S. Army Research Laboratory, Attn: AMSRD-ARL-SE-EM, 2800 Powder Mill Rd., Adelphi, MD 20783-1197
[2]U.S. Army Research Laboratory, Attn: AMSRD-ARL-WM-MA, 4600 Deer Creek Loop, Aberdeen Proving Ground, MD 21005-5069
[3]Laboratory for Physical Sciences, University of Maryland, 8050 Greenmead Drive, College Park, MD 20740
[*]robert.c.hoffman@us.army.mil

We describe for the first time the fabrication and poling of a 700 μm thick bulk piece of polymethylmethacrylate (PMMA) doped with various concentrations of Disperse Red 1 (DR1). Maker fringe analysis demonstrates that poling fields of 57 V/μm to 71 V/μm induce substantial ordering of the DR1 chromophores, resulting in birefringence and significant d_{33} and r_{33} values. The poling of thick sections of guest-host polymers will have applications for a range of electro-optic applications, including modulators and terahertz generation and detection.

Introduction

Much attention has been paid in recent years to the poling of thin-film doped polymers for electro-optics applications, mostly in the area of telecommunications switching devices. Poling is the spontaneous alignment of chromophores in an applied electric field that give rise to a host of useful electro-optical properties. When the aligned chromophores are locked in their new orientation within a polymer host, various electro-optic devices such as modulators can be constructed. However, there are applications where a larger thickness of guest-host polymer results in enhanced performance of the electro-optic device, such as in terahertz (THz) generation. The longer optical path length potentially available in bulk poled polymers may allow improved THz power, sensitivity and signal-to-noise ratio in both THz generators and detectors (*1–5*). Large-area Pockels cells constructed from electro-optic guest-host polymers may have uses as high-speed modulators and shutters, and any application requiring an appreciable d_{33} or r_{33} in a very thick film format could potentially benefit from bulk-poled guest-host polymer systems.

Optical poling (*6–10*) is a possible way to accomplish chromophore alignment of thick sections of guest-host polymer. The optical poling is accomplished using CW lasers of various types as well as femtosecond pulses. However, optical poling is generally limited to chromophores that undergo an optically-induced *cis-trans* isomerization such as DR1. While this may be of utility in aligning chromophores that undergo a *cis-trans* isomerization, it does not extend generally to all chromophores.

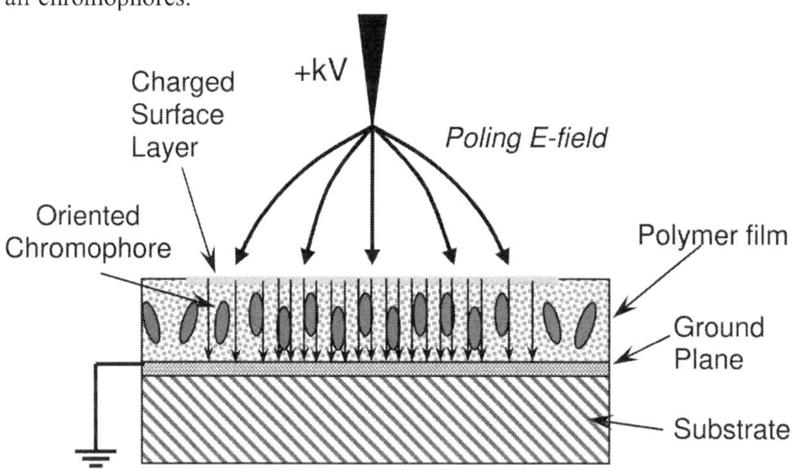

Figure 1. Normal configuration for corona poling showing chromophore orientation parallel to E-field induced by charged surface layer. (see color insert)

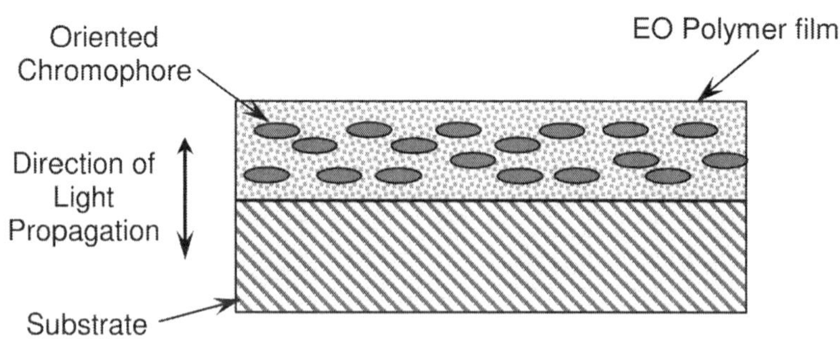

Figure 2. Desired chromophore orientation for EO device such as a Pockels cell (not experimentally practical). (see color insert)

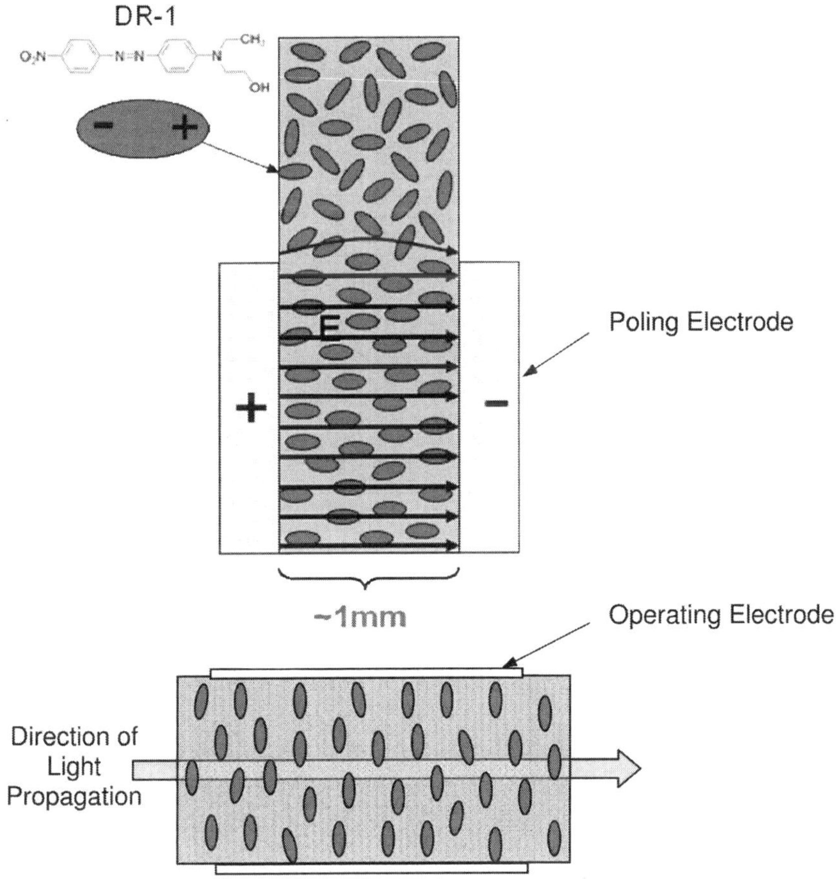

Figure 3. Top: bulk poling configuration showing alignment of chromophores with E-field; bottom: chromophore orientation for Pockels cell after bulk poling and subsequent reorientation. (see color insert)

In view of the shortcomings of optical poling, and the difficulties of processing and stacking extremely thin corona-poled films for bulk applications, we chose the more direct approach of high-voltage DC contact poling. This method has the advantage, in principle, of being able to pole any polar chromophore in a guest-host system.

The high-voltage corona poling configuration shown in Figure 1, while suitable for thin-film devices such as Mach-Zehnder modulators, is not suitable for fabrication of bulk devices requiring a thicker path length. The desired chromophore orientation for thicker devices is shown in Figure 2. However, poling of EO polymers of any thickness in this configuration is not practical; a more practical alternative is to pole thick sections of guest-host polymer and reorient them during subsequent fabrication so the poled chromophores are in the correct orientation, i.e., transverse to the direction of light propagation.

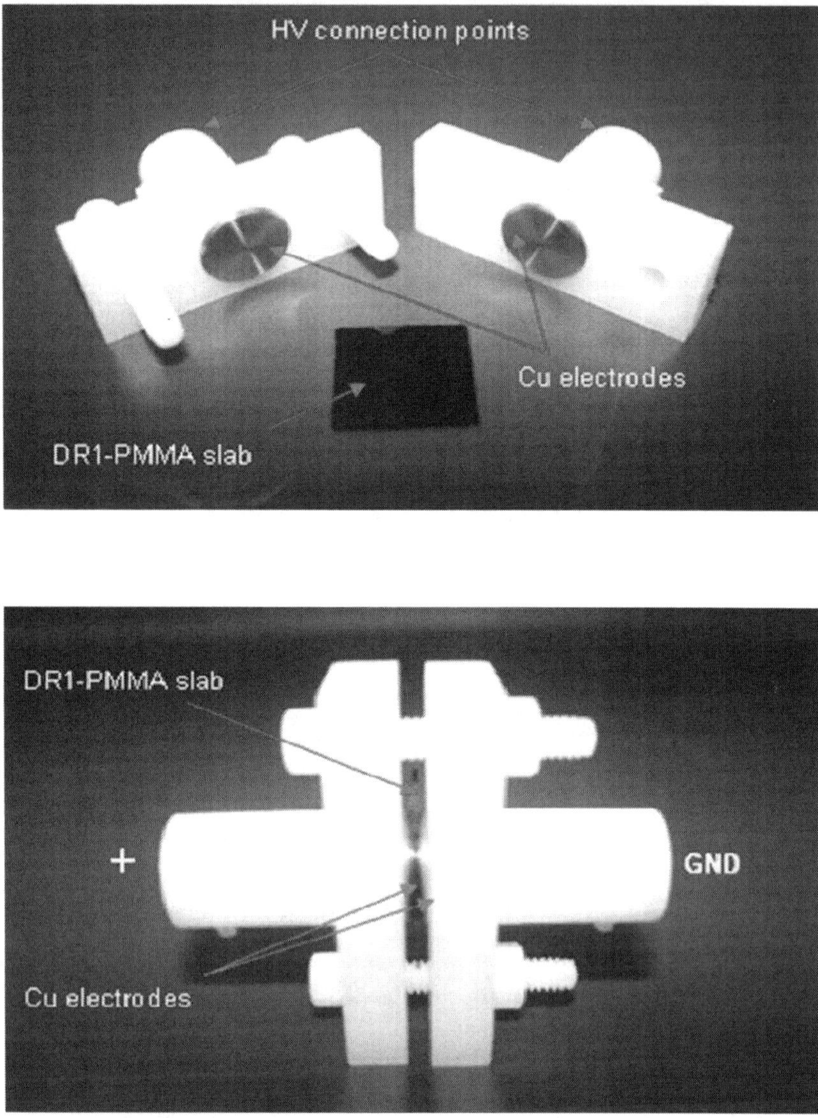

Figure 4. Electrode/sample configuration for poling bulk guest-host polymers; showing poling fixture open (top) and closed (bottom). (see color insert)

To this end, we investigated poling the guest-host polymer in the configuration (Figure 3, top) using contact electrodes. The as-poled "plaque" of guest-host polymer can then be cut, reoriented, and polished as shown in Figure 3, bottom. This allows a much longer path length, and opens a new pathway to a variety of bulk EO devices. Compared to contact poling of thin films, however, the larger electrode separation distance in slabs approaching 1 mm thickness necessitates the use of much higher poling voltages. These high voltages are potentially lethal and require extreme caution together with the use of a high-voltage power supply with current overlimit protection.

Experimental

Materials and Sample Preparation

The DR1 chromophore (Aldrich Chemical) was recrystallized from hot EtOH (co-precipitation) or used as received (micro-compounding). Mixing of the polymer guest-host systems was accomplished by two different methods: coprecipitation of the DR1 and PMMA from THF, or by the micro-compounder extrusion method.

In the co-precipitation method, a homogenous solution containing DR1 guest chromophore and the PMMA host polymer is precipitated by injection into a mutual non-solvent. The precipitates are then dried in an oven to remove traces of solvent or moisture that could adversely impact the dielectric breakdown properties of the guest-host polymer. This is followed by compression molding into a thick polymer slab that has smooth surfaces on both sides, allowing for convenient optical measurements.

The micro-compounder method begins with the macroscopic mixing of PMMA and the chosen chromophore. This is accomplished through the use of a vortex mixer or some other high-energy agitation. In the work described here, aDSM Xplore Micro-Compounder was used to perform the physical compounding tasks. The extruder barrel was heated to 250°C, and most processing was conducted at 100 rpm. A macroscopically mixed polymer blend was charged to the hopper and fed into the extruder. The extruder was fitted with two conical co-rotating screws and also possessed a recycle loop, allowing melt to recirculate through the system automatically. The general processing conditions consisted of the following: the screw rotation rate was 100 rpm, the barrel temperature was 250°C, and the processing cycle was approximately 10 minutes. A 16 g pre-mixed polymer charge resulted in approximately 13-15 g of polymer. The remainder of the mixed polymer adhered to screw or barrel. The blended extrudate was then pelletized and used for compression molding.

All sample plaques were prepared by compression molding at 180°C. Figure 4 shows a representative polymer plaque ready for poling. The polymer plaque is approximately 2.5 cm x 2.5 cm x 0.7 mm thick. The first experiments used two aluminum electrodes fabricated from electron microscopy sample mounts attached to the polymer sample using Kapton tape. The electrodes were polished flat to allow the electrode intimate contact with the polymer.

Two other methods of electrode attachment were attempted as well: temporary bonding of the aluminum electrode to a 100-200 nm, vacuum-deposited Au pad with Ag paste, and simple pressure contact of the aluminum stub against the polymer sample with no Au contact pad. The current preferred method of electrode attachment involves the use of a poling fixture of our own invention (Figure 4), which consists of two nylon blocks fabricated into a clamp-like device, secured by two nylon screws. Copper electrodes were inserted into the fixture through holes drilled into the clamp. The two screws allow one to adjust the tension against the polymer plaque, and to ensure that the copper electrodes are seated parallel to the faces of the plaque. Unless otherwise noted, all samples had 100-200nm gold pads deposited by vacuum evaporation. This was done to insure a good contact, as the silicone oil can penetrate between the Cu electrodes and the polymer sample in the absence of a deposited Au pad; the pads are removed with a dilute KI solution after poling.

High-voltage connections were made by securing the high-voltage leads from the high-voltage power supply to the poling fixture by set screws. Electrical contact was accomplished by compression of the exposed high-voltage lead against the end of the copper electrode. This was done so that the poling fixture can easily be removed from the silicone oil bath to access the sample. A much more satisfactory approach that is currently used allows us to make secure high-voltage connections via custom-made connectors at the end of each copper electrode. This insures that arcing does not occur between connections, and that the full voltage is applied to the sample.

Instrumentation and Poling Procedure

A Spellman 120 kV high-voltage DC power supply (Model SL120P60) was used to supply the high voltage required for poling. The power supply has an adjustable current overlimit setting that is designed to prevent continuous arcing. The overlimit setting was set at about 250μA. Once the limit is exceeded, an internal circuit breaker shuts off the high voltage. The power supply also has 0-10V proportional outputs to monitor both current and voltage. The poling fixture was connected to the power supply as described above and then immersed in Dow-Corning 710 silicone oil contained in a Teflon cup, or a Pyrex 1.5 liter cylindrical beaker. A Teflon cap is placed over the bath to prevent arcing to the walls of the oven. The entire bath/sample assembly was then placed in an oven for poling.

The original oven used to pole the first DR1 samples showing EO activity was a VWR Scientific vacuum drying oven modified for the experiment. The oven currently used is a Cascade Tek TFO-3 oven with a Watlow Series 981 controller. This oven allows a much more consistent and repeatable temperature profile. In both cases the oven was modified to allow the insulated high voltage lead to enter through the back of the oven. The orginal samples were poled without the benefit of a data collection system to closely monitor current and voltage on the sample. Subsequently, the voltage-current-temperature data were collected using a USB-operated DAQ device (Omega Scientific), and collected on a laboratory PC. However, high-voltage arcing events would disrupt the data collection process, and in some cases the arcing events would damage the data collection devices and shut down the PC inadvertently. To circumvent this problem, we now use wireless data transmission devices to transmit the data to a receiver attached to the PC. For wireless temperature data collection we use an Omega OM-CP-RFTC4000A transmitter; for voltage data we use an Omega OM-CP-RFVolt101A wireless transmitter. The data is transmitted every 30 seconds to a wireless RF receiver (Omega OM-CP-RFC101A). The wireless receiver is connected to the PC via an interface cable (Omega OM-CP-IFC110).

To accomplish poling of the original samples, the voltage was increased to 40-50 kV (57 V/μm to 71 V/μm) and the temperature ramped to T_g (~90°C) from ambient (25°C) in approximately 30 min, followed by a period of poling at T_g for 15min, then followed by a cool-down period with forced air to ambient over a period of 120-180 minutes. The high voltage was applied to the sample throughout the entire poling process. When the sample has cooled, the voltage is turned off and sample was removed and gently cleaned of the silicone oil with a mild solution of Alconox detergent. The gold pads were removed either by gentle rubbing or by etching with a solution of aqueous potassium iodide. The samples were stored at approximately minus 10°C until measurements were taken.

A representative poling profile of a DR1 sample using the wireless data collection system is shown in Figure 5. As expected, as the temperature nears T_g, the current passing through the sample peaks each time the voltage is increased, then the curent settles to a lower value before the next voltage increase.

Results and Discussion

The as-poled polymers were placed on a rotation stage between crossed polarizers to measure their poling-induced birefringence (Figure 6). The resulting birefringence data, shown in Figure 7, clearly indicate that the DR1-PMMA was poled under the conditions described. The birefringence decayed over a period of several days, an expected property of the low T_g DR1-PMMA guest-host system. Further confirmation of poling comes from an examination of second-harmonic generation (SHG) Maker-fringe data (*11*). The apparatus used in the analysis is shown in Figure 8.

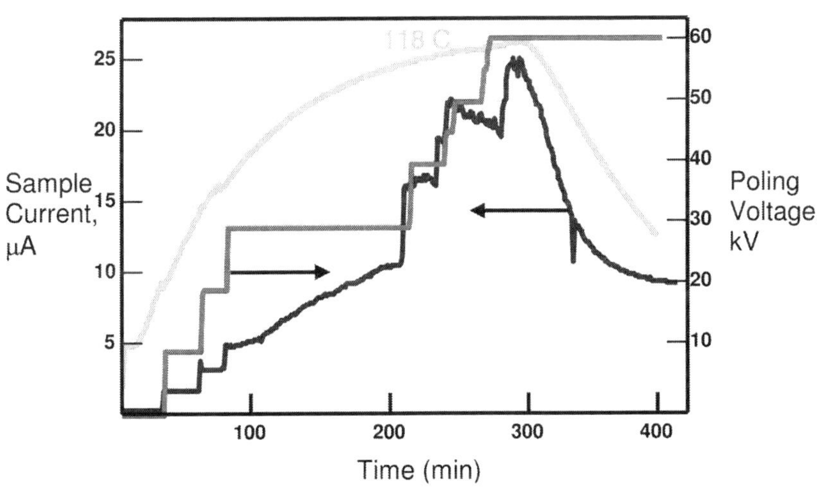

Figure 5. Representative temperature vs. time poling protocol. (see color insert)

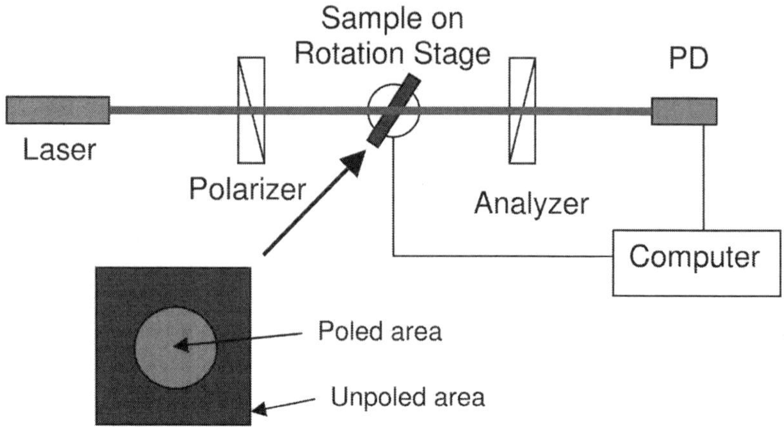

Figure 6. Apparatus for measuring birefringence. (see color insert)

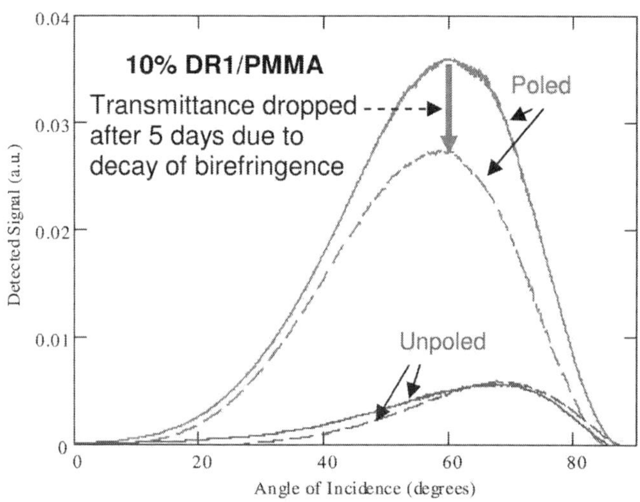

Figure 7. Poling-induced birefringence of 10% DR1-PMMA 700 μm slab. (see color insert)

In the SHG study, the sample was positioned at the waist of a lightly focused (150 mm lens) fundamental beam (1319nm) produced by a diode-pumped Q-switched Nd:YAG laser. The pulse width was 14 ns and the repetition rate 750Hz. The fundamental beam was polarized inside the laser cavity and a half-wave plate was used to control the polarization of the beam incident on the sample.

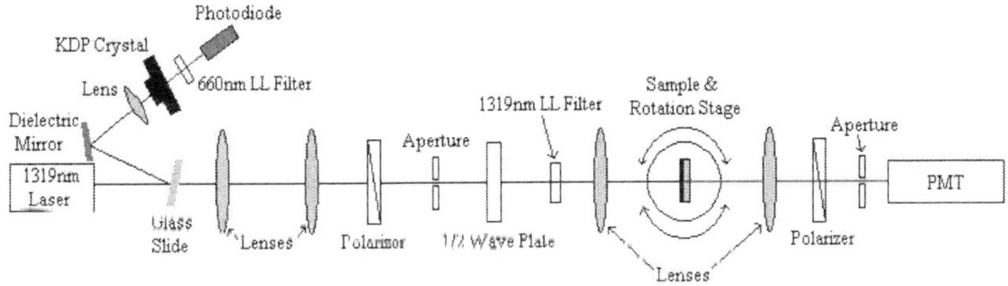

Figure 8. Maker fringe experiment to measure SHG d-coefficient. (see color insert)

For the data shown below, the fundamental beam was p-polarized and a polarizer after the sample was set for detection of the p-polarized component of the second harmonic output. The SHG signal was detected by a Hamamatsu R928 photomultiplier tube in conjunction with a Stanford Research SR250 boxcar averager. To account for laser power fluctuations, part of the fundamental beam was split off and sent through a KDP crystal to produce a second-harmonic reference signal. The sample was mounted on a computer-controlled Oriel rotation stage and data collected by computer at 1-2° increments of the angle of incidence. From the Maker fringe data at 1319 nm collected from a 10% DR1 coprecipitated sample, the largest measurable d_{33} was an estimated 9 pm/V (Figure 9). From this d_{33}-value we estimate an r_{33} at 1319 nm of 3.7 pm/V. The λ_{max} for the DR1-PMMA system appears at 478 nm and therefore we estimate a value of r_{33} = 6.1 pm/V at 810 nm. This compares favorably with other measurements of r_{33} observed in a DR1-copolymer system (7.5 pm/V) (*12*). This sample was observed to have an abundance of DR1 precipitates that obscured some of the fine structure in the Maker fringe scan.

Subsequent samples (Figure 10) were processed using the micro-compounder extrusion method; they show a very smooth surface morphology and far fewer particulates. The 10% DR1 samples prepared by this method show clear Maker fringes; the attenuated response compared to the aggregated 10% DR1 sample shown in Figure 9 was likely due to a high-resistance contact problem within the high-voltage poling fixture.

The 10% and 5% DR1 samples shown in Figure 11 exhibited low-contrast fringes due to precipitates, while the 2.5% DR1 sample showed no evidence of precipitates, resulting in high-contrast fringes. The 2.5% DR1 sample exhibits an SHG response proportionately much larger than either of the 10% and 5% DR1 samples. This is presumably due to the absence of precipitates in the 2.5% DR1 sample, resulting in more efficient poling. Other optical inhomogeneities were also observed; one could often observe color banding due to the variation in absorption (and therefore concentration) of the DR1 in the PMMA matrix. These effects were due to particulates and concentration gradients accumulated at the leading edge of the polymer melt as the pellets melted and filled the mold.

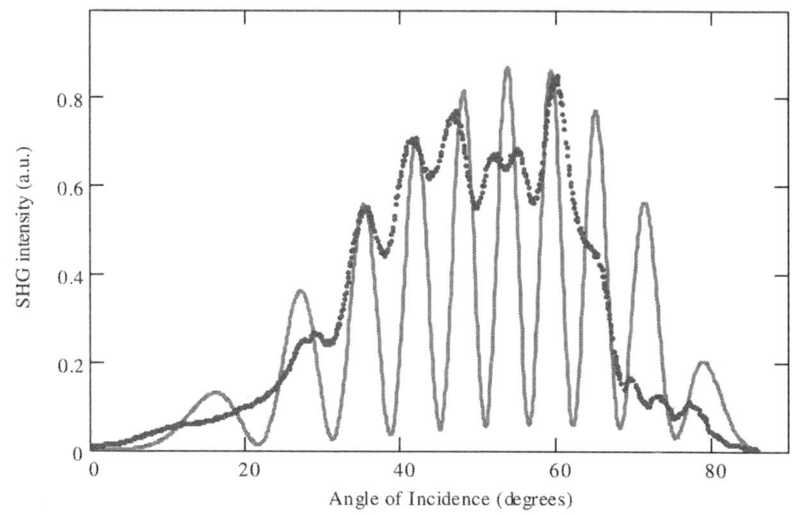

Figure 9. Projected (red) vs. observed (blue) Maker fringe fit for 10%DR1/PMMA (wt %). (see color insert)

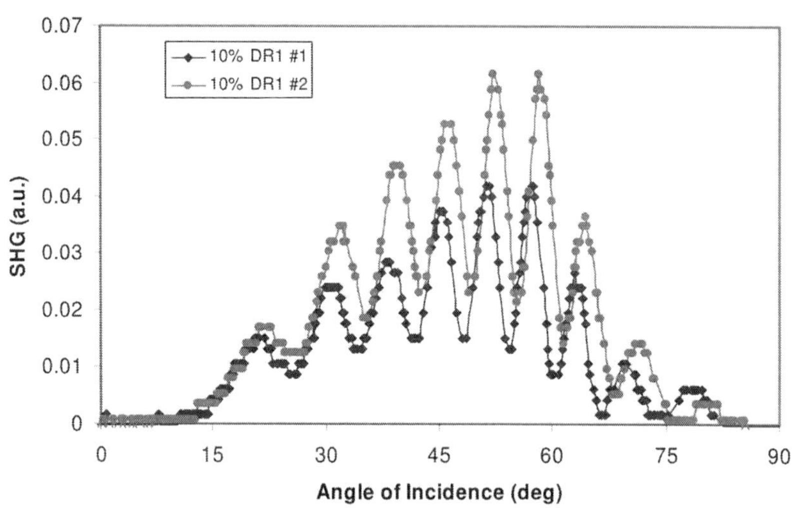

Figure 10. Maker fringe data for non-aggregated samples. (see color insert)

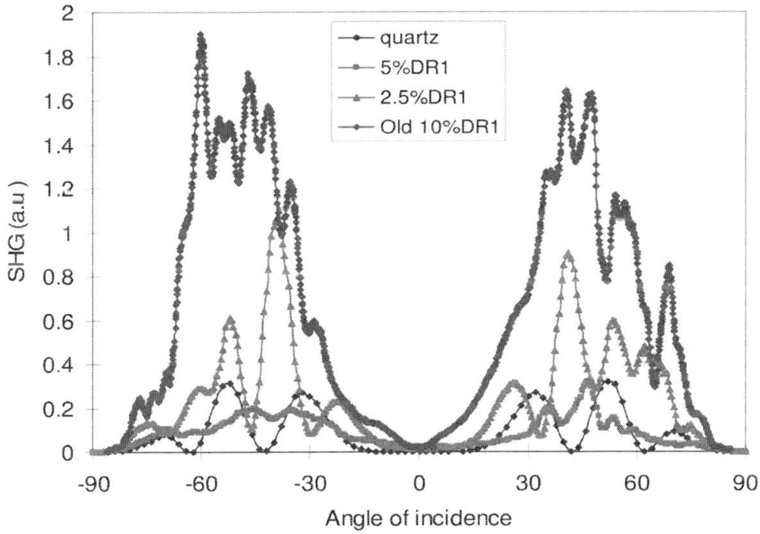

Figure 11. Non-aggregated 2.5% DR1 sample performance exceeds that of aggregated 5% and 10% DR1 samples. (see color insert)

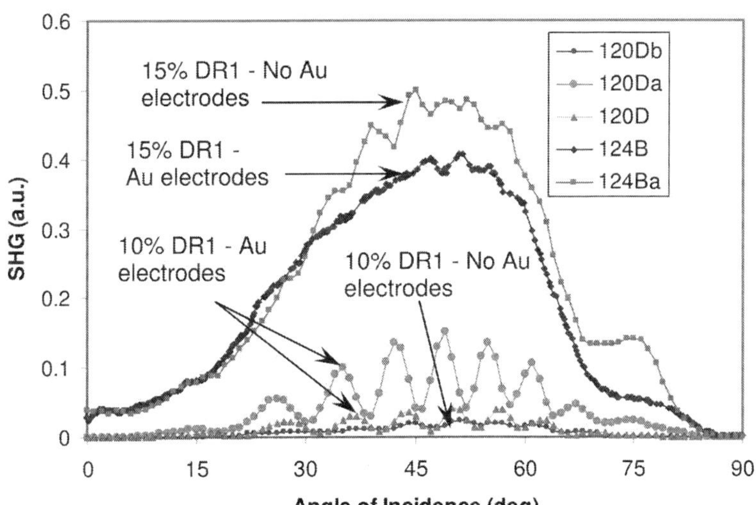

Figure 12. Maker-fringe data for 10% and 15 % DR1-PMMA samples prepared by micro-compounding. (see color insert)

To further limit impurities, we have begun to recrystallize the as-received DR1 to remove salt impurities. Furthermore, in comparing the first lot of samples that were made of co-precipitated DR1-PMMA material (~6 pm/V) with the later lots of micro-compounded DR1-PMMA material (less than 1 pm/V), it appears that every sample containing DR1 particulates has a higher r_{33} than would normally be expected for an aggregated sample. We expect to circumvent the problem of precipitates completely by using a copolymer in which the chromophore is chemically bonded to the polymer host, in addition to the use of the micro-compounding technique in conjunction with recrystallized dyes (to remove salts from commercially available materials) already shown to give good quality plaques.

Further differences can be seen in the type of electrodes used for poling, as shown in Figure 12. For example, sample 124B had gold pads whereas sample 124Ba did not, but sample 124Ba had a higher EO response. In this case the absence of gold electrodes appears to result in an enhancement, while in the 10% DR1 case it does not. At this point, there seems to be no clear advantage to using gold pads to assist in the poling process. If it is ultimately correct that gold pads are not necessary for poling, that will greatly simplify the fabrication of bulk EO devices.

Conclusions

In this paper we described the fabrication and poling of 700μm thick bulk pieces of polymethylmethacrylate (PMMA) doped with various concentrations of Disperse Red 1 (DR1). Maker fringe analysis demonstrated that under the poling fields currently available to us (57 V/μm to 71 V/μm), substantial ordering of the DR1 chromophores is occuring, resulting in birefringence and significant d_{33} and r_{33} values. An estimated value of r_{33} = 3.7 pm/V at 1319 nm was measured; this corresponds to an estimated value of r_{33} at 810 nm of 6.1 pm/V. We have shown that this compares favorably with other measurments of r_{33} observed in a DR1-copolymer system (7.5 pm/V at 800 nm).

The role of precipitates in the bulk poling process is not fully understood, as samples with DR1 precipitates had much more of an SHG response than would be expected of a poled electro-optic polymer laden with precipitates. We expect to circumvent this by switching to a chromophore-host copolymer in which the chromophore is chemically linked to the host polymer, as well as with improved melt processing of the conventional guest-host polymers.

Poling larger thicknesses of electro-optic polymers will result in enhanced performance of electro-optic devices in practical applications, such as those currently being investigated for THz generation. The thicker path length available in bulk poled polymers may potentially enable improved THz power, sensitivity and signal-to-noise ratio in both THz generators and detectors. Large-area Pockels cells constructed from electro-optic guest-host polymers will find new uses in high speed modulators and shutters, and any application requiring appreciable values of d_{33} or r_{33} will potentially benefit from bulk-poled guest-host polymers and copolymers.

Acknowledgments

Acknowledgements. The authors of this paper would like to thank Kim Olver for her help in preparing the samples; and Wei Zhang and Ron Moffitt of the Advanced and Applied Polymer Processing Institute (AAPPI) in Danville, VA.

Caution: The high voltage used for this experiment is potentially lethal; at a minimum, a high-voltage power supply with current overlimit protection is required.

References

1. Sinyukov, A.; Hayden, L. M. *Opt. Lett.* **2002**, *27*, 55.
2. Sinyukov, A.; Hayden, L. M. *J. Phys. Chem. B* **2004**, *108*, 8515.
3. Quon, O. P.; Kwon, S-J; Jazbinsek, M.; Brunner, F.; Seo, J-I; Hunziker, C.; Schneider, A.; Yun, H.; Lee, Y-S; Gunter, P. *Adv. Funct. Mater.* **2008**, *18*, 3242.
4. McLaughlin, C.; Hayden, L. M.; Polishak, B.; Huang, S.; Juo, J.; Kim, T-D; Jen, A. K-Y *Appl. Phys. Lett.* **2008**, *94*, 151107.
5. Hayden, L. M.; Sinyukov, A. M.; Leahy, M. R.; French, J.; Lindahl, P.; Herman, W. N.; Tweig, R. J.; He, M. *J. Polym. Sci., Part B: Polym. Phys.* **2003**, *41*, 2492.
6. Natansohn, A.; Rochon, P.; Barrett, C.; Hay, A. *Chem. Mater.* **1996**, *7*, 1612.
7. Xu, G.; Si, J.; Liu, X.; Yang, Q. G.; Ye, P.; Li, Z.; Shen, Y. *J. Appl. Phys.* **1999**, *85*, 681.
8. Apostoluk, A.; Nunzi, J.-M.; Boucher, V.; Essahlaoui, A.; Seveno, R.; Gundel, H. W.; Monnereau, C.; Blart, E.; Odobel, F. *Opt. Commun.* **2006**, *260*, 708.
9. Kitaoka, K.; Si, J.; Mitsuyu, T.; Hirao, K. *Appl. Phys. Lett.* **1999**, *75*, 157.
10. Kitaoka, K.; Matsuoka, N.; Si, J.; Mitsuyu, T.; Hirao, K. *Jpn. J. Appl. Phys.* **1999**, *38*, 1029.
11. Herman, W. N.; Hayden, L. M. *J. Opt. Soc. Am. B* **1995**, *12*, 416.
12. Nahata, A.; Shan, J.; Yardley, J. T.; Wu., C. *J. Opt. Soc. Am. B* **1993**, *10*, 1553.

Chapter 8

Photosensitive Polymeric Materials for Two-Photon 3D WORM Optical Data Storage and Microfabrication

Ciceron O. Yanez,[†] Carolina D. Andrade,[†] Sheng Yao,[†] Gheorghe Luchita,[†] Mykhailo V. Bondar,[§] and Kevin D. Belfield[*,†,‡]

[†]Department of Chemistry University of Central Florida, Orlando, FL 32816
[‡]CREOL, The College of Optics and Photonics, University of Central Florida, Orlando, FL 32816
[§]Institute of Physics, Academy of Sciences of Ukraine, Prospect Nauki, 46, Kiev-28, 03028
[*]Correspondence: belfield@mail.ucf.edu

Introduction

Since the first documented optical data storage (ODS) system was patented in the late 1920's the ODS capacity has been systematically increasing. This system, invented by the father of modern television John Logie Baird, was designed to store 15 minutes of poor quality TV by today's standards (*1*). Since then, researchers have come up with innovative new ways to attain ever smaller feature sizes used for writing, reading and processing digital information. The storage capacities of current ODS technologies, such as the compact disc (CD), digital video disc (DVD), and Blu-ray disc, depend on the bit density that can be achieved on the two-dimensional , 2D, recording surface. The bit densities in all of these technologies adhere to the Raleigh criterion, that, when applied to ODS, can be expressed as:

$$D \propto k \frac{S * NA^2}{\lambda^2} \quad (1)$$

where D is the storage capacity (usually expressed in bytes), S is the effective recording surface, NA is the numerical aperture of the scanning objective that is used for the writing process, and λ is the wavelength used for recording the information (*2*). These current ODS systems have increased their storage capacity by increasing the NA of the scanning objective, decreasing the wavelength of their source, or both. Further reduction of these two parameters beyond what is currently employed by the Blu-ray system (NA = 0.85, λ = 405 nm) does not seem feasible or is too expensive to be implemented commercially.

More recently, the increase of the storage capacity in ODS systems has relied on the development nonlinear optical effects both in the data recording and reading stages of the process. The storage capacity advantage that nonlinear ODS systems have over their linear counterparts is based on the quadratic dependence of two-photon absorption, 2PA, with respect to the intensity of incident light. This quadratic dependence of 2PA enables the photochemical processes that depend on this absorption to be confined to very small volumes. Hence, this nonlinear dependence leads to immense data storage capacity (*3*). Several efforts have been made to further develop both erasable (*4*) and non-ereasable (*3*, *5*, *6*) (write once read many, WORM) systems. The increased spatial resolution in nonlinear absorption processes enable true 3D ODS, and many recent efforts have proven that subdiffraction limit features can be recorded when the two-photon absorption properties of the ODS systems are exploited. A key step in this direction in the ODS field was the seminal work published by Rentzepis *et. al.* in 1989 (*7*). In this work two-photon writing, erasing, and reading was achieved by the photochromic interconversion of spyrobenzopyrans.

A fluorescence readout WORM ODS system was reported by this group where protonation of a 2PA fluorene dye, **1** (structure shown in Figure 1), in solution with a photoacid generator (PAG) was performed by both one- and two-photon excitation of photosensitive polymer films (*5*). Commercially available photoacid generators (PAGs) were used in this work. Because the 2PA cross sections of commercially available PAGs has been reported to be low (*8*), to further develop this system, more efficient 2PA PAGs were needed. Furthermore, in this effort, the potential for writing and reading in several layers by 2PA, without crosstalk was mentioned. Herein, we report the use of new photosensitive polymeric systems comprised of novel 2PA absorbing dyes **2-4**, and PAGs **5-7** in different polymer matrices. The chemical structures of these new dyes and PAGs are shown in Figure 1. In one of these new systems, comprised of a novel PAG, **5**, a new 2PA absorbing dye **2** in solution in a polymer matrix, we report a true three dimensional, 3D, WORM optical data storage system where the two-photon advantage is fully exploited. This improved ODS system allowed the two-photon recording and readout of voxels in five consecutive layers within the polymer matrix. The nonlinearity of the components ensured a crosstalk free system in which the 3D character of this 2PA ODS was truly realized.

Experimental

Materials

The syntheses of all PAGs and fluorescent dyes are reported elsewhere (*9–11*). The phosphorylated poly(VBC-co-MMA) used as polymer matrix was synthesized according to the literature (*12*). All solvents and monomers were used as received. The masks used during the photoexposure of prepared polymer films included TEM grids (600-mesh and 400-mesh hexagonal grids from Polysciences) and glass resolution targets (negative slide with the 1951 USAF test pattern from Edmund Scientific). Shell Epon® SU-8 (formaldehyde, polymer with (chloromethyl)oxirane and 4,4'-(1-methylethylidene)bis[phenol]), Sartomer K-126 (3,4-epoxycyclohexylmethyl 3,4-epoxycyclohexane carboxylate) were used as received. All PAGs used were synthesized according to the literature (*11*). Solvents were used as received. The glass resolution targets were purchased from Edmund Scientific (negative slide with the 1951 USAF test pattern).

One-Photon Recoding and Readout

One-photon recording was carried out by irradiating the photoreactive polymer with a Loctite 97034 light source equipped with a 200 W mercury lamp and an internal shutter to control the exposure times. The photomask was either projected or simply placed (as a contact mask) on the dry polymeric film. Output from the waveguide of the light source was focused into the condenser of an Olympus IX-81 confocal microscope. One-photon fluorescence images were recorded on this microscope that was equipped with a Hamamatsu EM-CCD C9100 digital camera. One-photon confocal fluorescence images for readout were taken using a FITC filter cube (Ex:477/50; DM: 507; Em:536/40) and a modified TRITC filter cube (Ex:525/40; DM: 555; Em:624/40) for the neutral and protonated forms of the dyes, respectively.

The films were solution cast onto 2.5 x 2.5 cm microscope glass cover slips. The film thickness of the polymer films was approximated by focusing on the glass surface and then on upper surface of the polymer and determining the distance the z stage had moved from one surface to the other.

Two-Photon Recording and Readout

Two-photon recording and readout were performed on a modified Olympus Fluoview FV300 laser scanning confocal microscopy system equipped with a broadband, tunable Coherent Mira Ti:sapphire laser (recording at 730 or 760 nm; readout at 860 nm, and 115 fs pulse width, 76 MHz repetition rate), pumped by a 10 W Coherent Verdi frequency doubled Nd:YAG laser. In two-photon writing, the exposure time and position was controlled by means of an electronic shutter and electronic stage, respectively, both from Thor Labs.

Spectra

Absorption spectra were recorded with an Agilent 8453UV–vis spectrophotometer. Steady-state fluorescence spectra were measured with a PTI Quantamaster spectrofluorimeter.

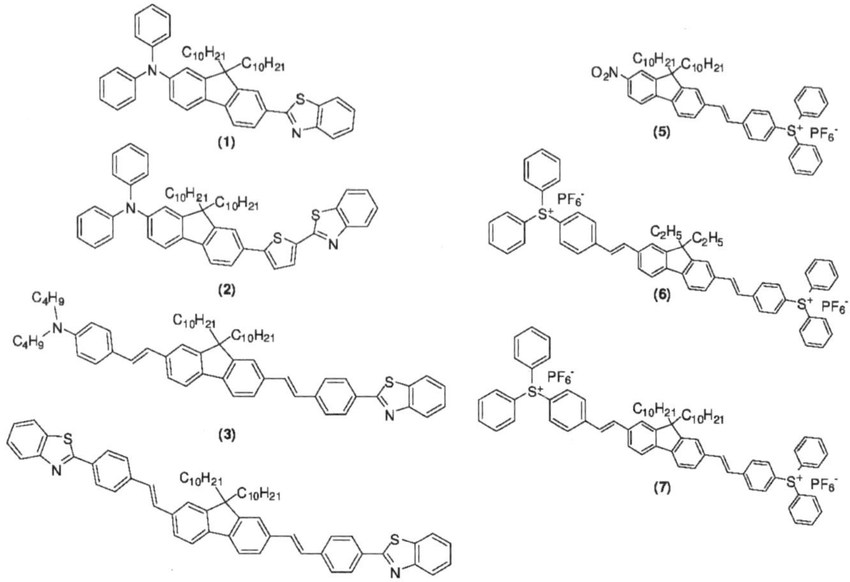

*Figure 1. 2PA absorbing dyes (**1-4**) and 2PA absorbing PAGs (**5-7**).*

Fluorescence Quantum Yield Measurements

Fluorescence quantum yields were determined relative to 9,10 diphenylanthracene in cyclohexane as a standard (*13*). Measurements were made in the photon counting regime of the PMT using an L-format configuration. The fluorescence spectra were corrected for the spectral dependence of the PMT. All measurements were performed at room temperature in 1 cm quartz cuvettes with dye concentrations on the order of 10^{-6} M.

Photoacid Quantum Yield Measurements

Steady-state photoacid quantum yields were measured by selectively exciting the PAG solutions at the desired wavelength using the monochromator of the PTI spectrofluorimeter. Irradiance of the incident radiation was measured with an Ophir Power Star power meter equipped with a UV 1.44 cm^2 detector head. Rhodamine B was used as a sensor of photoacid generation, observing that the change in optical density of the sulfonium salt didn't exceed 5% (*14*). Photoacid quantum yields were calculated by means of the following expression:

$$\Phi_{\lambda}^{H+} = \frac{\dfrac{\Delta OD^{555nm}}{\varepsilon_{555nm}^{RhB} * d * V_{sol}} N}{\Delta S * I_0 * \Delta t * (1 - 10^{-OD_0})} \qquad (2)$$

Where ΔOD^{555nm} is the change in optical density upon generation of rhodamine B measured at 555 nm, $\varepsilon^{RhB}{}_{555}$ is the extinction coefficient of rhodamine B at 555 nm in acetonitrile, d is the radiation path length (typically 1cm) of cuvette, ΔS is the area of solution irradiated (photon flux area), Δt is the time of irradiation, and OD_0 is the average absorption intensity (which did not change by more than 0.5% at any given time). The quantum yield of photoacid generation of triphenylsulfonium tetrafluoroborate was determined by this method ($\Phi^+ = 0.57$) and was comparable to the value reported in the literature (0.54) (*11*). PAG solutions were on the order of 10^{-5} M in acetonitrile.

General Procedures for 1PA Lithography and 2PA Microfabrication

Glass Functionalization

No. 1 glass microscope cover slips or 2.5 x 2.5 cm microscope glass slides were treated with a piranha solution (7:3 sulfuric acid:30% hydrogen peroxide) for 10 min at 60 °C and dried under a stream a nitrogen. The etched slides were placed in a 30% solution of 3-aminotrimethoxysilane in ethanol and dried again with nitrogen.

Spin Coating

Epon SU-8 and Sartomer K-126 (in an 8:2 ratio) were mixed with 1% weight of 6 and a minimal amount of propylene carbonate. The homogeneous mixture was dissolved (30%) in cyclopentanone. The resulting mixture was filtered through a 0.45 μm glass filter. Once filtered, the resin was spin coated on the substrates (1. 500 rpm, 10 sec and 2. 1000 rpm, 30s). The samples were dried in a vacuum oven over night at 80 °C.

Exposure and Developing

The resin was exposed to broadband UV radiation through (one-photon) and 760 nm (two-photon) for a number of different exposure times. Post-exposure baking was carried out on a hotplate at 90 °C (10 min). The resist was developed by carefully washing away the unexposed resist with propylene glycol methyl ether acetate, followed by isopropanol. The slide was then dried under nitrogen.

1PA and 2PA Polymerization

1PA and 2PA polymerizations were performed on the same respective systems described for ODS.

Discussion

Optical Data Storage

The System

Writing in the ODS system, originally reported by this group, was achieved by monoprotonation of the benzothiazole fluorene dye 1 when exposed to one- or two-photon irradiation of the adequate wavelength in presence of a commercially available PAG (*5*). Irradiation of localized volumes within the polymer matrix induced photodecomposition of the sulfonium salt, generating hexafluorophosphoric acid. Because the pK_b of the benzothiazole amine ($pK_b \approx 13$) (*15*) is lower than that of the diphenyl amine functionality ($pK_b \approx 19$) (*16*), the benzothiazole amine was the first to be protonated upon production of superacid. The monoprotonated species of the 2PA dye appeared red shifted in the absorption spectra (Scheme 1, 1a) due to the increase of electron withdrawing character of the protonated benzothiazole, resulting in emission at longer wavelengths than that of the unexposed, neutral form of the dye (*5*). Unprotonated 1 and protonated 1a could be interrogated for readout, enabling positive and negative image formation when addressing the exposed polymer at shorter or longer wavelengths, respectively. Longer wavelength readout had the advantage that the photoacid was unaffected at these wavelengths, resulting in nondestructive readout.

To evaluate the versatility of this ODS system, several of its components were studied, namely PAG efficiency, fluorescent dye efficiency, and polymer matrix.

The 2PA Fluorescent Dyes

To asses whether using systems with more basic, aliphatic amines (instead of phenyl amines) would still allow one to write and readout, dye **3** was characterized and used. Spectrophotometric absorption studies of this dye in solution with dipheyliodonium hexafluorophosphate showed an initial, quick blue shift of the absorption λ_{max} of this dye after 3 s of irradiation (Figure 2, **3a**), due to the protonation of the dibutylamine, as demonstrated by ^{1}H NMR and COSY studies. This quick blue shift was followed by a progressive red shift, due to the formation of species **3b**, affording recording and readout of images from both channels. As can be seen in the inset in Figure 2, the highest concentration in solution of the protonated species **3b** was reached after 15 s of irradiation. Over exposure led to an extreme blue shift of this dye, resulting in a species with an absorption λ_{max} of approximately 330 nm. Interrogating this blue shifted species would result in destructive readout because either the absorption or fluorescence spectra would overlap with the spectrum of the PAG, inducing acid generation. Furthermore, the 2PA wavelength would be below the output window (700-1000 nm) of the Ti:sapphire laser. Despite the limitations inherent to using aliphatic amine dyes, the versatility of the system allowed recording and readout while both channels were observed by 1PA fluorescence readout (Figure 5).

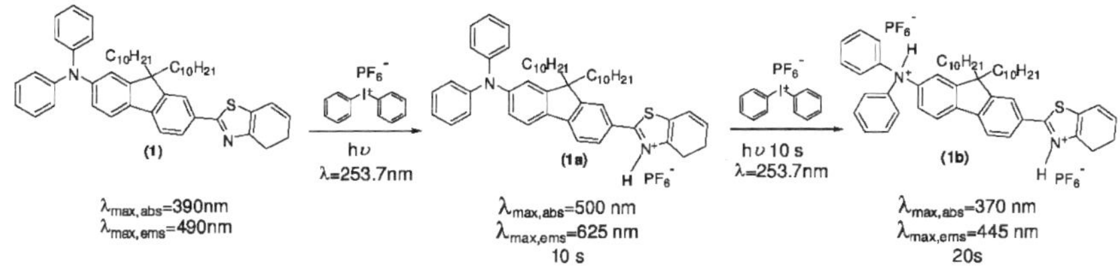

Scheme 1. Protonation of PAG (1) in ACN solution with increase of photoacid concentration upon exposure.

The behavior of dyes containing aromatic amines as a donor and benzothiazole as an electron-acceptor, such as **1** and **2**, is more predictable because the benzothaizole amine, having a higher pKb, will be the first to protonate once the acid is generated upon irradiation, yielding species **2a**. The solution studies of dye **2** (Figure 3 and Scheme 2) proved that overexposure resulted in a milder blue shift upon the generation of species **2b**. This effect gave the system greater tolerance to excessive irradiation time and power, as species **2a** and **2b** can both be interrogated without affecting systems in which conjugated PAGs are used. Furthermore, the absorption profile of species **2a** extended beyond 600 nm. This ultimately enabled effective 2PA readout at longer wavelengths (860 nm) where the 2PA of the PAG was negligible.

The symmetrical A-π-A system of dye **4** was quickly converted to a D-π-A chromophore upon the generation of photoacid (as shown in Figure 4 and Scheme 3), generating species **4a**. The clear transitions in solution studies reveal the protonation of a first benzothiazole group, generating a red shift due to the increased electron-withdrawing properties of the protonated benzothiazole, followed by the protonation of the second benzothaizole and consequent blue shift, forming species **4b**. This dye however presented solubility problems within the polymer matrix, at concentrations higher than 1.5% (w/w) the dye molecules crystallized, generating scattering domains that significantly affected the optical quality of the polymer film.

The Polymer Matrix

Three different polymer matrices were used as supports. Phosphorylated poly(VBC-*co*-MMA) is a highly transparent, robust matrix in which all of the dyes and PAGs were very soluble except **4**. However, this matrix required the use of solvents to help dissolve all of the system components and allow one to cast the polymers onto the glass slide support. These solvents must be low vapor pressure solvents that ensure that a bubble free film will remain after solvent drying. Ultimately, this translated into film drying procedures that required greater than 24 h and involved a series of steps.

To expedite the film drying process, creation of a polymer matrix from liquid monomers that would dissolve the 2PA dye and PAG was developed. Two polymer substrates were made from the photocrosslinking **7** and **8** in different proportions with **9**, in the presence of a free radical photoinitiator (Irgacure 184) and doped with the PAG-dye mixture. Even though thiolene crosslinked polymer matrices are formed by a free radical initiated mechanism, these reactions are typically not very oxygen-sensitive since they involve chain transfer termination by reaction with the thiol SH. Furthermore, many different mechanical and visco-elastic properties of the resulting polymer can be obtained by tailoring the structure of the monomers (*17*). Special care was taken in choosing the adequate free radical photoinitiator λ_{max} so as to not generate photoacid during the formation of the polymer film. The films were made in 3 to 10 min and images were successfully recorded by both channels (Figure 5). The FITC filter cube of the confocal microscope was equipped with an excitation bandpass filter (452-502 nm), a dichroic mirror (507 nm) and an emission bandpass filter (496-556 nm) that allowed fluorescence readout of the unprotonated species of the dye where the resist remained unexposed (Figure 5 A and C). The modified TRITC filter cube, equipped with an excitation bandpass filter (505-545 nm), a dichroic mirror (555 nm) and an emission bandpass filter (604-644 nm), was used to capture the fluorescence of the exposed protonated form of **3** (Figure 5 B and D).

The 2PA Photoacid Generators

The synthesis of the PAGs was carried out by the previously described methods. Since commercially available PAGs are not conjugated enough to generate photoacid by 2PA in the near-IR and because their 2PA cross sections are low, the PAGs used in this system were designed to have to have high 2PA cross sections at wavelengths accessible to a Ti:sapphire laser system (700-1000 nm). It has been observed that symmetrical and unsymmetrical conjugated systems with two electron-donating groups (D) or electron accepting subtituents (A) tend to have significant 2PA properties and high 2PA cross sections (*18–20*). Based on the high thermal and photophysical stabilities of the florenyl π-system (*21*), fluorene was chosen as the core structure when designing the PAGs. Because fluorene can be readily substituted in its 2-, 7-, and 9-positions a stilbene or thiophene motif was introduced (2- and/or 7-positions) to increase the photoacid generation efficiency per molecule by incorporating two sulfonium salt motifs per molecule in an A-π-A architecture, **6** and **7**. However, the high fluorescence of these molecules, as demonstrated by their high fluorescence quantum yield values, hindered the efficiency of photoacid generation (Table 1).

The direct photolysis of triarylsulfonium salts has been reported to occur mainly from the singlet state. However, sensitization studies have proven that triplet triarylsulfonium salts are also labile (*22*). Consequently, we incorporated a nitro group in the architecture of PAG **5** to induce the internal heavy atom effect, quench the fluorescence of the PAG by induction of intersystem crossing, and increase the photoacid quantum yields. The photoacid quantum yield values of PAG **5** showed a significant improvement in photoacid generation efficiencies when compared to PAG **7**, indicating that that the introduction of groups that favor intersystem crossing by means of the internal heavy atom effect is a viable mechanism for increasing the photoacid quantum yield of highly fluorescent triarylsulfonium salt PAGs.

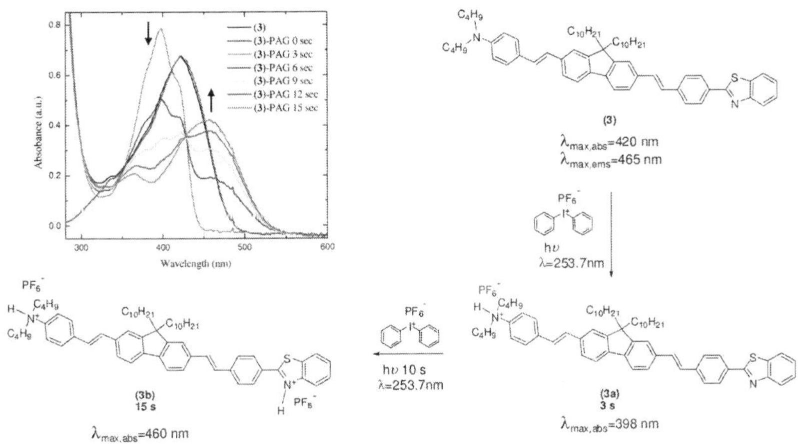

Figure 2. Spectral change of ***3*** *in ACN solution with diphenyliodonium hexafluorophophate (PAG) upon irradiation at 253.7 nm. Equililibria of species* ***3, 3a****, and* ***3b*** *at irradiation times ranging from 0-15 s. (see color insert)*

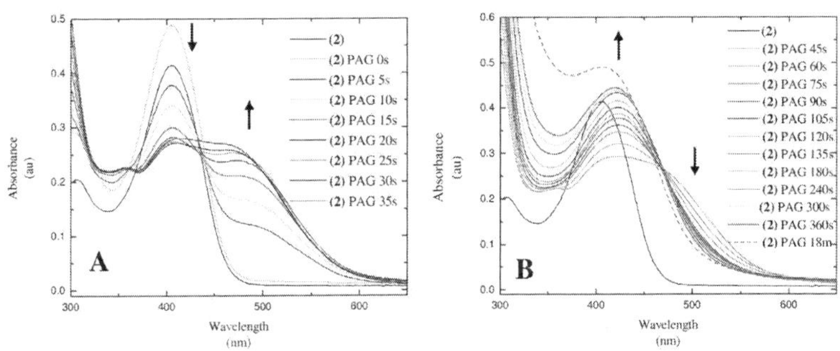

Figure 3. Spectral change of ***2*** *in ACN solution with diphenyliodonium hexafluorophophate (PAG) upon irradiation at 253.7 nm.* ***A)*** *Equililibria of species* ***2*** *and* ***2a*** *at 0-35 s of irradiation.* ***B)*** *equililibria of species* ***2a*** *and* ***2b*** *at 45 s-18 min of irradiation. (see color insert)*

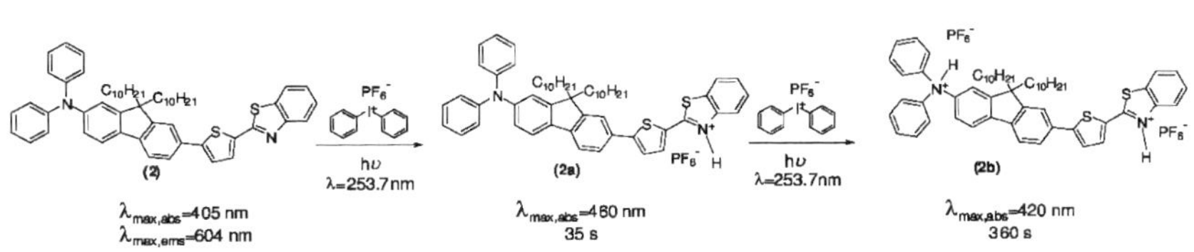

Scheme 2. Sucesive protonation of 2PAfluorescent dye ***2*** *in ACN solution with increase of photoacid concentration upon exposure.*

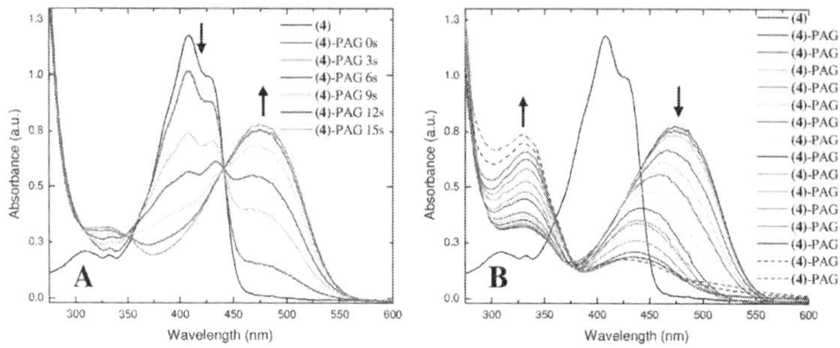

*Figure 4. Spectral change of **4** in ACN solution with diphenyliodonium hexafluorophophate (PAG) upon irradiation at 253.7 nm. **A)** equililibria of species **4** and **4a** at 0-15 s of irradiation. **B)** equililibria of species **4a** and **4b** at 18-59 s of irradiation. (see color insert)*

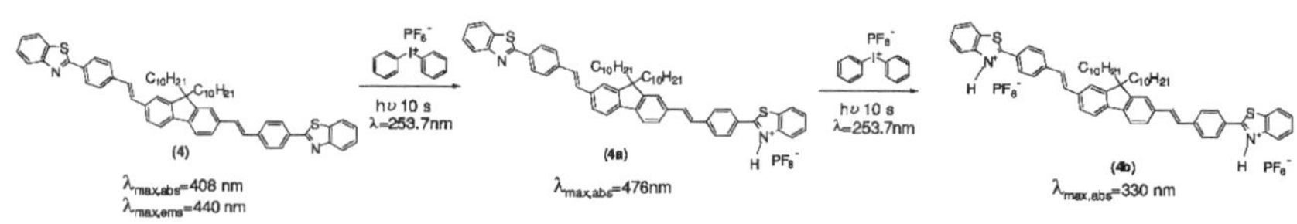

*Scheme 3. Sucesive protonation of 2PA dye **4** in ACN solution with increase of photoacid concentration upon exposure.*

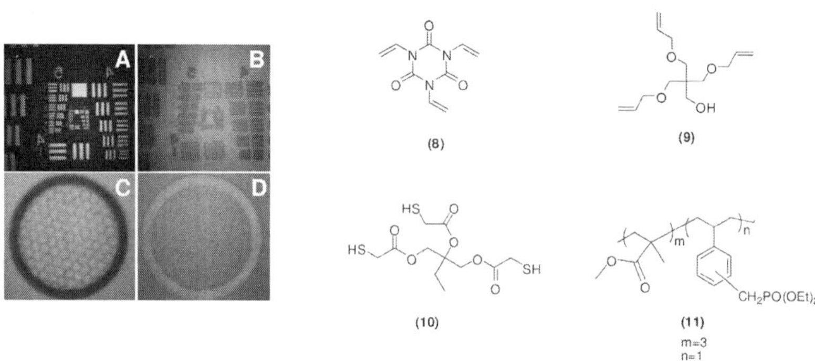

*Figure 5. One-photon writing and readout images with contact (A and B) and projection photomasks (C and D). Composition: **6** 12%; **3** 1.3%; **8** 25%; **9** 25%; **10** 50%. Exposed at 253.7 nm (3 min to form film). Later exposed at 300-400 nm (40 s for image recording).*

The photacid quantum yields were determined by a steady state method (*14*). The solutions of the PAGs in acetonitrile were selectively irradiated at the desired wavelength with an excitation source of a spectrofluorimeter. Rhodamine B was used as a sensor for photacid generation as reported by Scaiano *et. al.* (*14*) Special care was taken in observing a photo-decomposition conversion no greater than 5% in order keep secondary photoproduct acid generation to a minimum.

One-Photon vs. Two-Photon Writing and Readout

The advantage of using two-photon absorbing dyes is clearly evidenced in the initial two-photon readout experiments, Figure 6, in which the images were recorded by projecting the image of a 600 mesh TEM hexagonal grid onto the photoreactive polymer surface using a 200 W mercury lamp as the light source. After recording the image by 1PA, the fluorescent image was collected by both one 1PA and 2PA. The two-photon, upconverted fluorescence image (readout) proved to be much more forgiving to film defects when compared to the one photon readout. Furthermore the inherent nonlinearity (Figure 6b) of the system allowed better contrast and sharper images than one-photon readout (Figure 6a). Scanning across the indicated line on the one photon readout images, afforded the fluorescence intensity scan (Figure 6c). A threshold can be established over which any fluorescence signal can be translated into a digital signal. The full width at half maximum (FWHM) of the fluorescence intensity scan peaks was used to estimate the minimum size feature for one-photon recording/two-photon readout (ca. 3 μm).

The advantage of using 2PA for writing and readout in all three dimensions of the polymer matrix was demonstrated by a simple experiment in which voxels were recorded by 2PA in five layers using 730 nm fs irradiation to induce photoacid generation via 2PA of PAG **5**. The electronic shutter was programmed to open for 50 ms at 10 μm intervals along the x and y axes. After recording (60x, 1.4 NA oil immersion objective) an x-y layer (plane) of voxels, the microscope objective was moved ≈12 μm to focus deeper into the polymer matrix, generating another set of voxels. To ensure that the entire depth of the polymer matrix was used, this process was repeated until the objective focal point had cleared the polymer matrix. After recording, the voxels were readout by both one- and two-photon excitation. 2PA upconverted fluorescence readout was performed by scanning consecutive 0.4 μm layers of the polymer film at 860 nm in order to excite the protonated dye **2** (Figure 8B). Similarly, one-photon readout was performed by taking a series of confocal fluorescence pictures of the recorded volume at consecutive focal points of the objective separated by a distance of 0.4 μm (Figure 7A). The one-photon confocal fluorescence images were taken using a modified TRITC filter cube (Ex:525/40; DM: 555; Em:624/40). One and two-photon fluorescence intensity vs. distance (μm) graphs were then recorded for each voxel layer in order to compare the readout quality (signal-to-noise) obtained by each of the readout methods (Figures 7B and 8C, respectively). Clearly, in one-photon readout, the fluorescence signal-to-noise ratio is affected significantly as one focuses deeper into the polymer film. In the case of the two-photon readout experiment, where the two-photon advantage is fully exploited, the fluorescence signal readout shows well defined voxels, and comparable signal-to-noise ratio for all five layers through out the entire polymer matrix. Furthermore, the nonlinearity of the system ensured a crosstalk free system (Figure 8A) in which

Table 1. Quantum Yields of Photoacid Generation and Fluorescence of PAGs

PAG	Φ_{H+}	Φ_F
5	0.3[a] ;0.16[b]	*0.10*
7	0.03[b]	0.80

[a] 350 nm. [b] 400 nm in acetonitrile.

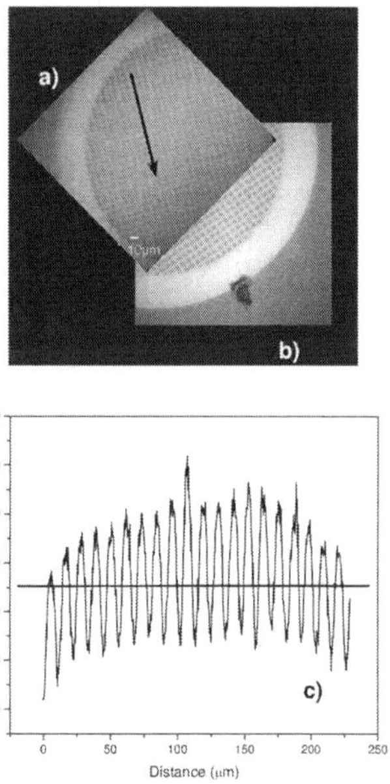

*Figure 6. Readout images using a TEM 600 mesh projection mask of a mixture of **1** and **6** in phosphorylated poly(VBC-co-MMA). (a) one-photon (FITC filter cube, 40x objective, 180 s recording exposure time, 20 ms readout time), (b) two-photon (40x objective, readout: 800 nm, 2mW, PMT band pass filters 510IF-550RIF).(c) Fluorescence intensity plot of 1PA fluorescence along arrow on Figure 6a.*

the 3D character of this ODS was truly evidenced, where the advantages of two-photon writing and readout were clearly demonstrated, resulting in increased storage density capacity (up to $\approx 5x10^{11}$ bytes/cm^3).

Two-Photon CROP Microfabrication

1PA Lithographic Studies

The photolysis of triarylsulfonium salts has been known to generate Brønsted acids. Typically the counter ion of these salts is a heavy atom halide salt (commonly AsF_6^-, SbF_6^-, PF_6^-, BF_4^-). In these cases, the acid generated from this photolysis is a superacid. The superacid anions are very bulky and quite poor nucleophiles, leading to their efficiency as photoinitiators for cationic ring

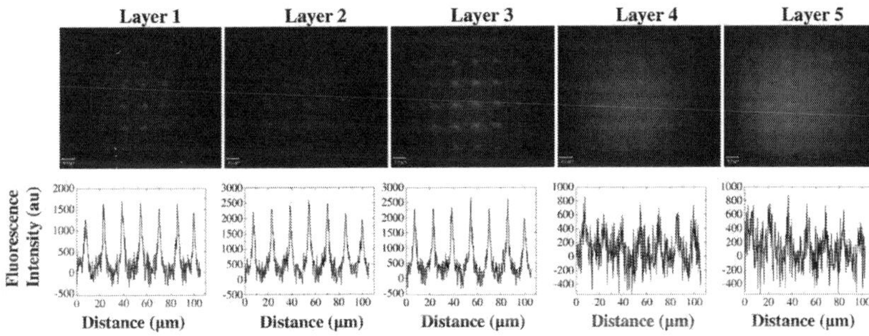

*Figure 7. Photosensitive polymeric system used for 3D, two-photon optical data storage. Composition: **2** (1%), **5** (5%), **11** (94%). Two-photon writing was performed at 730 nm (2.4 mW), 200 fs, 60 ms exposure/voxel with a 60x, 1.4 NA oil immersion objective. One-photon readout (upper) was performed, layer by layer (~0.4 µm/layer), with a modified TRITC filter cube (Ex:525/40; DM: 555; Em:624/40), with same objective used for writing. Layers 1-5 show significant reduction of the signal to noise ratio (lower) in the fluorescence intensity scan for all the layers. Scale bar: 10 µm. (see color insert)*

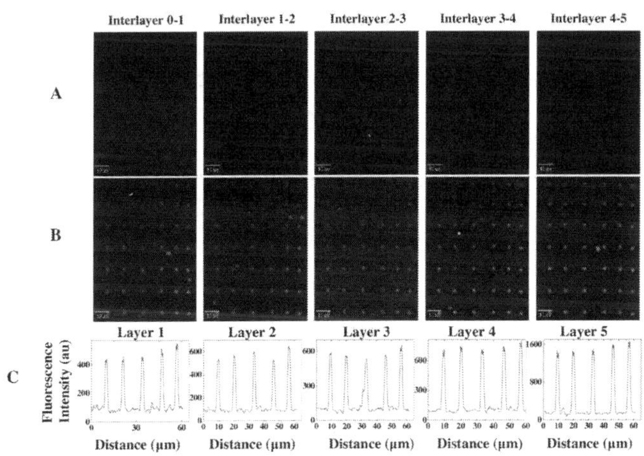

*Figure 8. Photosensitive polymeric system used for 3D, two-photon optical data storage. Composition: **2** (1%), **5** (5%), **11** (94%). Two-photon writing was performed at 730 nm (2.4 mW), 200 fs, 60 ms exposure/voxel with a 60x, 1.4 NA oil immersion objective. Two-photon readout was performed, layer by layer (~0.4 µm/layer), at 860 nm (7 mW), 200 fs, with same objective used for writing. (A) Blank interlayers. (B) Two-photon readout of the five layers. (C) Fluorescence intensity scan of all layers showing excellent signal to noise ratio throughout all five recorded layers of data. Note that because there is virtually no fluorescence signal in between the layers (A), the system is crosstalk free. Scale bar: 10 µm. (see color insert)*

opening polymerization (CROP) of epoxy resins. Among the many products of decomposition of the arylsulfonium salts, the most abundant are diarylsulfides (Figure 9). Since 6 and 7 have two sulfonium functionalities per molecule, they will likely form both symmetrical and unsymmetrical aryl sulfide photoproducts. Symmetrical diarylsulfide 6a is the synthetic precursor of 6, and was necessarily characterized during the synthesis of the sulfonium salt (*11*). Sulfide 6a was found to possess a fluorescence quantum yield of 1.0. When polymerization is carried out, this diarylsulfide that is formed from the corresponding sulfonium salt upon photolysis, is entrapped in the polymer matrix, and remains as such after development. The high fluorescence quantum yield of 6, 6a, and other similar photoproducts then allows one to observe highly fluorescent structures after development (Figure 10). This is potentially useful to monitor defects in microstructures that would otherwise be difficult to detect by other means.

The resist was deposited onto a glass cover slip by means of spin coating, and left to dry overnight (see details in the Experimental section). In the one-photon fabrication experiments, the resist system, Epoxy system Epon SU-8 and Sartomer K-126 (in an 8:2 ratio) and PAG **6** was exposed to UV-vis radiation from a high pressure mercury lamp focused through a liquid waveguide projected onto the stage of a confocal microscope. The microscope condensor helped to further focus the light through the photomask (USAF resolution mask) onto the surface of the resist. Once exposed, the resist was developed, and the same microscope was used to record DIC and confocal fluorescence images. In the DIC images of the structure resulting from the contact photomask experiments (Figure 10A) clear, undistorted structures were obtained with up to ~3.5 μm size features.

2PA Microfabrication

Finally, in order to evaluate whether these PAGs would induce two-photon CROP of epoxy resins for 3D microfabrication, a mixture of SU-8 and PAG **5** or SU-8 and PAG **7** (5% w/w of the PAG, relative to SU-8, in cyclopentanone) were spin coated on a glass coverslip. The stage was programmed to fabricate a pattern consisting of six 20 x 20 μm squares. This pattern was repeated at 6 μm intervals within the epoxy resin by moving the motorized microscope objective, creating open boxes or cubes. After inscribing the latent pattern into the resist, the sample was postexposure baked then developed. The result was a free standing set of open cubes (Figure 11). The 3D image was reconstructed from the layer-by-layer confocal fluorescence image data (as described above for the 1PA readout experiments in ODS), and overlaying them sequentially. The result of this superimposition is shown in Figure 11C.

Conclusions

Of the different polymer matrices that were used as supports, the phosphorylated poly(VBC-*co*-MMA) was a highly transparent, robust matrix in

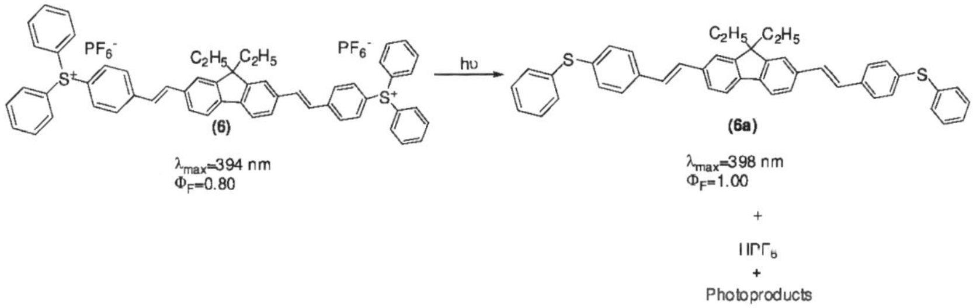

Figure 9. Photolysis of PAG 6.

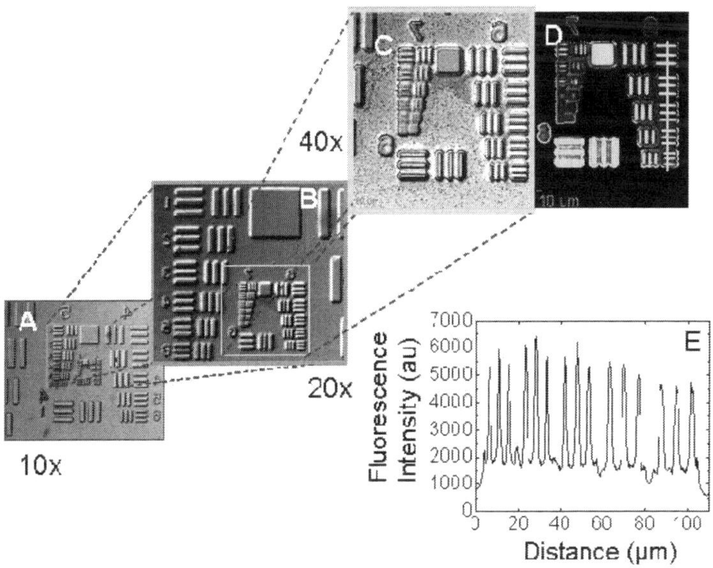

Figure 10. DIC images of microstructure generated by one-photon contact lithography using USAF target resolution masks. Polymerization exposure time: 300 s. A) 10x, 0.3 NA objective; 62 ms exposure time for DIC image. Scale bar: 100 μm; B) 20x, 0.45 NA objective; 250 ms exposure time for DIC image. Scale bar 50 μm. C) 40x, 0.6 NA objective; 62 ms exposure time for DIC image. Scale bar 50 μm. D) confocal fluorescence image (40x, 0.6 NA objective; 165 ms exposure time). Scale bar 10 μm. E) fluorescence intensity along the line shown in D. (see color insert)

which the dyes and PAGs were quite soluble. The photoinduced formation of a thiol-ene polymer matrix was, however, a feasible alternative in which one can easily obtain recording media of different mechanical and optical propeties.

The system proved to be versatile enough to tolerate a wide range of pK_b of the 2PA absorbing dyes. A wide variety of architectures can be used as the dye, as long as these dyes red shift upon protonation by an acid, possess high 2PA cross sections, along with high fluorescence quantum yields. The structures of the 2PA PAGs proved to be very useful in this system, and is of considerable

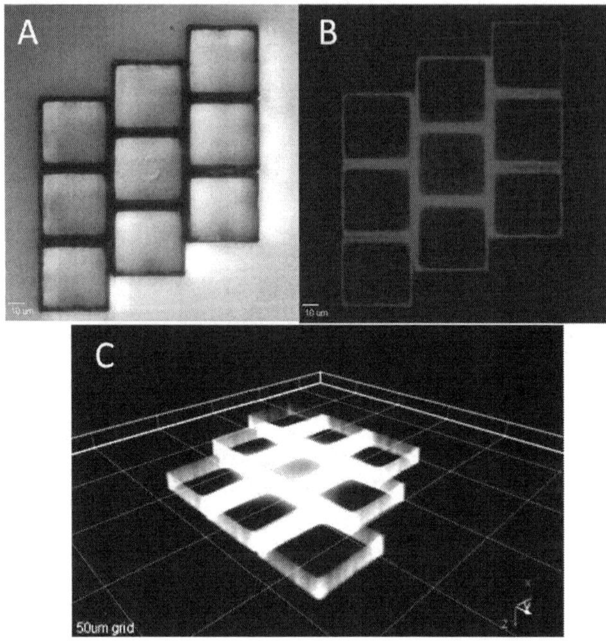

Figure 11. SU-8 with **5** *(5% w/w). Free standing structure resulting from scanning of 730 nm pulsed (210 fs, 76 MHz, 5 mW) Ti:sapphire laser irradiation. Two-photon microfabrication using a 60x, 1.4 NA oil immersion objective. One-photon readout was performed, layer by layer (~0.4 μm/layer), with a modified TRITC filter cube (Ex:525/40; DM: 555; Em:624/40), using a 40x, 0.45 NA objective. (A) DIC image, scale bar: 10 μm. (B) Confocal fluorescence image (one layer), scale bar: 10 μm. (C) 3D image reconstruction was done by overlaying all readout layers using SlideBook 4.1 (surface mode). Scale: 50 μm grid.*

value in systems in which the photoacid, generated by 2PA, is desired (i.e., 2PA microfabrication). These PAGs represent an important contribution to the very few sulfonium salt structures that have proven to generate photoacid upon 2PA (*23*, *24*).

The 2PA WORM ODS system, consisting of fluorene dye **2** (1%), PAG **5** (5%), and polymer matrix **11** (94%), proved to be resilient to overexposure. Furthermore, the inherent nonlinearity of 2PA components enabled a crosstalk free system in which the 3D character of the ODS was demonstrated. The advantages of two-photon writing and readout were clearly evidenced, affording a substantial storage density capacity (up to ca. 10^{11} bytes/cm^3).

A testament to the versatility of the new 2PA PAGs was their use in two-photon 3D microfabrication via cationic ring-opening polymerization of an epoxy resin system. Due to the nature of the PAG photoproduct, the resulting structures were readily visualized by fluorescence imaging.

Acknowledgments

The authors wish to acknowledge the National Science Foundation (ECS-0621715, CHE-0832622, and ECS-0524533) for support of this work.

References

1. Baird, J. L. Improvements in or relating to the recording of views of objects or scenes, or optical images or the like. British Patent GB289104, 1928.
2. van de Nes, A. S. ; Braat, J. M.; Pereira, S. F. High-density optical data storage. *Rep. Prog. Phys.* **2006** (69), 2323–2363.
3. Strickler, J. H.; Webb, W. W. 3-dimensional optical-data storage in refractive media by 2-photon point excitation. *Opt. Lett.* **1991**, *16* (22), 1780–1782.
4. Corredor, C. C.; Huang, Z. L.; Belfield, K. D. Two-photon 3D optical data storage via fluorescence modulation of an efficient fluorene dye by a photochromic diarylethene. *Adv. Mater.* **2006**, *18* (21), 2910–2914.
5. Belfield, K. D.; Schafer, K. J. A new photosensitive polymeric material for WORM optical data storage using multichannel two-photon fluorescence readout. *Chem. Mater.* **2002**, *14* (9), 3656–3662.
6. Dvornikov, A. S.; Rentzepis, P. M. Novel organic ROM materials for optical 3D memory devices. *Opt. Commun.* **1997**, *136* (1-2), 1–6.
7. Parthenopoulos, D. A.; Rentzepis, P. M. 3-Dimensional Optical Storage Memory. *Science* **1989**, *245* (4920), 843–845.
8. Schafer, K. J.; Hales, J. M.; Balu, M.; Belfield, K. D.; Van Stryland, E. W.; Hagan, D. J. Two-photon absorption cross-sections of common photoinitiators. *J. Photochem. Photobiol., A* **2004**, *162* (2-3), 497–502.
9. Belfield, K. D.; Schafer, K. J.; Mourad, W.; Reinhardt, B. A. Synthesis of new two-photon absorbing fluorene derivatives via Cu-mediated Ullmann condensations. *J. Org. Chem.* **2000**, *65* (15), 4475–4481.
10. Belfield, K. D.; Morales, A. R.; Kang, B. S.; Hales, J. M.; Hagan, D. J.; Van Stryland, E. W.; Chapela, V. M.; Percino, J. Synthesis, characterization, and optical properties of new two-photon-absorbing fluorene derivatives. *Chem. Mater.* **2004**, *16* (23), 4634–4641.
11. Yanez, C. O.; Andrade, C. D.; Belfield, K. D. Characterization of novel sulfonium photoacid generators and their microwave-assisted synthesis. *Chem. Commun.* **2009**, 827–829.
12. Belfield, K. D.; Wang, J. X. Modified horner-emmons reaction of polymeric phosphonates: Versatile synthesis of pendant stilbene-containing polymers. *J. Polym. Sci., Part A: Polym. Chem.* **1995**, *33* (8), 1235–1242.
13. Lakowicz, J. R. *Principles of Fluorescence Spectroscopy*; Kluwer Academic Publishers: New York, 1999.
14. Pohlers, G.; Scaiano, J. C.; Sinta, R.; Brainard, R.; Pai, D. Mechanistic studies of photoacid generation from substituted 4,6-bis(trichloromethyl)-1,3,5-triazines. *Chem. Mater.* **1997**, *9* (6), 1353–1361.

15. Dey, J. K.; Dogra, S. K. Solvatochromism and prototropism in 2-(aminophenyl)benzothiazoles. *Bull. Chem. Soc. Jpn* **1991**, *64* (10), 3142–3152.
16. Arnett, E. M.; Quirk, R. P.; Burke, J. J. Weak bases in strong acids. III. Heats of ionization of amines in fluorosulfuric and sulfuric acids. New general basicity scale. *J. Am. Chem. Soc.* **1970**, *92* (5), 1260–1266.
17. Belfield, K. D.; Crivello, J. V. *Photoinitiated Polymerization*; Oxford University Press: Washington, DC, 2003.
18. Denk, W.; Strickler, J. H.; Webb, W. W. 2-Photon laser scanning fluorescence microscopy. *Science* **1990**, *248* (4951), 73–76.
19. He, G. S.; Bhawalkar, J. D.; Zhao, C. F.; Park, C. K.; Prasad, P. N. Two-photon-pumped cavity lasing in a dye-solution-filled hollow-fiber system. *Opt. Lett.* **1995**, *20* (23), 2393–2395.
20. Ehrlich, J. E.; Wu, X. L.; Lee, I. Y. S.; Hu, Z. Y.; Rockel, H.; Marder, S. R.; Perry, J. W. Two-photon absorption and broadband optical limiting with bis-donor stilbenes. *Opt. Lett.* **1997**, *22* (24), 1843–1845.
21. Belfield, K. D.; Bondar, M. V.; Przhonska, O. V.; Schafer, K. J. One- and two-photon photostability of 9,9-didecyl-2,7-bis(*N*,*N*-diphenylamino)fluorene. *Photochem. Photobiol. Sci.* **2004**, *3* (1), 138–141.
22. Dektar, J. L.; Hacker, N. P. Photochemistry of Triarylsulfonium Salts. *J. Am. Chem. Soc.* **1990**, *112* (16), 6004–6015.
23. Zhou, W. H.; Kuebler, S. M.; Braun, K. L.; Yu, T. Y.; Cammack, J. K.; Ober, C. K.; Perry, J. W.; Marder, S. R. An efficient two-photon-generated photoacid applied to positive-tone 3D microfabrication. *Science* **2002**, *296* (5570), 1106–1109.
24. Zhou, W. H.; Kuebler, S. M.; Carrig, D.; Perry, J. W.; Marder, S. R. Efficient photoacids based upon triarylamine dialkylsulfonium salts. *J. Am. Chem. Soc.* **2002**, *124* (9), 1897–1901.

Chapter 9

Fabrication of High-Performance Optical Devices Using Multiphoton Absorption Polymerization

Linjie Li,[1] George Kumi,[1] Rafael R. Gattass,[1] Erez Gershgoren,[1] Wei-Yen Chen,[2] P.-T. Ho,[2] Warren N. Herman,[3] and John T. Fourkas[*,1,4,5,6]

[1]Department of Chemistry and Biochemistry, University of Maryland, College Park, MD 20742
[2]Department of Electrical and Computer Engineering, University of Maryland, College Park, MD 20742
[3]Laboratory for Physical Sciences, University of Maryland, College Park, MD 20740
[4]Institute for Physical Science and Technology, University of Maryland, College Park, MD 20742
[5]Center for Nanophysics and Advanced Materials, University of Maryland, College Park, MD 20742
[6]Maryland NanoCenter, University of Maryland, College Park, MD 20742
[*]fourkas@umd.edu

We report the fabrication of high-performance optical devices using multiphoton absorption polymerization (MAP). MAP has been used to create supported, waveguide-based optical devices, such as microring resonators, from both acrylic and epoxy polymers. The devices exhibit performance that is competitive with or superior to that of supported polymeric devices fabricated with conventional techniques. By taking advantage of the three-dimensional fabrication capabilities of MAP, it is also possible to create novel and compact waveguide-based architectures that promise to have unique functionality for applications in areas that include optical logic and sensing.

Introduction

Multiphoton absorption polymerization (MAP) is a technique that is attracting increasing attention in the realm of micro- and nanofabrication (*1–5*). In this method, photopolymerization is driven by excitation of a photoinitiator via the simultaneous absorption of two or more photons of laser light. Because the photons must be present in the same place at the same time for this absorption to occur, the excitation probability depends nonlinearly upon the laser intensity. As a result, excitation (and therefore polymerization) can be constrained to occur solely within the focal volume of a microscope objective. By scanning the position of the focal point, three-dimensional polymeric structures of arbitrary complexity can be created. After fabrication, any unexposed prepolymer resin is washed away with a solvent, leaving a free-standing, polymeric structure on the substrate.

One area in which MAP has found numerous applications is photonics. Examples of devices that have been fabricated include waveguide couplers (*6*), unsupported waveguide-based devices including Mach-Zehnder interferometers (*7*), microlenses (*8*), distributed-feedback dye lasers (*9*) and photonic crystals (*10–12*). However, in most of these examples the device performance was limited or additional materials processing had to be performed after fabrication of an original polymeric structure using MAP.

We have recently reported the fabrication of high-performance, supported, waveguide-based devices using MAP (*13*). These devices were fabricated from acrylic polymers. We demonstrated the creation of supported microring resonators that at communications wavelengths exhibited performance on par with that of some of the best supported polymeric devices created with conventional fabrication techniques.

Here we will review our work with acrylic waveguide-based devices created with MAP and then will discuss new results obtained with epoxy-based polymers. The higher refractive index of the epoxy polymers will lead to further enhancements in device performance. Finally, we also discuss new, 3-D architectures that will provide improved performance and new functionalities in devices created using MAP.

Experimental

Fabrication Setup

Our fabrication setup is based on a commercial Ti:sapphire laser (Coherent Mira 900-F) with a repetition rate of 76 MHz and a pulse duration of roughly 100 fs. The laser is generally tuned to 800 nm. The laser enters an inverted microscope (Zeiss Axiovert 100) through the reflected light port and is reflected to the objective with a dichroic mirror. Samples were mounted on a 3-D piezo stage (Physik Instrumente) controlled by a LabView program.

Acrylic Resins

A typical acrylic prepolymer resin (*14*) was composed of two polyacrylic monomers, dipentaerythritol pentaacrylate (Sartomer) and tris (2-hydroxy ethyl) isocyanurate triacrylate (Sartomer), at weight percentages of 54% and 43%, respectively. The photoinitator employed was Lucirin TPO-L (BASF) at 3% by weight, which has a modest two-photon absorption cross section at 800 nm (*15*) and has proven to be an effective two-photon initiator (*14*). The resin was mixed thoroughly prior to use. The threshold value of average power for fabrication in this resin is on the order of 2 mW.

Polydimethylsiloxane (PDMS) was used as a low-index substrate. Sylgard 184 (Corning) was mixed in a 10:1 mass ratio (prepolymer:curing agent) and was centrifuged to remove any bubbles. The PDMS resin was placed on a glass substrate that had been cleaned in an oxygen plasma. A piece of silicon wafer whose native oxide layer had been treated with (tridecafluoro-1,1,2,2-tetra-hydrooctyl)dimethylchlorosilane was pressed against the PDMS using a 200 μm spacer. The PDMS was then cured at 110 °C for 20 minutes and the spacer and silicon wafer were removed. The PDMS was again exposed to an oxygen plasma and then was reacted with (3-acryloxypropyl)trimethoxysilane in order to promote adhesion of the polymeric structures. Using a razor blade mounted on a micrometer, two parallel slices were made in the PDMS at a spacing of 250 μm, and all of the PDMS except that between the slices was peeled away.

For MAP fabrication, the substrate was immersed in the prepolymer resin. Fabrication was performed with a 100×, 1.45 NA oil-immersion objective (Zeiss α PLAN-Fluar). Waveguides were fabricated in successive slices with a spacing of 100 nm at a fabrication power of less than 2 mW at the sample. When fabrication was complete, successive washes in dimethyl formamide and ethanol were used to rinse away the unexposed resin.

Epoxy Resins

The epoxy resin used was SU8 2025 (MicroChem). Fused silica capillaries with a square cross section were used as substrates due to their lower refractive index. The substrate was cleaned to remove any debris and then treated with an oxygen plasma for 3 minutes. After plasma treatment, the capillary was placed in a solution of 10% aminopropyl trimethoxy silane and 90% methanol by volume. After 4 hours in the solution, the substrate was rinsed with methanol for 1 min, sonicated in methanol for 1 min, and then dried at 100 °C for 1 hour.

Surface-treated substrates were spin coated with a ~80 μm film of SU8 resin. The film was pre-baked in an oven at 65 °C for 5 min. It was then baked for 12 min in a 95 °C oven and then cooled down for 5 min in a 65 °C oven prior to being exposed. Waveguide structures were fabricated with a 40×, 1.3 NA objective at an average laser power of 8 mW measured prior to the objective (which had ~60% transmission). After exposure, the sample was post-baked for 5 min, 12 min and

5 min at temperatures of 65 °C, 95 °C and 65 °C respectively. The samples were then developed with MicroChem SU8 Developer for 10 min.

Device Characterization

All devices were characterized using an Agilent 8164A Lightwave Measurement System, which served as both the light source and the measurement system. Tapered, single-mode fibers were used to couple light into and out of the waveguide devices. Measurements were made at wavelengths ranging from 1530 nm to 1560 nm at a temperature of 21 °C.

Results

Acrylic Devices

Here we summarize some of the important results from our previously-reported studies (*13*). MAP is able to create acrylic waveguides with very low surface roughness. Although acrylic polymers have some degree of vibrational overtone absorption in the communications region of the spectrum (*16*), tests performed on waveguide structures of different lengths showed that propagation losses were too small to measure over distances of hundreds of microns. As shown in Figure 1, these waveguides support high-quality modes.

Microring resonators were fabricated in an "IOU" configuration, as shown schematically in Figure 2. Light was introduced into an input port on a straight waveguide (the "I"). The through port was at the other end of the input bus. Light couples into the microring (the "O") evanescently through a linear region in which the microring and the input bus are parallel. The separation between the input bus and the microring can be controlled precisely down to a spacing of ~200 nm during fabrication. The typical spacing of adjacent lines in a waveguide is 100 nm. Light in the microring is coupled into the ouput bus (the "U") through a second linear coupling region. The coupled signal exits the output bus at the drop port. The other end of this bus serves as the add port.

A set of typical devices is shown in Figure 3. Many devices can be fabricated on a single substrate, and parameters such as the length of coupling region and the radius of curvature of the rings can be changed readily. Fabrication of a typical IOU structure takes 20 minutes or less, so many structures can be fabricated and tested within a reasonable time period.

Representative through-port and drop-port data for an acrylic IOU device are shown in Figure 4. All of the resonances in the through-port spectrum are of similar depth, and exhibit an extinction of greater than 20 dB. This performance rivals that of the best supported polymeric devices that have been reported using conventional fabrication techniques (*17*, *18*). The resonances in the drop-port spectrum are also of similar heights. Other IOU devices that we have fabricated have exhibited drop-port spectra with considerably higher finesse (*13*).

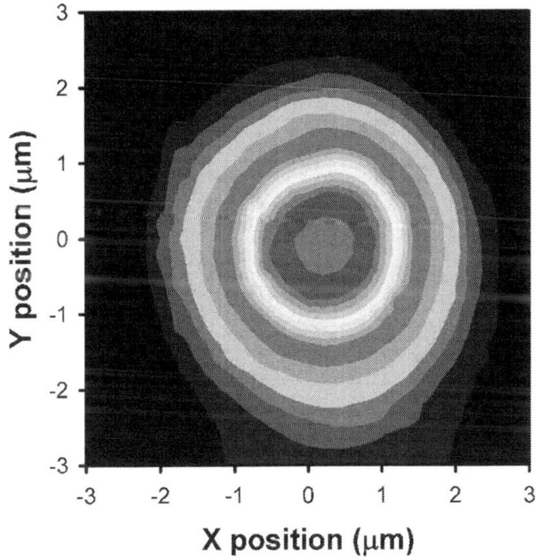

Figure 1. Contour map of a laser mode after propagation through an acrylic waveguide fabricated with MAP.

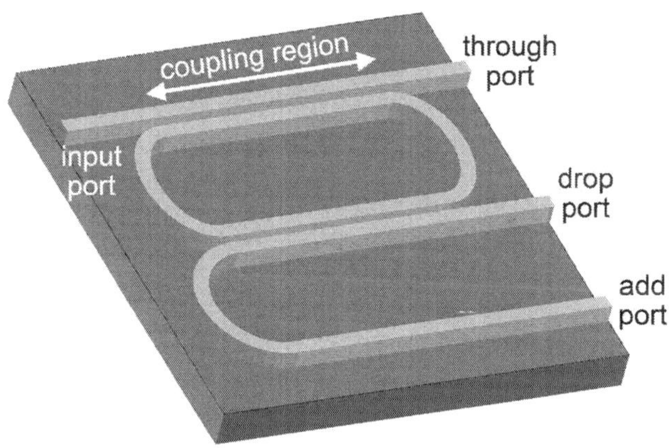

Figure 2. Schematic illustration of the layout of an IOU microring device.

Epoxy Devices

Using SU8 as the fabrication medium for MAP offers a number of advantages as compared to acrylic prepolymer resins. First, SU8 is a solid during fabrication, making it possible to avoid any issues with sagging during the fabrication of complex, 3-D structures. Second, SU8 also has a higher refractive index than typical acrylic resins, which is a favorable property for creating waveguide-based devices. Third, the minimal refractive index change of SU8 upon exposure makes it possible to create complex structures without distortions due to the fabrication beam passing through exposed regions of the resin. Finally, the

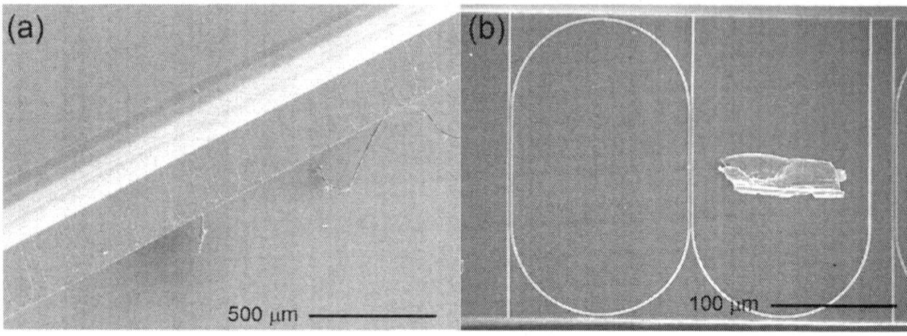

Figure 3. (a) PDMS substrate with a series of IOU devices with different designs. (b) Closeup of a single IOU device.

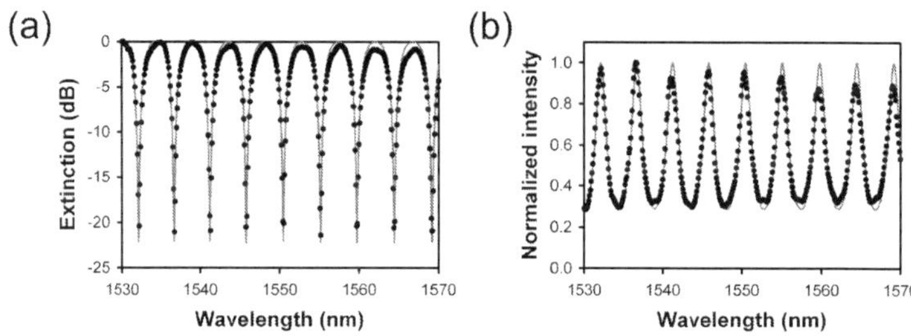

Figure 4. (a) A through-port extinction spectrum and (b) a drop-port transmission spectrum for a representative acrylic IOU device along with fits (solid lines). (Reproduced from reference (13). Copyright 2008 Wiley-VCH).

cationic polymerization of SU8 involves species that are less reactive than radicals, and may therefore be more amenable to the incorporation of other functional materials within structures.

SU8 also has a number of disadvantages compared to acrylic resins. For instance, SU8 is considerably more expensive than typical acrylic monomers. The processing of SU8 is also considerably more complex than that of acrylic resins. In addition, the adhesive properties of SU8 are generally not as good as those of acrylic resins. Finally, the low index change of SU8 upon exposure makes locating and leveling the interface between the photoresist and the substrate a difficult task.

Whether it is preferable to use acrylic resins or SU8 for MAP depends on the application being pursued. The advantageous properties of SU8 are attractive for the creation of waveguide-based devices, and so we have pursued its use in this application. For instance, representative IOU devices fabricated with SU8 are shown in Figure 5. The feature sizes and gaps that we can attain in the material are comparable to or smaller than those possible in acrylic resins, leading to even greater flexibility in the design and implementation of waveguide-based devices.

Through-port and drop-port spectra for a representative IOU device created from SU8 are shown in Figure 6. Both spectra demonstrate narrow resonances with similar intensities.

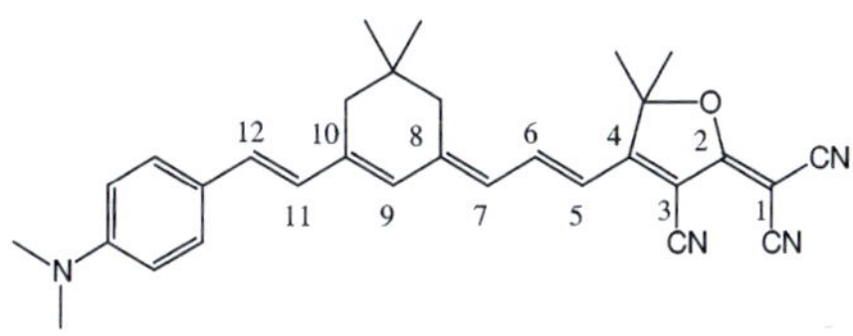

Figure 1. The CLD dye containing the isophorone and furan ring-locking units.

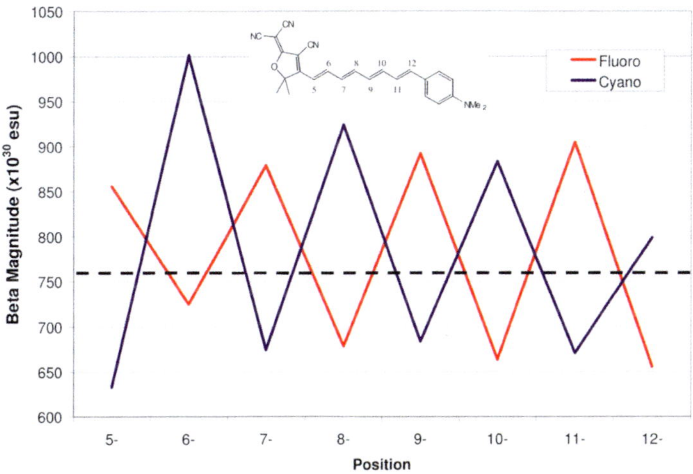

Figure 2. Calculated total magnitude of β versus position for the trans-polyene dye. Black dashed line is beta for the unsubstituted dye, 760 × 10^{-30} *esu. The IUPAC name of this dye scaffold is 2-{3-cyano-4-[8-(4-(dimethylamino)phenyl)-octa-13,5,7-tetraenyl]-5,5-dimethyl-5H-furan-2-ylidene}-malononitrile.*

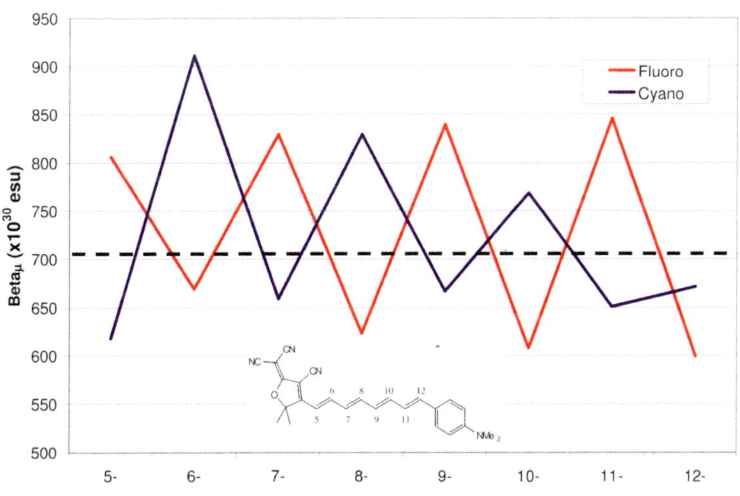

Figure 4. Calculated β_μ *versus methine position for fluorine and cyano mono-substitutions. The black dashed line is* β_μ *for the unsubstituted dye. Taken from reference .*

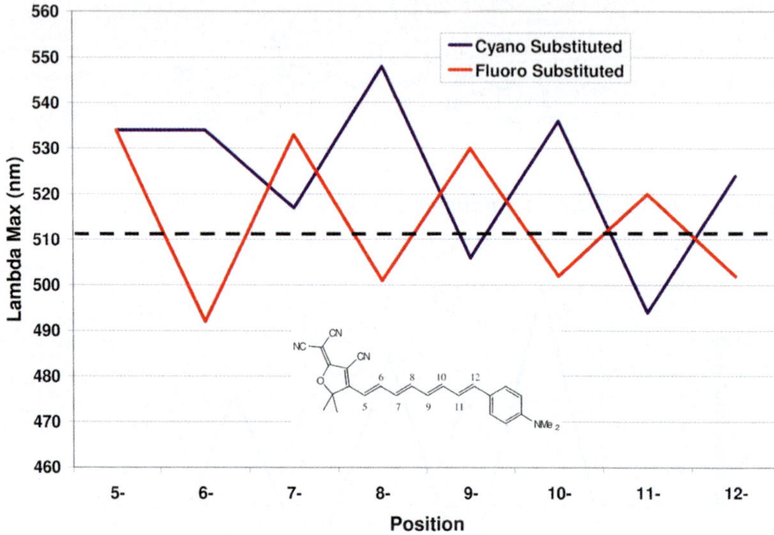

Figure 5. Calculated λ_{max} versus position for the trans-polyene dye. The black dashed line is the calculated λ_{max} for the unsubstituted dye (512 nm). The BH&B functional was used for the λ_{max} calculation.

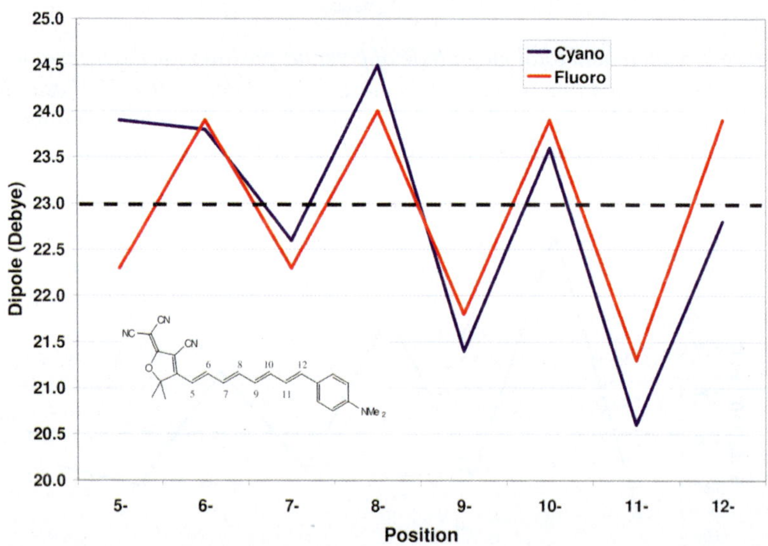

Figure 6. Calculated ground state dipole moment versus position for the trans-polyene (black dashed line is the dipole for the unsubstituted dye).

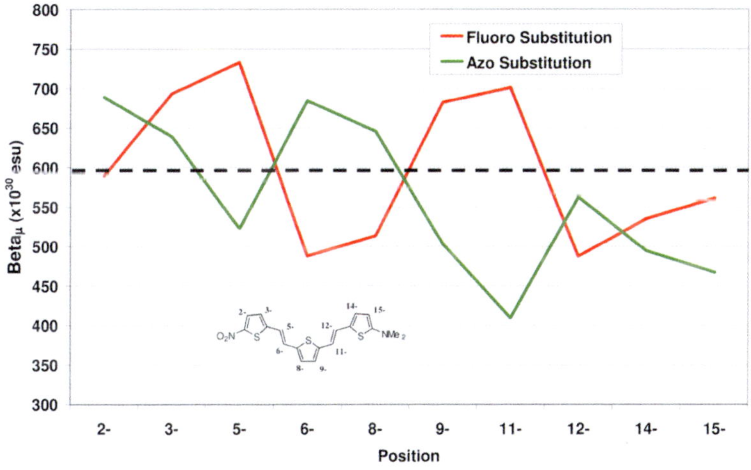

Figure 7. The calculated β_μ versus position for fluorine and azo mono-substituted dimethyl-[5-(2-{5-[2-(5-nitro-thiophen-2-yl)-vinyl]-thiophen-2-yl}-vinyl)-thiophen-2-yl]-amine. The black dashed line is β_μ for the bare dye (600×10^{-30} esu).

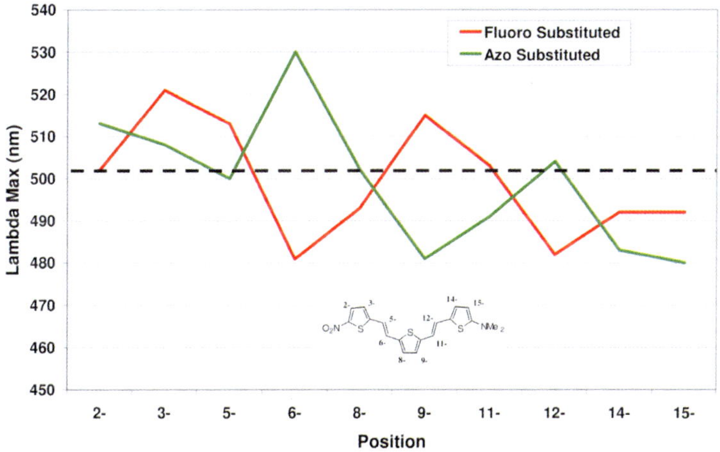

Figure 8. Calculated λ_{max} versus position for fluoro and azo mono-substituted dimethyl-[5-(2-{5-[2-(5-nitro-thiophen-2-yl)-vinyl]-thiophen-2-yl}-vinyl)-thiophen-2-yl]-anime. The black dashed line is λ_{max} for the bare dye (502 nm). The BH&B functional was used for the λ_{max} calculation.

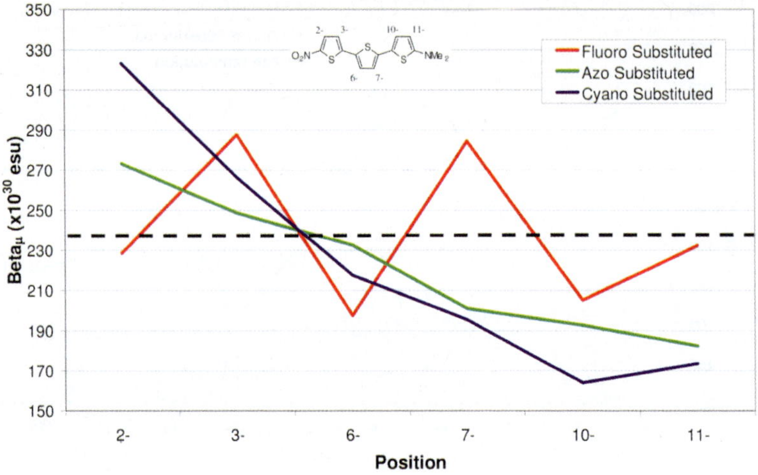

Figure 9. DFT-calculated β versus position for fluoro, azo and cyano mono-substitutions on 12-dimethylamino-1-nitrotristhiophene. The black dashed line is β_μ for the bare dye (237 × 10^{-30} esu).

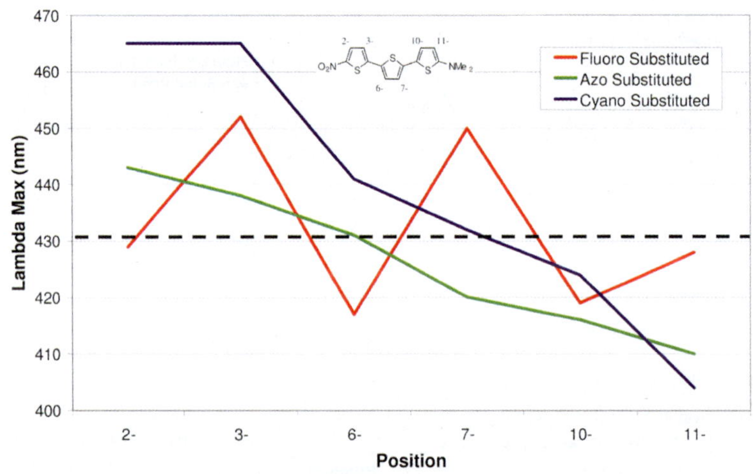

Figure 10. Calculated λ_{max} versus position for fluoro, azo and cyano substituted 12-dimethylamino-1-nitrotristhiophene. The black dashed line is λ_{max} for the bare dye (431 nm). The BH&B functional was used for the λ_{max} calculation.

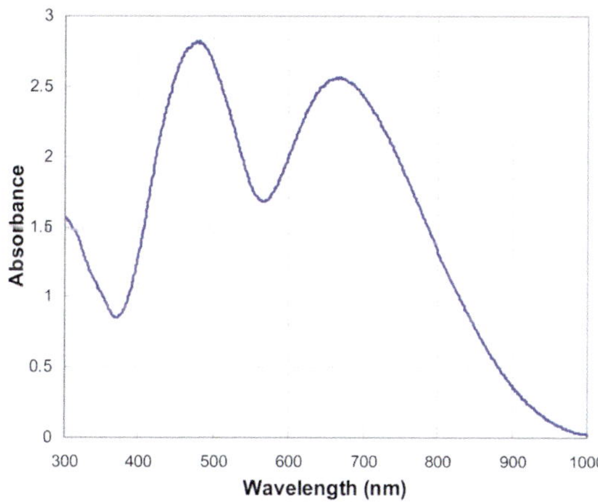

Figure 1. Absorption Spectrum of 1.9 um thick PMMA-DR1/CLD42A layer on glass substrate. The absorption spectrum shows the characteristic absorption peak of DR1 centered around 480nm, and then a second peak centered around 660nm that is contributed to CLD42A. The height of the absorption peaks are directly related to the chromophore concentration in the film, and in this case the solid concentration of CLD42A is 29%.

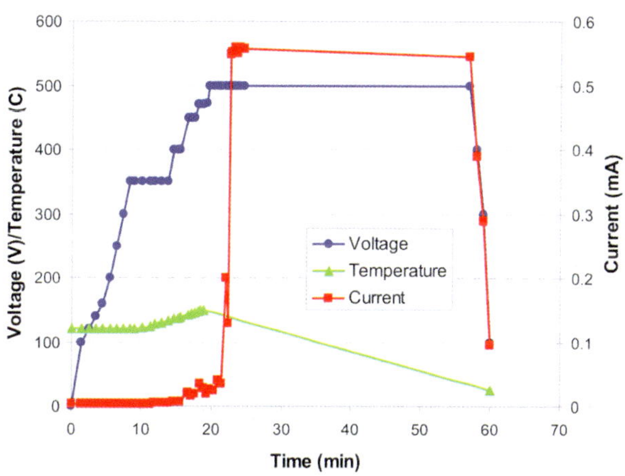

Figure 2. The poling schedule of PMMA-DR1/CLD42A film (1.5 μm). Blue curve is for the applied electric voltage, red curve is for the current flowing through the film, and green curve is for the temperature of the sample.

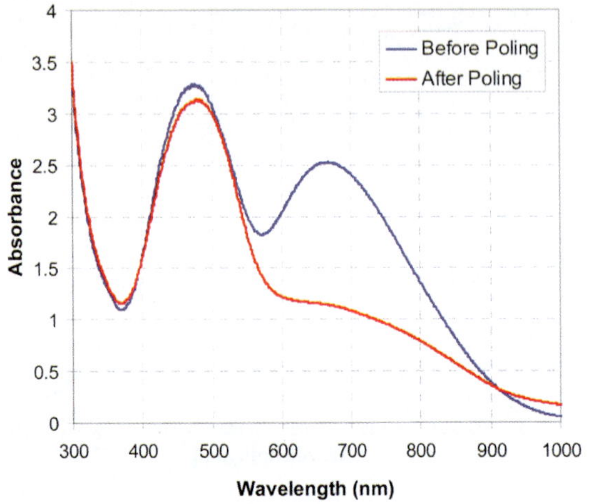

Figure 3. The absorption curve of PMMA-DR1/CLD42A film (1.5 μm) before (blue) and after (red) poling for normally incident light. As the CLD molecules are poled to be aligned normal to the film plane, light propagating along the film normal will not be absorbed by the CLD molecules, as is indicated by the disappearance of the absorption peak around 660nm. The absorption spectrum measured here also clearly indicates that the DR1 chromophores are not poled using this poling schedule, and additional poling would be required to pole the DR1 chromophores.

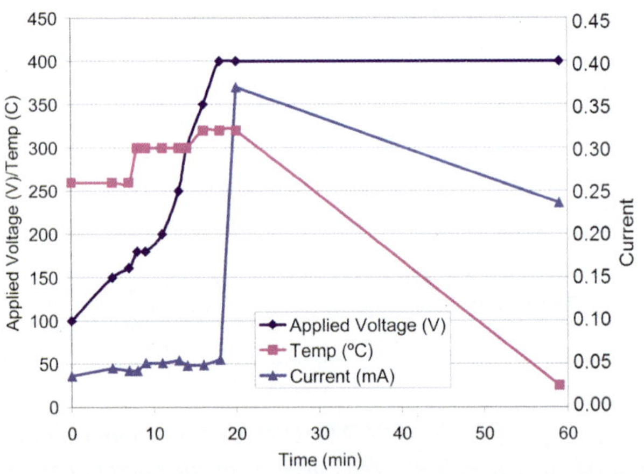

Figure 4. Poling schedule for a 15% DCM in polyimide film. The glass transition of the polyimide is at 325°C, forcing a much higher poling temperature than that for the PMMA/DR1 CLD. The polymer film is 3.8 μm thick.

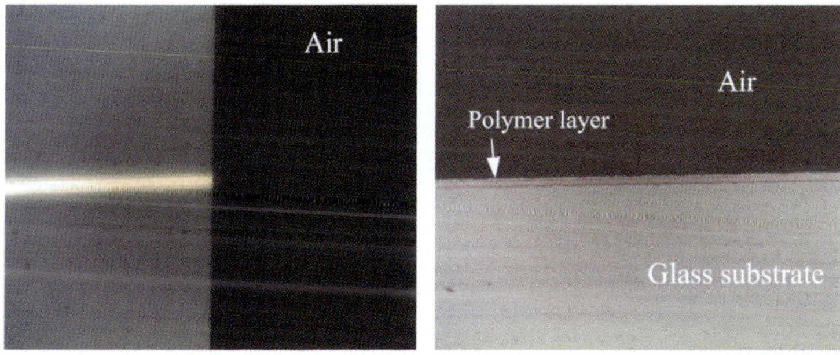

(a) (b)

Figure 5. Microscope image of polyimide SPP WG. (a) top-view of the end face of SPP WG. The 5 µm width silver stripe is cleared seen through the polyimide layer. (b) cross-sectional view of the end face. Both pictures show that careful dicing of the polymer on top of the glass substrate enabled a clear cut and at the same time polishing of the end face of the polymer waveguide, which is crucial to the end fire coupling of the light wave into the waveguide.

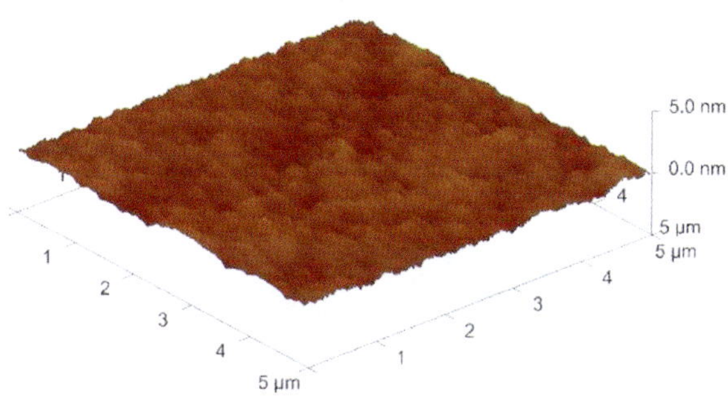

Figure 6. The AFM image of PMMA/DR1 film on ITO coated glass substrate. The image shows that the coated film is smooth to nanometer scales.

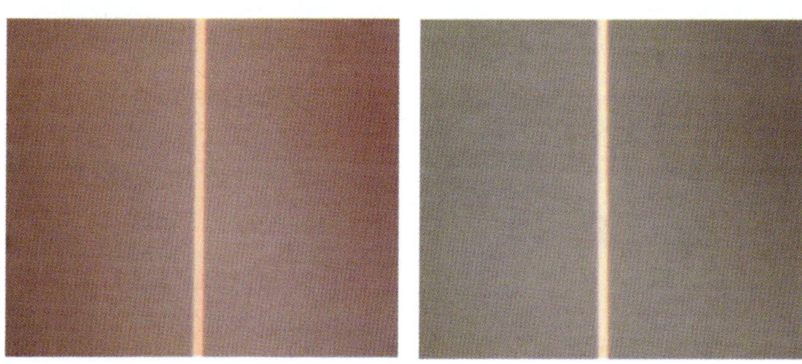

(a) (b)

Figure 7. Gold strip deposition results using shadow mask patterning on PMMA-DR1 (a) and on PMMA-DR1/CLD42A (b). The gold stripes are 10µm wide.

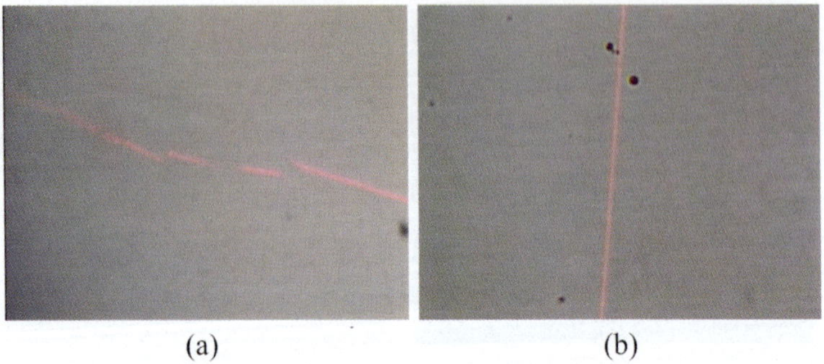

Figure 8. *Microscope image of gold strip in PMMA-DR1/CLD42A film (a) without and (b) with PAA protective layer between two NLO films. Red line is 10 μm wide gold stripe. The gold strip is broken when it is sandwiched straightforwardly between the two electrooptic polymer films. Thin layers, ~100 nm, of crosslinked PAA protective layer is deposited after the deposition of the gold strip to protect the gold strip from the electrooptic films.*

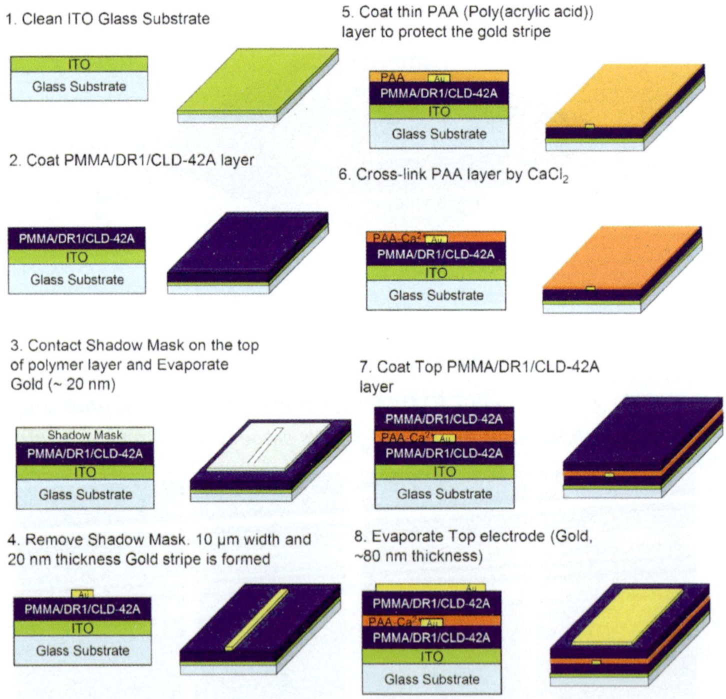

Figure 9. *The illustration for the fabrication procedure of active SPP WG fabrication with NLO.*

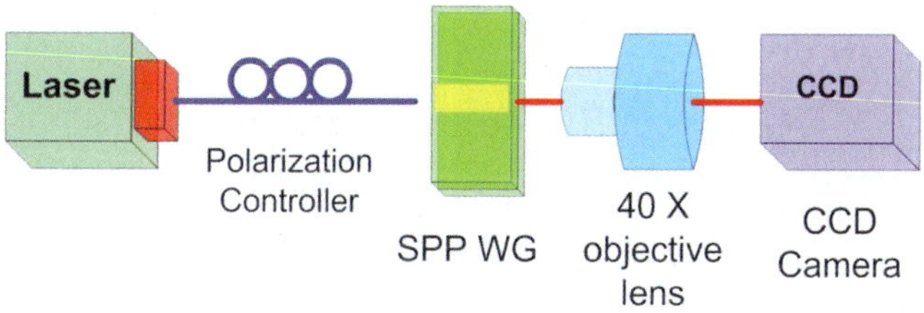

Figure 10. The near-field of SPP WG measurement setup.

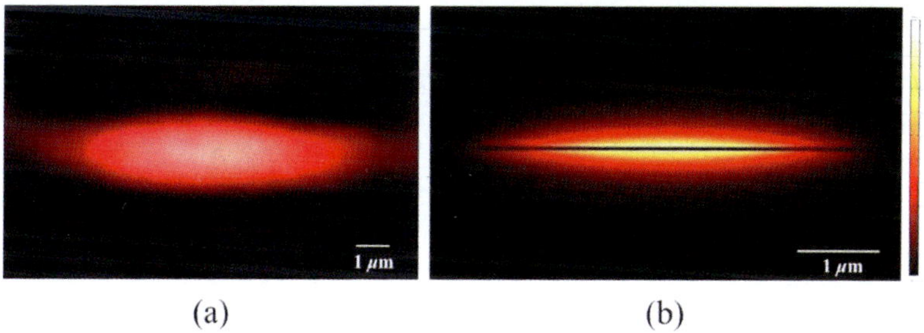

Figure 11. Experimentally measured near field image at output of SPP WG (a). The passive SPP waveguide is comprised of a silver stripe that is 5 μm wide and 20 nm thick with polyimide as claddings. TEM00 single mode HeNe laser is coupled to the waveguide by endfire coupling technique and the image was taken at output end of the 2 mm SPP waveguide. (b) numerically simulated SPP WG mode pattern.

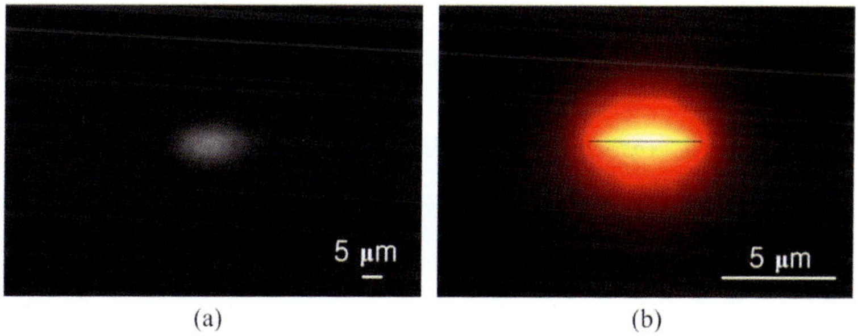

Figure 12. Experimentally measured near field image at output of SPP WG with 1550 nm laser input (a). The image was taken at output end of the 2 mm SPP waveguide by the infrared camera. (b) numerically simulated SPP WG mode pattern at 1550 nm.

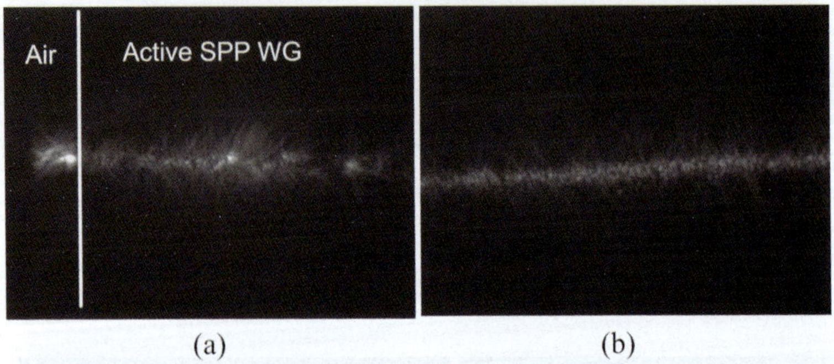

Figure 13. Infrared laser light guided through the active SPP WG. (a) top view of input of WG (b) light scattering through metal stripe in the active SPP WG.

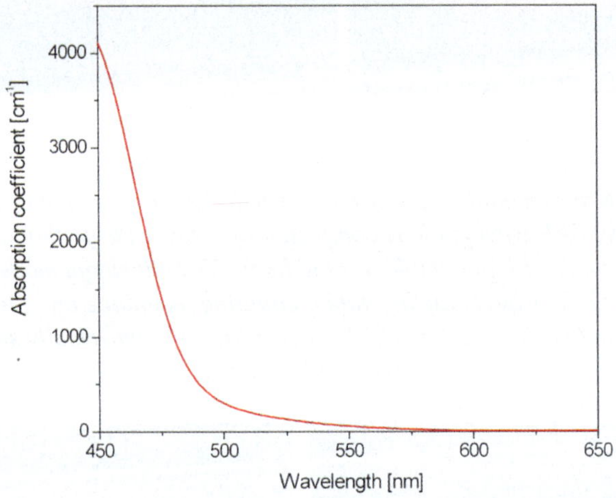

Figure 2. Absorption of the PR material in the visible region.

Figure 3. Sample sizes from 2 × 2 in² to large 6×6 in².

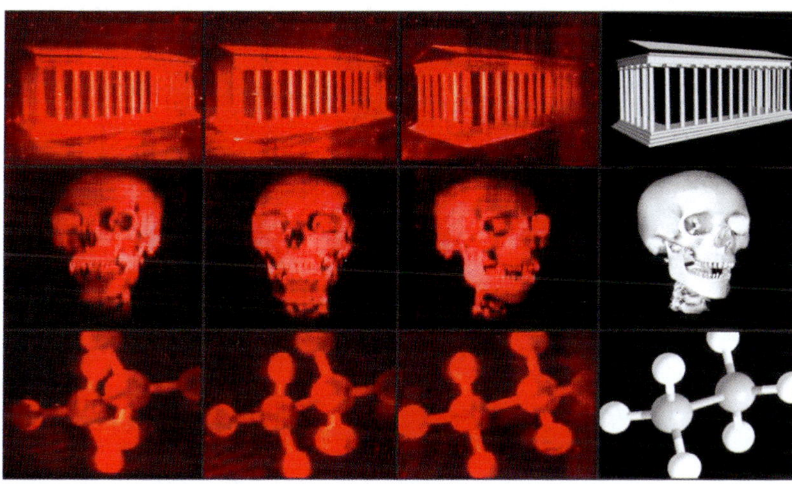

Figure 4. Holographic images of an ionic temple, human skull, and ethane molecule produced in a photorefractive polymer composite film and captured with a digital camera moving in the horizontal plane.

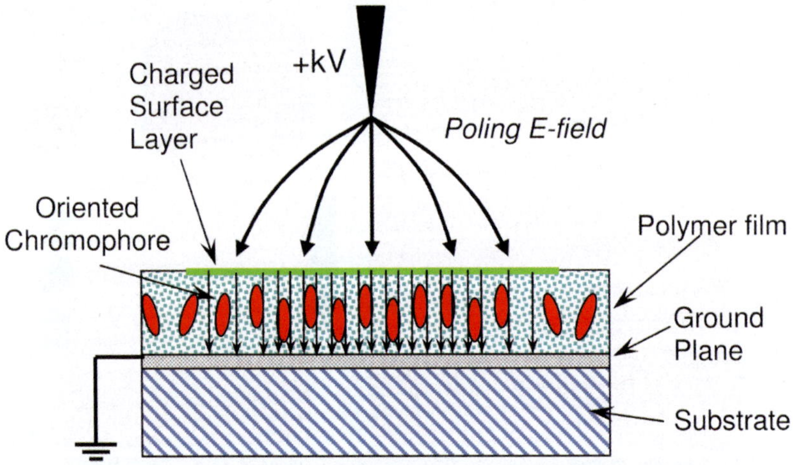

Figure 1. Normal configuration for corona poling showing chromophore orientation parallel to E-field induced by charged surface layer.

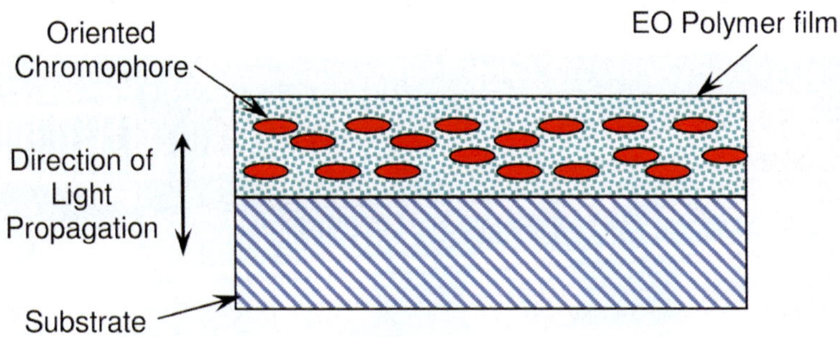

Figure 2. Desired chromophore orientation for EO device such as a Pockels cell (not experimentally practical).

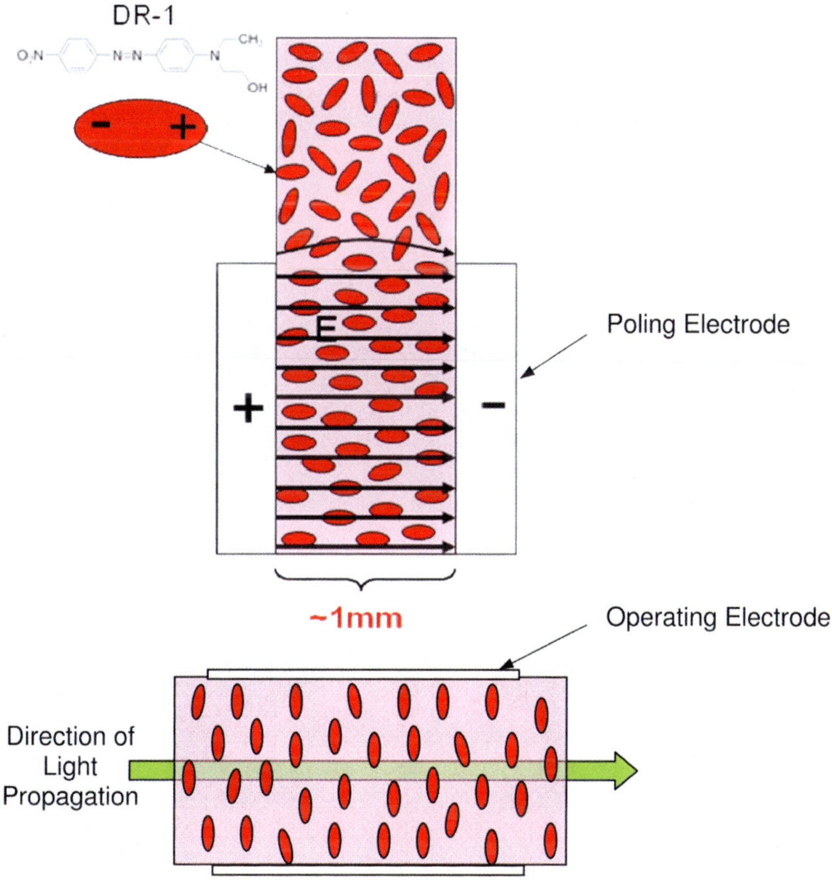

Figure 3. Top: bulk poling configuration showing alignment of chromophores with E-field; bottom: chromophore orientation for Pockels cell after bulk poling and subsequent reorientation.

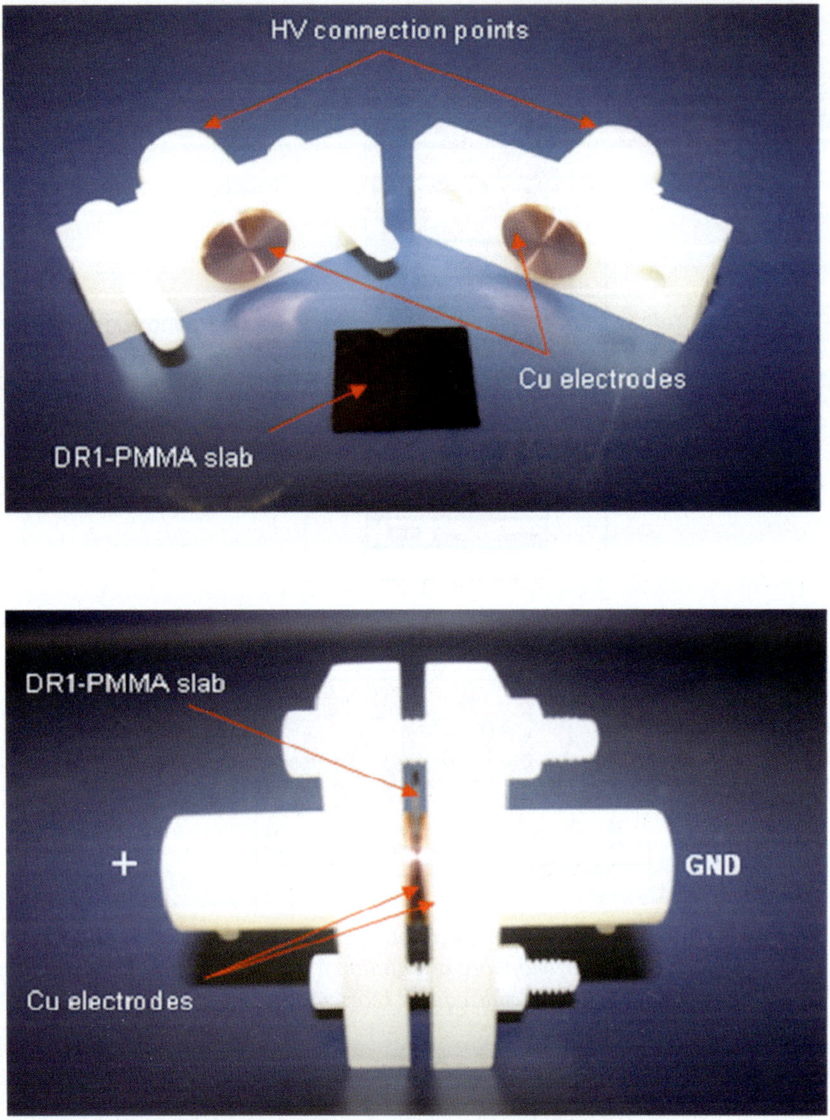

Figure 4. Electrode/sample configuration for poling bulk guest-host polymers; showing poling fixture open (top) and closed (bottom).

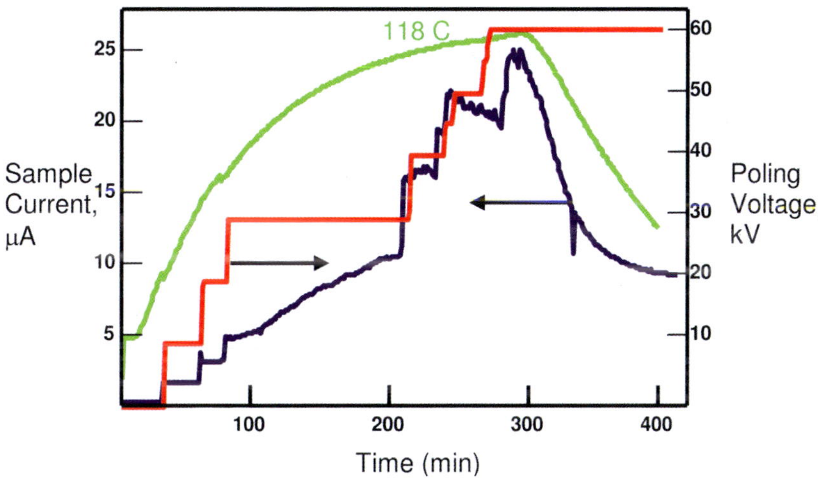

Figure 5. Representative temperature vs. time poling protocol.

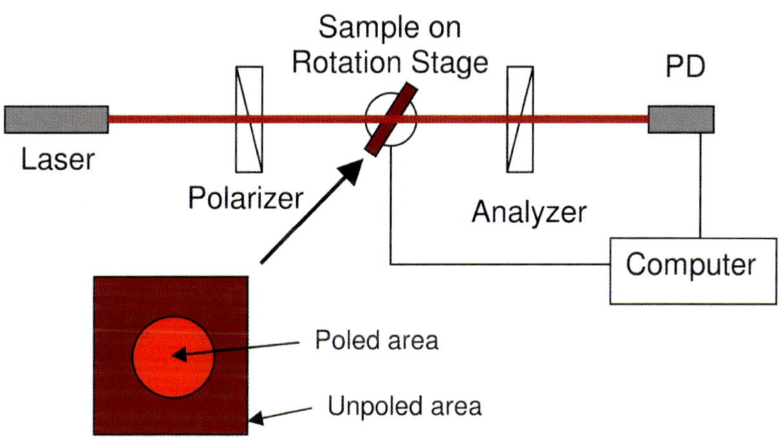

Figure 6. Apparatus for measuring birefringence.

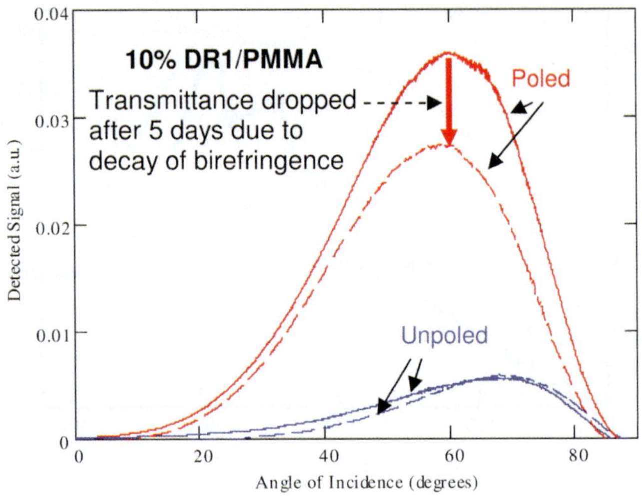

Figure 7. Poling-induced birefringence of 10% DR1-PMMA 700 μm slab.

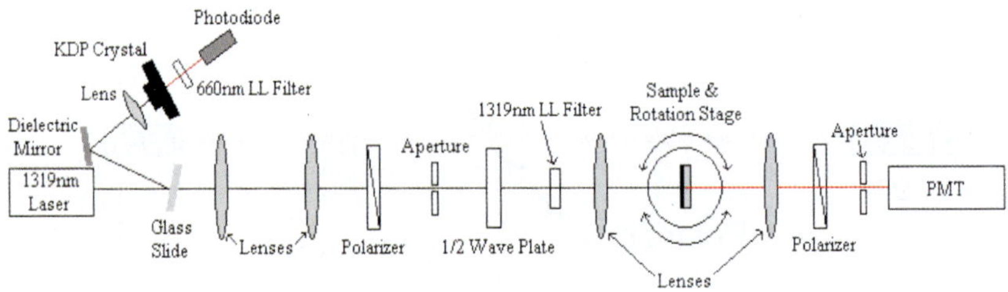

Figure 8. Maker fringe experiment to measure SHG d-coefficient.

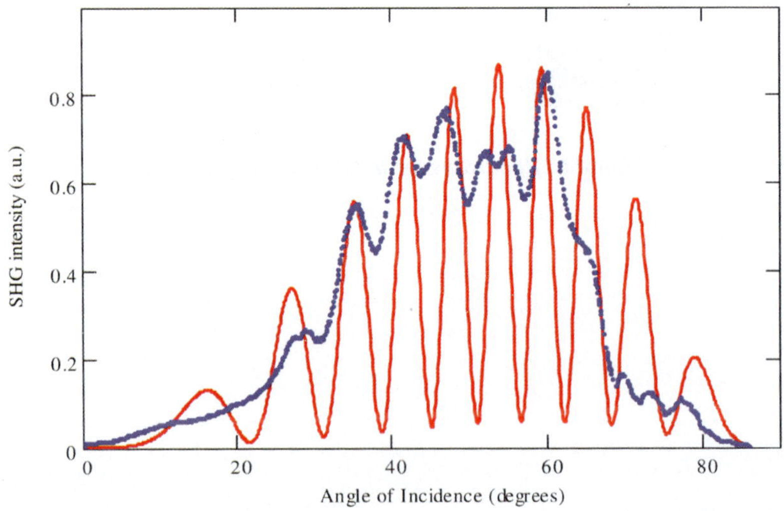

Figure 9. Projected (red) vs. observed (blue) Maker fringe fit for 10%DR1/PMMA (wt %).

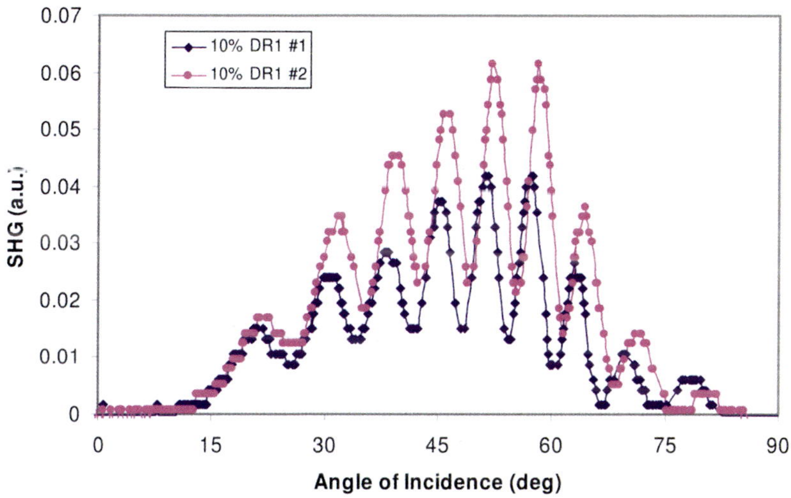

Figure 10. Maker fringe data for non-aggregated samples.

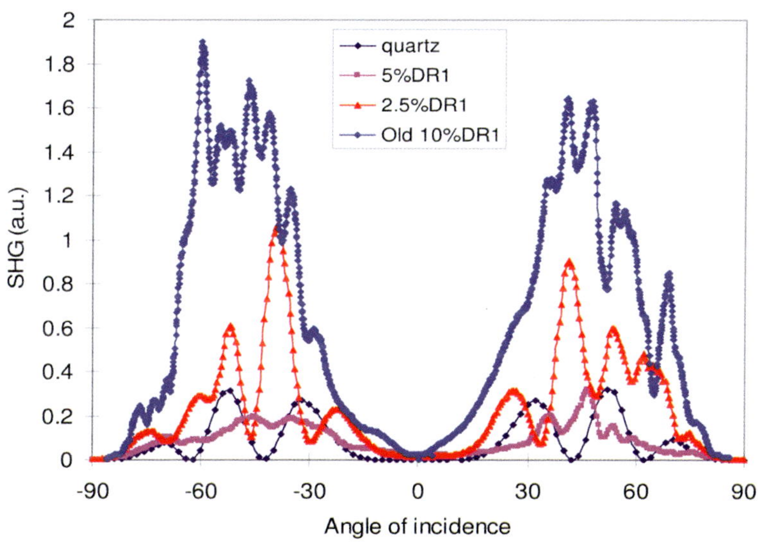

Figure 11. Non-aggregated 2.5% DR1 sample performance exceeds that of aggregated 5% and 10% DR1 samples.

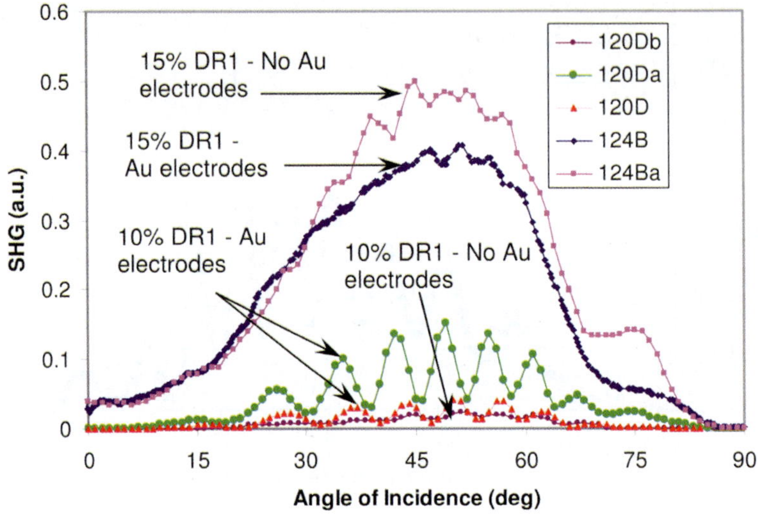

Figure 12. Maker-fringe data for 10% and 15 % DR1-PMMA samples prepared by micro-compounding.

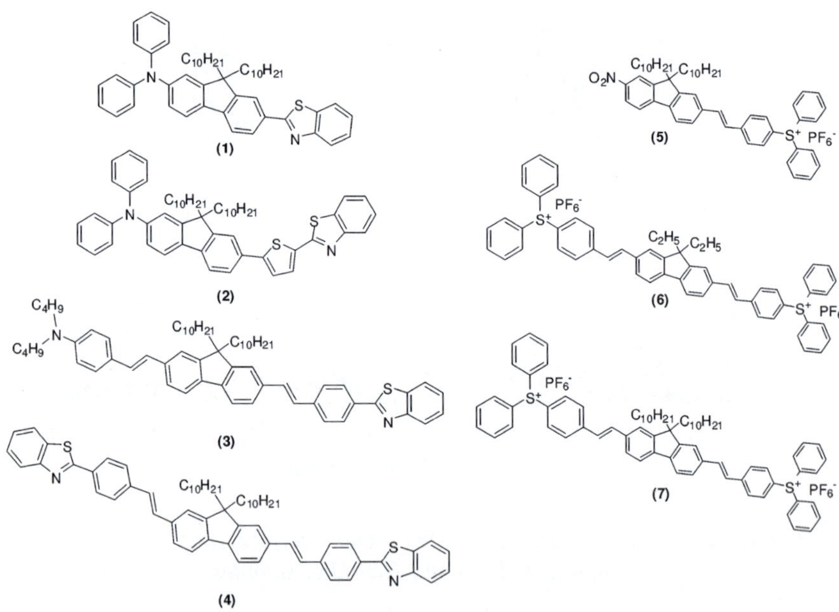

*Figure 1. 2PA absorbing dyes (**1-4**) and 2PA absorbing PAGs (**5-7**).*

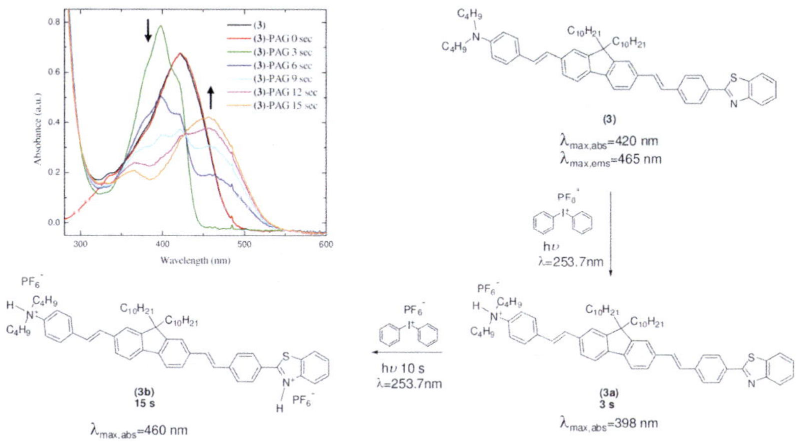

*Figure 2. Spectral change of **3** in ACN solution with diphenyliodonium hexafluorophophate (PAG) upon irradiation at 253.7 nm. Equililibria of species **3**, **3a**, and **3b** at irradiation times ranging from 0-15 s.*

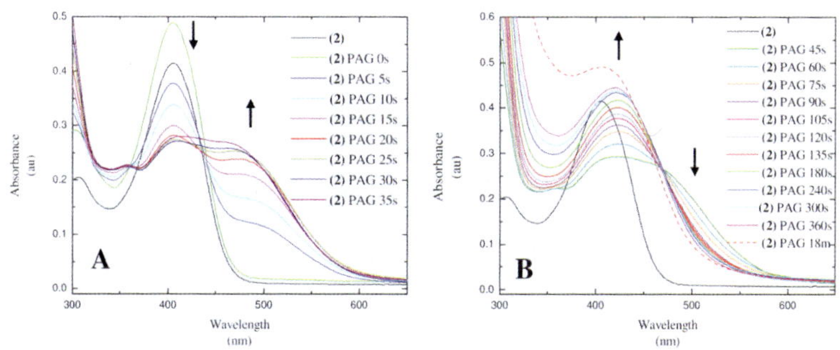

*Figure 3. Spectral change of **2** in ACN solution with diphenyliodonium hexafluorophophate (PAG) upon irradiation at 253.7 nm. **A)** Equililibria of species **2** and **2a** at 0-35 s of irradiation. **B)** equililibria of species **2a** and **2b** at 45 s-18 min of irradiation.*

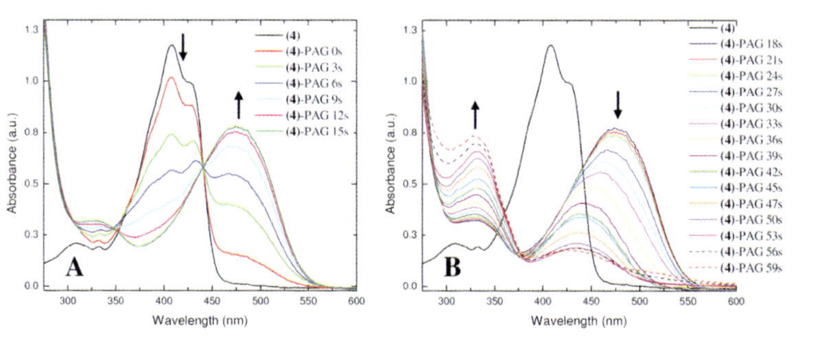

*Figure 4. Spectral change of **4** in ACN solution with diphenyliodonium hexafluorophophate (PAG) upon irradiation at 253.7 nm. **A)** equililibria of species **4** and **4a** at 0-15 s of irradiation. **B)** equililibria of species **4a** and **4b** at 18-59 s of irradiation.*

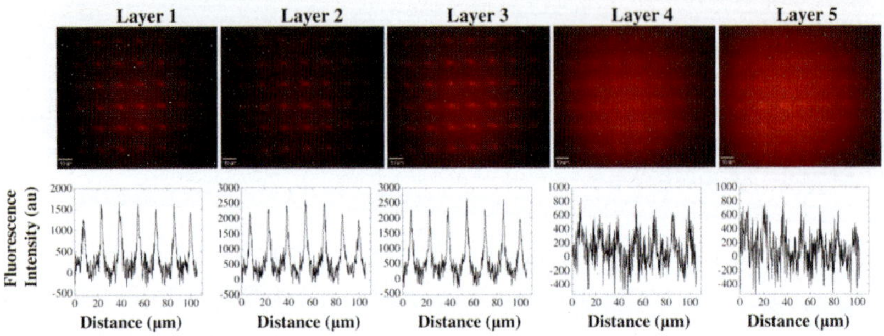

Figure 7. Photosensitive polymeric system used for 3D, two-photon optical data storage. Composition: ***2*** *(1%),* ***5*** *(5%),* ***11*** *(94%). Two-photon writing was performed at 730 nm (2.4 mW), 200 fs, 60 ms exposure/voxel with a 60x, 1.4 NA oil immersion objective. One-photon readout (upper) was performed, layer by layer (~0.4 µm/layer), with a modified TRITC filter cube (Ex:525/40; DM: 555; Em:624/40), with same objective used for writing. Layers 1-5 show significant reduction of the signal to noise ratio (lower) in the fluorescence intensity scan for all the layers. Scale bar: 10 µm.*

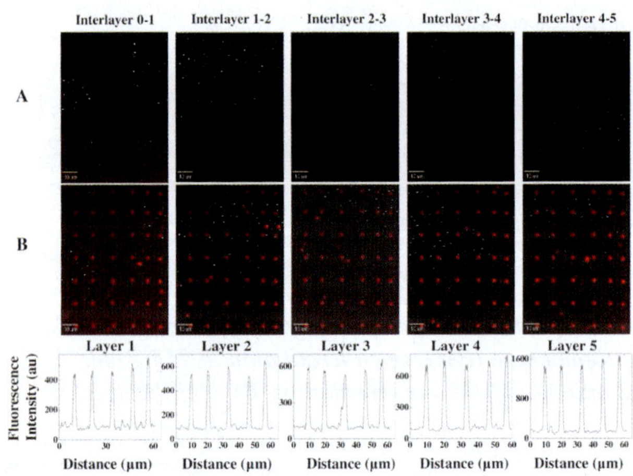

Figure 8. Photosensitive polymeric system used for 3D, two-photon optical data storage. Composition: ***2*** *(1%),* ***5*** *(5%),* ***11*** *(94%). Two-photon writing was performed at 730 nm (2.4 mW), 200 fs, 60 ms exposure/voxel with a 60x, 1.4 NA oil immersion objective. Two-photon readout was performed, layer by layer (~0.4 µm/layer), at 860 nm (7 mW), 200 fs, with same objective used for writing. (A) Blank interlayers. (B) Two-photon readout of the five layers. (C) Fluorescence intensity scan of all layers showing excellent signal to noise ratio throughout all five recorded layers of data. Note that because there is virtually no fluorescence signal in between the layers (A), the system is crosstalk free. Scale bar: 10 µm.*

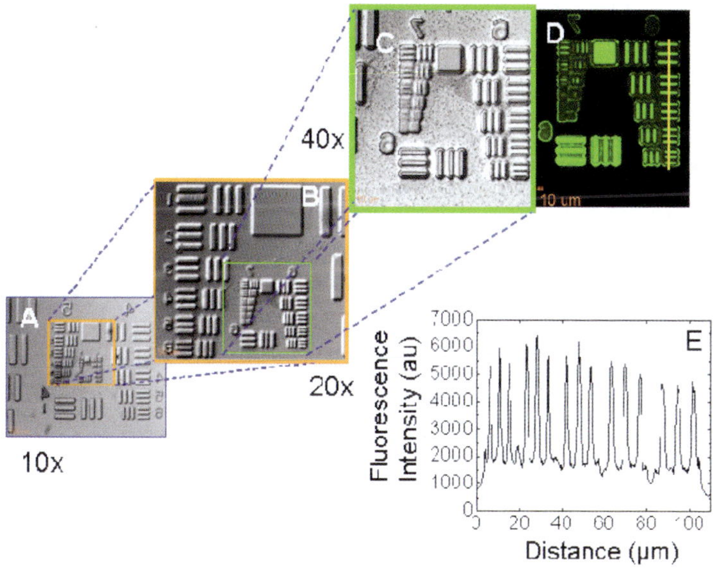

Figure 10. DIC images of microstructure generated by one-photon contact lithography using USAF target resolution masks. Polymerization exposure time: 300 s. A) 10x, 0.3 NA objective; 62 ms exposure time for DIC image. Scale bar: 100 µm; B) 20x, 0.45 NA objective; 250 ms exposure time for DIC image. Scale bar 50 µm. C) 40x, 0.6 NA objective; 62 ms exposure time for DIC image. Scale bar 50 µm. D) confocal fluorescence image (40x, 0.6 NA objective; 165 ms exposure time). Scale bar 10 µm. E) fluorescence intensity along the line shown in D.

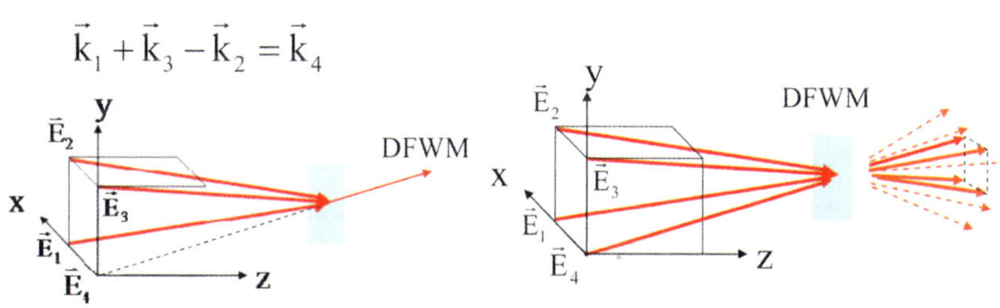

Figure 1. Schematic illustration of pump beams in a DFWM experiment in the BOXCARS configuration a) for measuring the magnitude of $\chi^{(3)}$ using three pump beams and monitoring generation of the fourth beam; b) for measuring the phase of $\chi^{(3)}$ using four equal pump beams and varying the input phase of beam 4 (several non-phase matched beams are also shown).

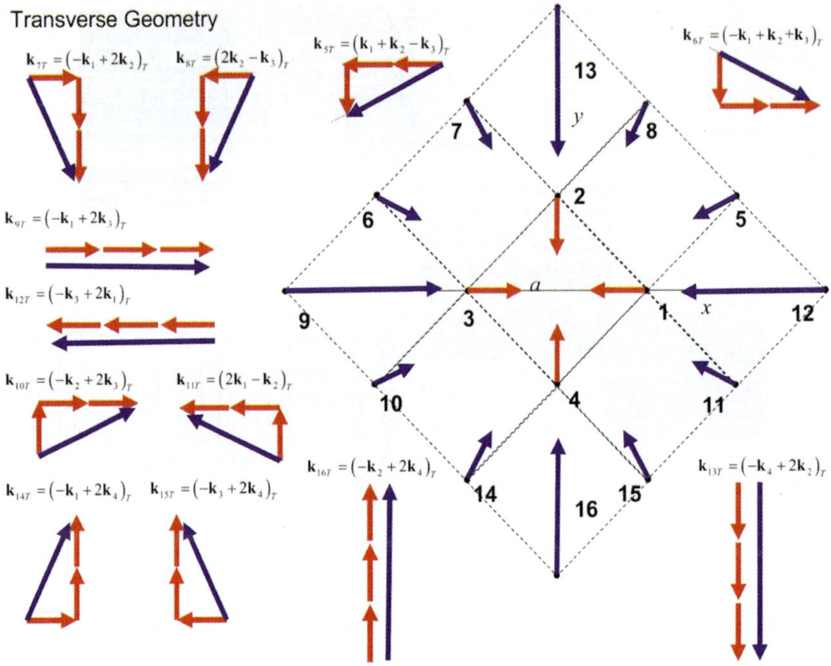

Figure 2. Relation of the transverse components of the 12 phase-mismatched beams to the four primary beams.

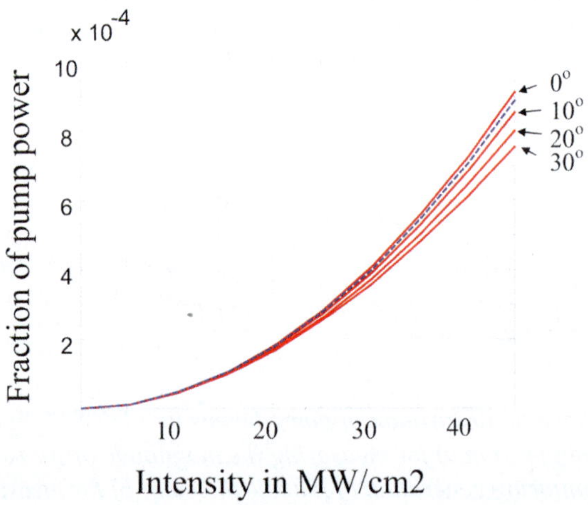

Figure 3. The blue dashed line is a plot of Eq.(14) for nonlinear material with $|\chi^{(3)}|=10^{-10}$esu, n=2,and L=100 μm, compared to the split-step method numerical calculations with four different phase values of φ_χ.

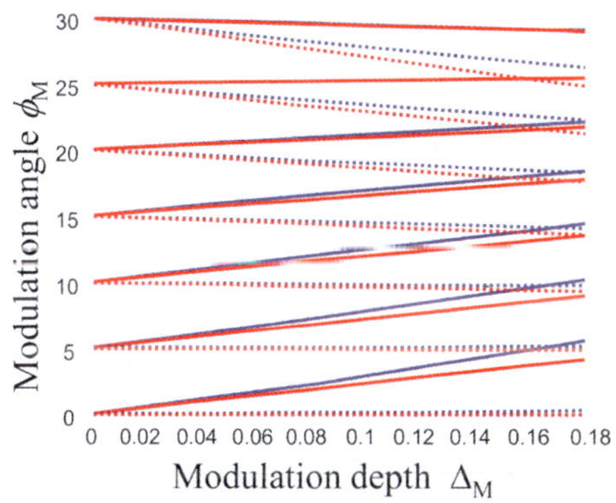

Figure 4. Simulation plots of the modulation phase vs. modulation depth for φ_χ= 0, 5°, 10°, 15°, 20°, 25°, 30°. For each φ_χ is plotted the approximate analytical expression(red) and the numerical solution of the differential equations(blue). The dotted lines are for four beams only and the solid lines include the phase mismatched beams.

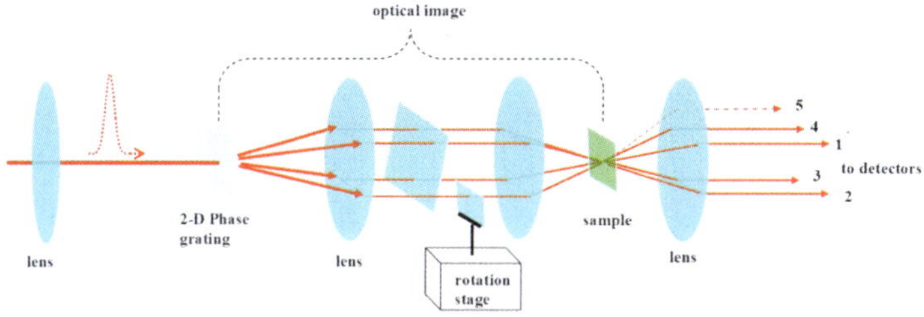

Figure 5. Optical setup for the measurement of the phase of $\chi^{(3)}$.

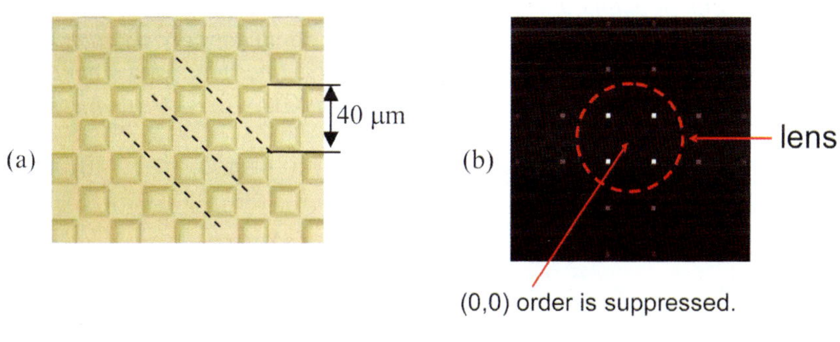

Figure 6. (a) Phase mask for multiple diffraction beam generation. (b) Far field diffraction pattern from the grating.

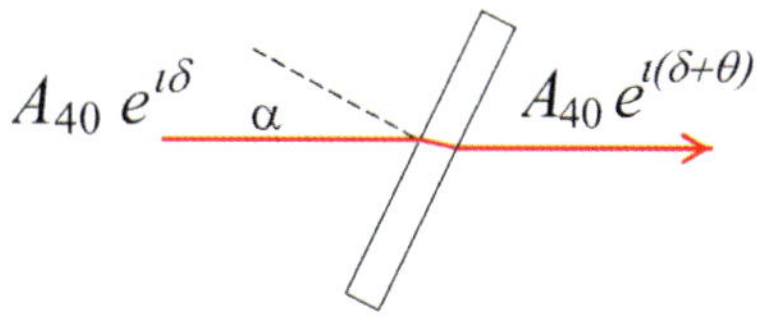

Figure 8. Depiction of the change in phase of beam #4 from a tilted plane slab.

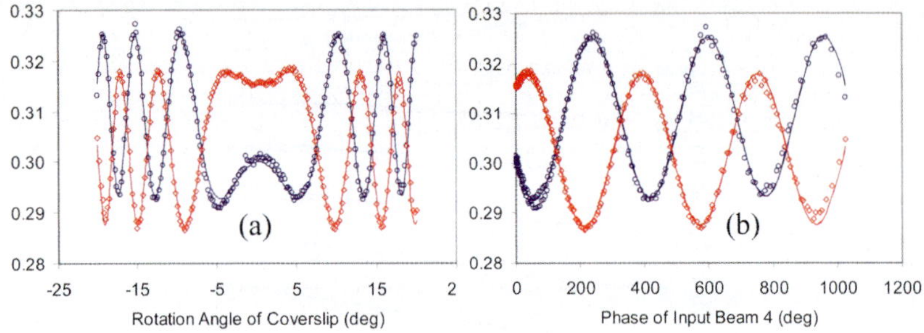

Figure 9. Power (a.u.) for beam #4 (red) and beam #1(blue) plotted vs. coverslip rotation angle (a) and vs. phase θ of beam #4 input to the sample. (b).

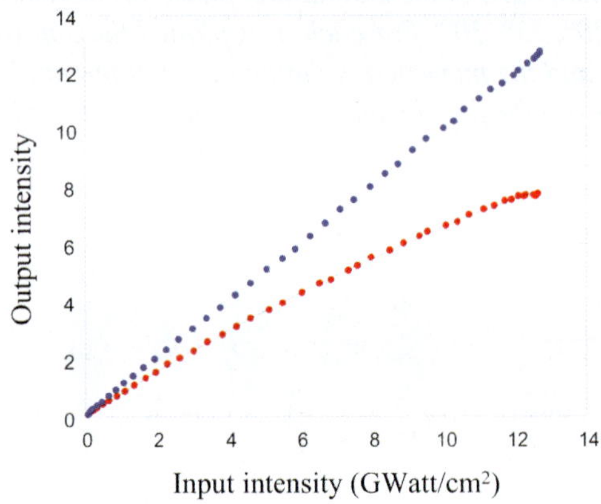

Figure 11. Plot of output intensity vs. laser input intensity. Blue dots correspond to output without the sample and red dots with the nonlinear polymer sample. The black line is the calculated transmission with 21% linear loss and a nonlinear loss coefficient of 2.1 cm/GW.

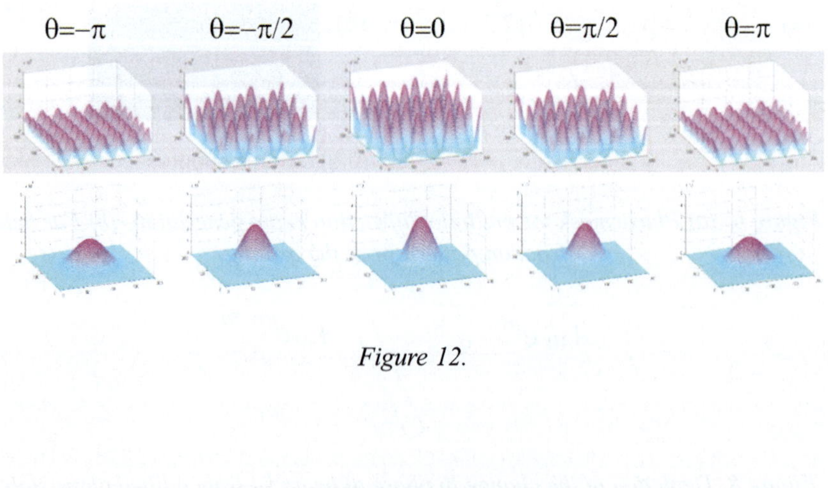

Figure 12.

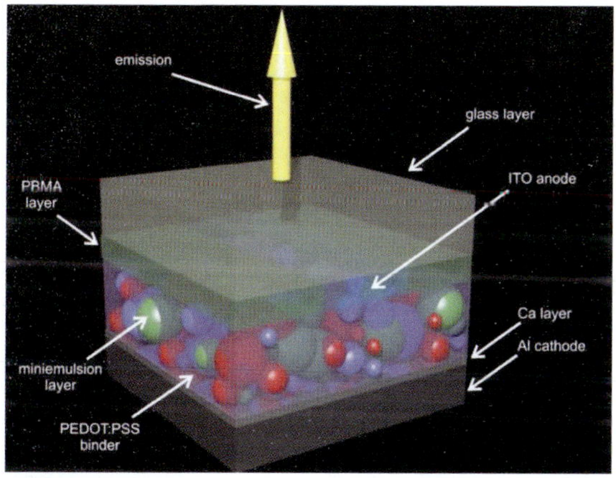

Figure 1. Schematic illustration depicting the construction of a single pixel colloidally-based organic light emitting device. A glass substrate with a templated layer of indium tin oxide serves as the anode of the device. In order to planarize the surface and prevent regions of _eld enhancement, a ca. 150 nm layer of poly(butyl methacrylate) is spin-coated onto the anode and removed from the ITO region. Emissive colloids, mixed with a binding poly(3,4-ethylenedioxythiophene):poly(styrenesulfonate) dispersion, are deposited onto this electrode assembly. The cathode is formed by the evaporation of a ca. 30 nm thick Ca layer onto the colloids, over which a thicker Al layer is evaporated.

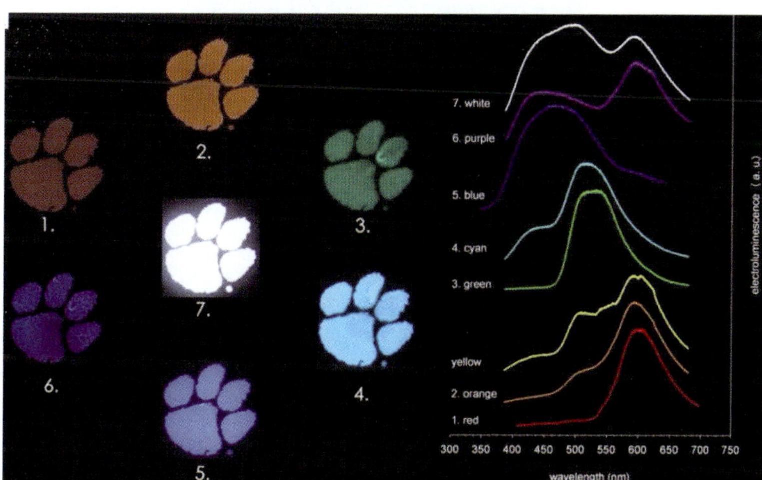

Figure 2. Optical images (1-7) of colloidally-based organic light emitting devices templated into a 12.7 mm wide tiger paw insignia with varying RGB dye ratios (left side of the figure) and the corresponding electroluminescence spectra (right side of the figure). In addition, the electroluminescence spectrum of a yellow emitting device is presented.

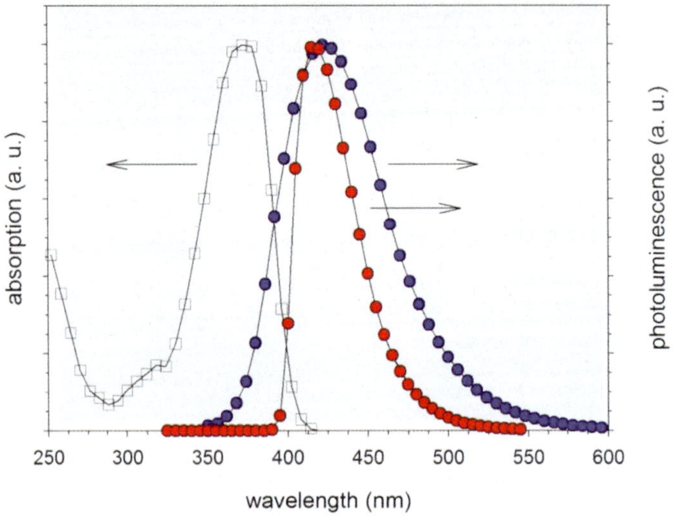

Figure 4. The absorption (2) and photoluminescence (excitation at 375 nm; red) spectra of the Coumarin 1 (B) dye in chloroform. The photoluminescence spectra (excitation at 343 nm; blue) of a colloid composed of only PVK and tBu-PBD in a ca. 63/37 weight ratio. All curves have been scaled for clarity.

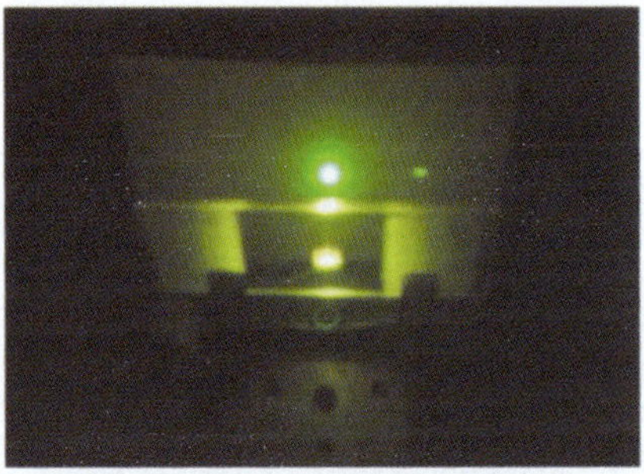

Figure 8. Lasing R6G-DFB laser pumped by a 530 nm pulse generated from a YAG-third harmonic-operated OPO. The pump was set at 18 degrees from the sample surface normal. The small green spot on the right is the pump residue.

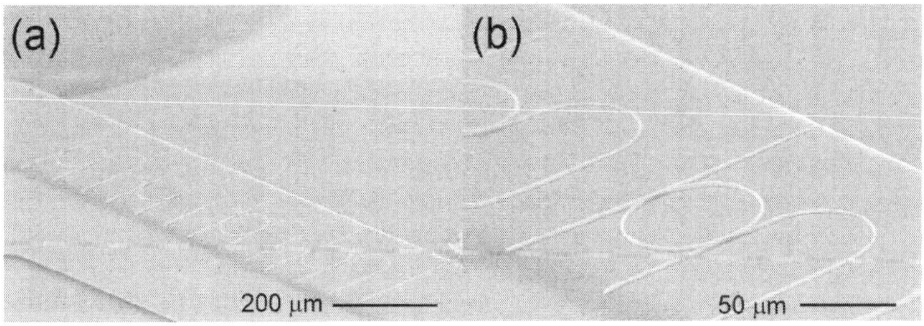

Figure 5. Two different views of IOU devices fabricated on a silica capillary using SU8.

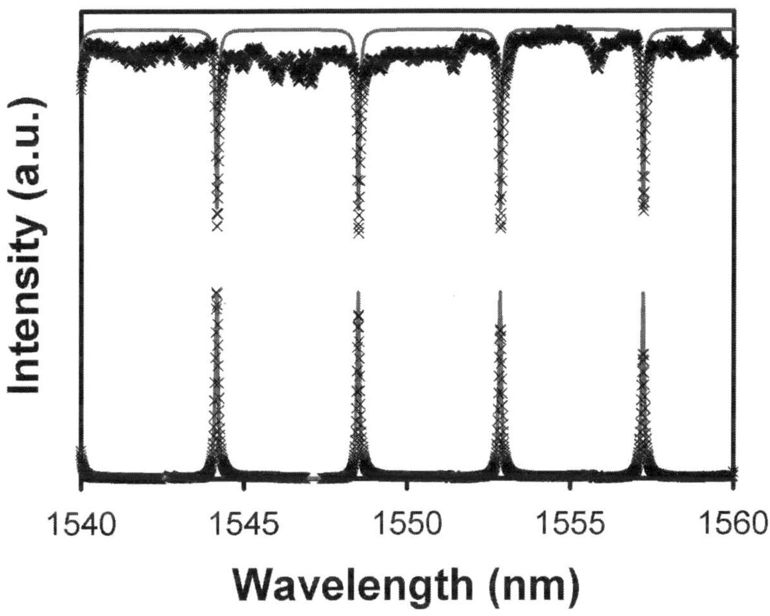

Figure 6. Through-port spectrum (bottom) and drop-port spectrum (top) of a representative IOU device fabricatedin SU8, along with fits (solid lines).

New Architectures

The 3-D fabrication capability of MAP makes possible the creation of waveguide-based devices with unique functionality. Some examples of novel waveguide architectures and devices that we have created with MAP are shown in Figure 7.

The refractive index contrast between our polymer waveguides and the substrates is high enough to allow us to create high-performance devices, but is low enough that effects such as bending losses limit device designs. However, it is straightforward to use MAP to create waveguides that are on thin "pedestals." So long as the pedestals are designed appropriately, the waveguides act as if they are effectively air-clad (*19*, *20*). Examples of such waveguide designs are shown in Figure 7. The "Γ" shaped waveguide shown in Figure 7a has the pedestal offset to one side of the waveguide to prevent any mode leakage. However, so long as the pedestal is thin enough, "T" waveguides such as the one shown in Figure 7b also exhibit high-quality, single-mode performance.

Also shown in Figure 7 are two examples of 3-D devices that would be difficult to fabricate with conventional techniques. Vertical microrings such as the one shown in Figure 7c save space on the substrate and have different coupling to the bus waveguides than to coplanar microrings because the buses couple to one another through the ring. As shown in Figure 7d, MAP can also be used to create microrings that are fixed above bus waveguides. Extending this design to architectures such as stacks of microrings will be a straightforward process that will lead to compact devices with new functionalities.

Conclusions

MAP shows great promise as a technique for the fabrication of waveguide-based optical devices. Here we have demonstrated high-performance devices created with both acrylic and epoxy polymers, and have demonstrated new architectures that will lead to further improvements in device performance and add new functionality.

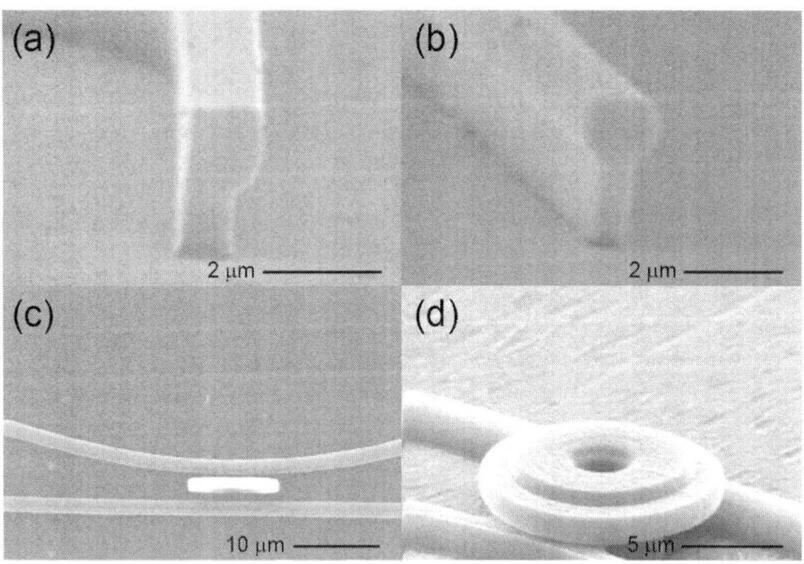

Figure 7. Examples of unique architectures that can be created readily with MAP. (a) A waveguide with a "Γ" cross section. (b) A waveguide with a "T" cross section. (c) A standing-ring waveguide device. (d) A device in which the microring is held above the bus waveguides.

References

1. LaFratta, C. N.; Fourkas, J. T.; Baldacchini, T.; Farrer, R. A. *Angew. Chem., Int. Ed.* **2007**, *46*, 6238.
2. Li, L. J.; Fourkas, J. T. *Mater. Today* **2007**, *10*, 30.
3. Maruo, S.; Fourkas, J. T. *Laser Photonics Rev.* **2008**, *2*, 100.
4. Marder, S. R.; Bredas, J. L.; Perry, J. W. *MRS Bull.* **2007**, *32*, 561.
5. Rumi, M.; Barlow, S.; Wang, J.; Perry, J. W.; Marder, S. R. Two-Photon-Absorbing Materials and Two-Photon-Induced Chemistry. In *Photoresponsive Polymers I*; Springer-Verlag: Berlin, 2008; Vol. 213, pp 1–95.
6. Jia, B. H.; Li, J. F.; Gu, M. *Aust. J. Chem.* **2007**, *60*, 484.
7. Klein, S.; Barsella, A.; Leblond, H.; Bulou, H.; Fort, A.; Andraud, C.; Lemercier, G.; Mulatier, J. C.; Dorkenoo, K. *Appl. Phys. Lett.* **2005**, *86*, 211118.
8. Guo, R.; Xiao, S.; Zhai, X.; Li, J.; Xia, A.; Huang, W. *Opt. Express* **2006**, *14*, 810.
9. Yokoyama, S.; Nakahama, T.; Miki, H.; Mashiko, S. *Thin Solid Films* **2003**, *438*, 452.
10. Cumpston, B. H.; Ananthavel, S. P.; Barlow, S.; Dyer, D. L.; Ehrlich, J. E.; Erskine, L. L.; Heikal, A. A.; Kuebler, S. M.; Lee, I. Y. S.; McCord-Maughon, D.; Qin, J. Q.; Rockel, H.; Rumi, M.; Wu, X. L.; Marder, S. R.; Perry, J. W. *Nature* **1999**, *398*, 51.
11. Deubel, M.; Von Freymann, G.; Wegener, M.; Pereira, S.; Busch, K.; Soukoulis, C. M. *Nature Mater.* **2004**, *3*, 444.
12. Tetreault, N.; von Freymann, G.; Deubel, M.; Hermatschweiler, M.; Perez-Willard, F.; John, S.; Wegener, M.; Ozin, G. A. *Adv. Mater.* **2006**, *18*, 457.
13. Li, L.; Gershgoren, E.; Kumi, G.; Chen, W. Y.; Ho, P. T.; Herman, W. N.; Fourkas, J. T. *Adv. Mater.* **2008**, *20*, 3668.
14. Baldacchini, T.; LaFratta, C.; Farrer, R. A.; Teich, M. C.; Saleh, B. E. A.; Naughton, M. J.; Fourkas, J. T. *J. Appl. Phys* **2004**, *95*, 6072.
15. Mendonca, C. R.; Correa, D. S.; Baldacchini, T.; Tayalia, P.; Mazur, E. *Appl. Phys. A: Mater. Sci. Process.* **2008**, *90*, 633.
16. Eldada, L.; Shacklette, L. W. *IEEE J. Sel. Top. Quantum Electron.* **2000**, *6*, 54.
17. Kim, D. H.; Im, J. G.; Lee, S. S.; Ahn, S. W.; Lee, K. D. *IEEE Photonics Technol. Lett.* **2005**, *17*, 2352.
18. Poon, J. K. S.; Huang, Y. Y.; Paloczi, G. T.; Yariv, A. *IEEE Photonics Technol. Lett.* **2004**, *16*, 2496.
19. Armani, D. K.; Kippenberg, T. J.; Spillane, S. M.; Vahala, K. J. *Nature* **2003**, *421*, 925.
20. Chen, W. Y.; Grover, R.; Ibrahim, T. A.; Van, V.; Herman, W. N.; Ho, P. T. *IEEE Photonics Technol. Lett.* **2004**, *16*, 470.

Chapter 10

Phase Scan Technique for Measuring Phase of Complex $\chi^{(3)}$ of Nonlinear Polymer Thin Films

Weilou Cao, Yi-Hsing Peng, Yongzhang Leng, Chi H. Lee, Warren N. Herman, and Julius Goldhar*

Laboratory for Physical Sciences and Department of Electrical and Computer Engineering, University of Maryland, College Park, MD 20740
***jgoldhar@umd.edu**

A practical technique for absolute measurement of the complex phase of the third-order nonlinear optical susceptibility $\chi^{(3)}$ in thin films was developed using forward Degenerate Four Wave Mixing (DFWM) in the BOXCARS configuration. An apparent variation of the complex phase of $\chi^{(3)}$ with pump laser intensity was observed in highly nonlinear polymers. Detailed numerical models of the DFWM process were used to simulate the nonlinear interaction and it was found that this apparent intensity dependence can be explained by relatively efficient generation of phase mismatched beams. A procedure for calculating the correct phase of the nonlinear susceptibility from the data is presented and the effects of beam quality, both temporal and spatial, are discussed. It is shown that this technique does not depend on knowledge of the magnitude of $\chi^{(3)}$ and is absolute in the sense that it does not require a separate reference material, such as fused silica.

Introduction

Highly nonlinear polymer waveguide devices offer the possibility of major advances in optical signal processing and development of parametric amplifiers. For fiber optic communications they could be used for optical switching, gating and routing signals. It has been recently demonstrated in nonlinear optical fiber based devices, that phase sensitive parametric amplifiers can be built with noise figure below 3dB and that they could be used for regeneration of signals in DPSK format (*1–3*). However, these applications require either long lengths of fibers or high power coherent pump lasers. Also, in most nonlinear optical fibers, there is a severe limitation on operating regimes due to other nonlinear processes such as SBS (stimulated Brillouin scattering) and SRS (stimulated Raman scattering). The key to successful implementation of these concepts is identification of polymers that have complex third order nonlinear susceptibility, $\chi^{(3)}$ with sufficiently large magnitude and a small imaginary component, which corresponds to nonlinear loss.

Development of new nonlinear materials requires a characterization set-up which is able to quickly evaluate the nonlinear properties of polymers without the many processing steps involved in device fabrication. Measurements need to be made in thin films with many orders of magnitude shorter interaction lengths than what would be used in actual devices where the materials eventually would be used. We also need precise knowledge of the phase φ_χ of the complex susceptibility $\chi^{(3)}=|\chi^{(3)}|\exp(i\varphi_\chi)= \chi^{(3)}{}_R+i\chi^{(3)}{}_{Im}$. The real part corresponds to the intensity dependent index of refraction and the imaginary part is responsible for two photon absorption and possibly excited state absorption in some materials. Figures of merit assigned to different material media for nonlinear optical devices directly depend on the phase of $\chi^{(3)}$ (*4*). We have found that measurement of the phase of highly nonlinear polymers turns out to be a non-trivial task and there can be a significant variation in measured values depending on measurement conditions. In particular, we have observed a substantial dependence of the measured phase on pump intensity. In this work we explain the origin of the apparent intensity dependence and provide a procedure for absolute measurement of the phase of the complex susceptibility $\chi^{(3)}$.

There are two well-known conventional techniques for measurement of both the magnitude and the phase of $\chi^{(3)}$. They are the Degenerate Four Wave Mixing (DFWM) method (*5–8*) and the Z-scan technique (*9*). Z-scan is an elegant technique which, because of its simplicity, is most commonly used for material characterization. In simple DFWM experiments, the measured signal is proportional to $|\chi^{(3)}|^2$, but the phase of $\chi^{(3)}$ is not available. In more sophisticated DFWM experiments, it has been shown (*10*) to be possible to measure the phase of $\chi^{(3)}$ but implementation is significantly more difficult because it involves interferometric alignment and synchronization of multiple optical beams. Ref. (*8*) introduced a method that allows one to compare both the magnitude and the phase of all $\chi^{(3)}$ tensor elements of a given thin film to those of a thick substrate. Recently, we demonstrated a simple optical arrangement for phase sensitive detection of DFWM to characterize the real and imaginary parts of $\chi^{(3)}$ using an imaged 2-D phase grating (*11*) based on the folded BOXCARS geometry (*12*).

In DFWM experiments, three or four pump beams of the same wavelength, in our case 1550 nm, overlap in the thin polymer film to generate additional beams which draw energy from the pump beams. In order to measure the magnitude of $\chi^{(3)}$ we can send in the three pump beams and measure the fraction of the pump power converted to the phase matched fourth beam, as illustrated in Figure 1a. This approach has been extensively investigated previously (*10*, *13*), and will not be discussed in detail here.

In DFWM, the wave vector of a generated beam may or may not be equal to the wave vector of an optical beam which has the same wavelength and satisfies all boundary conditions. If it is not, the generated beam will be reduced in power by a phase-mismatch factor $F=[\sin(x)/x]^2$ over that for a phase-matched beam (*5*). Here $x = |\Delta k|L/2$, where Δk is the difference of the wave vectors of the generated phase-matched and phase-mismatched beams and L is the thickness of the thin polymer film. Appearance of these non-phase matched beams can result in small apparent variation of the magnitude of $\chi^{(3)}$ with intensity by causing additional pump depletion. It also can have a very large effect on the measurement of the phase of $\chi^{(3)}$ as will be described below.

For the measurement of the phase of $\chi^{(3)}$, the fourth reference beam is introduced as illustrated in Figure 1b. Additional, non-phase matched, beams will also be created in the sample by the nonlinear interaction. The experiment involves precise variation of the initial optical phase of beam #4 with respect to that of the other pump beams at the input to the sample. This input phase variation causes a change in the relative power output of all the pump beams and changes drastically the efficiency for generation of the phase mismatched beams. By measuring these effects and comparing them to a theoretical model, we are able to obtain a precise measure of the phase of $\chi^{(3)}$. The presence of the phase mismatched beams significantly complicates the analysis of DFWM interactions in the sample. However, we found that it is essential to include them in the calculations in order to obtain correct values of the measured physical parameters.

Theory

We first review forward DFWM from the point of view of plane waves and then discuss the effects of beam pulse shape in space and time. In the presence of two to four pump beams, the nonlinear four wave mixing interaction in the material results in four phase-matched output beams and up to 12 additional non-phase matched beams.

Plane Wave Approach

The plane wave equations for the electric fields $\mathbf{E}_j$ for these 16 beams in Gaussian units are (*6*)

$$\nabla\times\nabla\times\mathbf{E}-\left(\frac{\omega}{c}\right)^2\ddot{\boldsymbol{\varepsilon}}\cdot\mathbf{E}=\left(\frac{\omega}{c}\right)^2 4\pi\mathbf{P}^{NL}, \tag{1}$$

$$\mathbf{E}=\sum_j \mathbf{u}_j A_j(z)e^{i\mathbf{k}_j\cdot\mathbf{r}}, \quad j=1,\ldots,16 \tag{2}$$

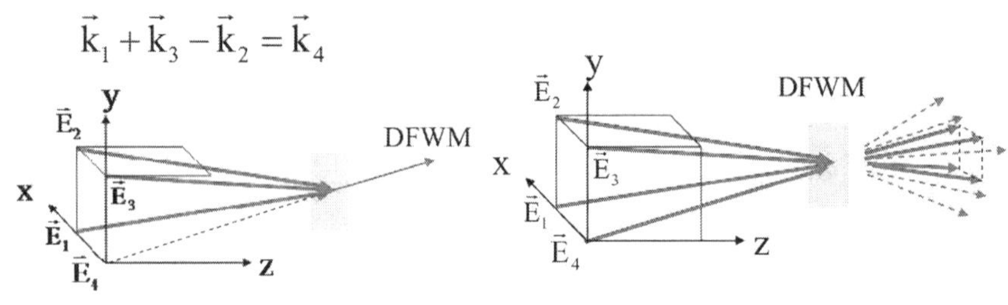

Figure 1. Schematic illustration of pump beams in a DFWM experiment in the BOXCARS configuration a) for measuring the magnitude of $\chi^{(3)}$ using three pump beams and monitoring generation of the fourth beam; b) for measuring the phase of $\chi^{(3)}$ using four equal pump beams and varying the input phase of beam 4 (several non-phase matched beams are also shown). (see color insert)

where **u** and **k** are the field polarization and propagation vectors, respectively, and $\mathbf{P}^{NL}$ is the total material nonlinear polarization given by

$$\mathbf{P}^{NL} = 3\chi^{(3)} \vdots \left[\left(\sum_{k=1}^{16} \mathbf{u}_k A_k e^{i\mathbf{k}_k \cdot \mathbf{r}} \right)^2 \sum_{l=1}^{16} \mathbf{u}_l A^*_l e^{-i\mathbf{k}_l \cdot \mathbf{r}} \right]. \tag{3}$$

The complex susceptibility $\chi^{(3)}$ is in general a 4th rank tensor and we have only included the 4-wave mixing terms that are frequency matched to $+1\omega$. In the following, we will consider the case that all the field polarization vectors $\mathbf{u}_j$ are equal and, for simplicity, that $\chi^{(3)}$ has only one independent tensor component. In addition, the A_j are assumed to depend only on the propagation direction z, but the phases $\mathbf{k}_j \cdot \mathbf{r}$ still depend in general on x, y, z. We can then integrate over x and y and make use of the fact that for sufficiently large a the integral $\frac{1}{2a}\int_{-a}^{a} e^{i\Delta k_x x}dx = \mathrm{sinc}(\Delta k_x a)$ is zero unless $\Delta k_x=0$. Also using the slowly varying envelope approximation (*6*) together with $\mathbf{k}_j \cdot \mathbf{u}_j = 0$ we obtain

$$\frac{\partial A_j}{\partial z} = \frac{2\pi i k_0^2}{k_{jz}} \frac{1}{4a^2} \int_{-a}^{a}\int_{a}^{a} \mathbf{u}_j \cdot \mathbf{P}^{NL} e^{-i\mathbf{k}_j \cdot \mathbf{r}} dxdy, \tag{4}$$

where $k_0 = \omega/c = 2\pi/\lambda$, , $k_j = |\mathbf{k}_j|$, $\varepsilon = n^2$, and $k_j = k_0 n$ with n being the refractive index. The term $\mathbf{u}_j \cdot \mathbf{P}^{NL} e^{-i\mathbf{k}_j \cdot \mathbf{r}}$ contains a great many terms in general, with each term containing a phase mismatch factor $\exp(i\Delta\mathbf{k} \cdot \mathbf{r})$. Only the first four beams are phase matched in the sense that $\mathbf{k}_4 + \mathbf{k}_2 + \mathbf{k}_1 + \mathbf{k}_3$. The propagation vectors for these four primary beams inside the film can be written

$$\begin{aligned} \mathbf{k}_1 &= k_0 n(-\sin\alpha, 0, \cos\alpha), & \mathbf{k}_2 &= k_0 n(0, -\sin\alpha, \cos\alpha), \\ \mathbf{k}_3 &= k_0 n(\sin\alpha, 0, \cos\alpha), & \mathbf{k}_4 &= k_0 n(0, \sin\alpha, \cos\alpha), \end{aligned} \tag{5}$$

where $\sin\alpha = \sin\alpha_0/n$ and α_0 is the angle of incidence of the beams onto the sample.

For the remaining beams, the propagation vectors can be separated into longitudinal and transverse components as $\mathbf{k}_j = k_{jz}\hat{\mathbf{z}} + \mathbf{k}_{jT}$ and the relation of these transverse components to the four primary beams is shown in Figure 2. Once the transverse components are obtained, the longitudinal components can be determined from Figure 2 and Eq.(5) together with the requirement that $|\mathbf{k}_j| = k_0 n$. From this requirement, one finds that beams 5,6,7,8,10,11,14, and 15 have the same z-component, *viz.*, $k_0 n K_a \cos\alpha\hat{\mathbf{z}}$ with $K_a = \sqrt{1-5\sin^2\alpha}/\cos\alpha$, and, similarly, the z-component for beams 9,12, 13, and 16 is $k_0 n K_b \cos\alpha\hat{\mathbf{z}}$ with $K_b = \sqrt{1-9\sin^2\alpha}/\cos\alpha$. The phase mismatches are thus of the form $\Delta\mathbf{k}\cdot\mathbf{r} = k_0 n(1 - K_{a,b})\cos\alpha z$ for beams 5 through 16.

When all terms in $\mathbf{P}^{NL}$ are explicitly displayed, Eq.(4) giving the differential equation for the complex amplitude of the field of beam j takes the form

$$\frac{\partial A_j}{\partial z} = i\kappa_j \left[\sum_{p=1}^{15} \sum_{q=p+1}^{16} \sum_{s=1}^{16} A_p A_q A_s^* M_{pqsj} e^{i(k_{zp}+k_{zq}-k_{zs}-k_{zj})z} + \frac{1}{2} \sum_{p=1}^{16} \sum_{s=1}^{16} A_p{}^2 A_s^* M_{ppsj} e^{i(2k_{zp}-k_{zs}-k_{zj})z} \right], \tag{6}$$

where

$$\begin{aligned} M_{pqsj} &= \left(\frac{1}{4a^2}\right) \int_{-a}^{a} \int_{-a}^{a} e^{i(k_{xp}+k_{xq}-k_{xs}-k_{xj})x} e^{i(k_{yp}+k_{yq}-k_{ys}-k_{yj})y} dxdy \\ &= \mathrm{sinc}\left[\left(k_{xp}+k_{xq}-k_{xs}-k_{xj}\right)a\right] \mathrm{sinc}\left[\left(k_{yp}+k_{yq}-k_{ys}-k_{yj}\right)a\right] \end{aligned} \tag{7}$$

Here $\kappa_j = 12\pi k_0 \chi^{(3)}/nC_j \cos\alpha$, $\chi^{(3)} = \mathbf{u}\cdot\boldsymbol{\chi}^{(3)}\vdots\mathbf{uuu}$, and C=1 for beams 1 through 4 , $C=K_a$ for beams 5,6,7,8,10,11,14, 15, and $C=K_b$ for beams 9,12, 13, and 16. Eq.(7) implies that, for a given beam j, the only terms on the right hand side of Eq.(6) that contribute are those with phase-matched transverse components, i.e., those for which $(\mathbf{k}_p + \mathbf{k}_q - \mathbf{k}_s - \mathbf{k}_j)_T = 0$.

Split-Step Approach

Another conventional approach to simulation of nonlinear optical interactions involves division of the spatial domain into discrete small cells to form a volume mesh or grid with forward propagation accomplished using the split step method (*14*). We implemented this approach using a simple MATLAB program. In the transverse plane, the optical electric field is calculated on a $(2N+1)\times(2N+1)$ grid with typical N~100. The grid dimensions $2D\times2D$ were typically D=0.4 mm. Experimentally, our typical beam size had radius on the order of 0.1-0.2 mm so

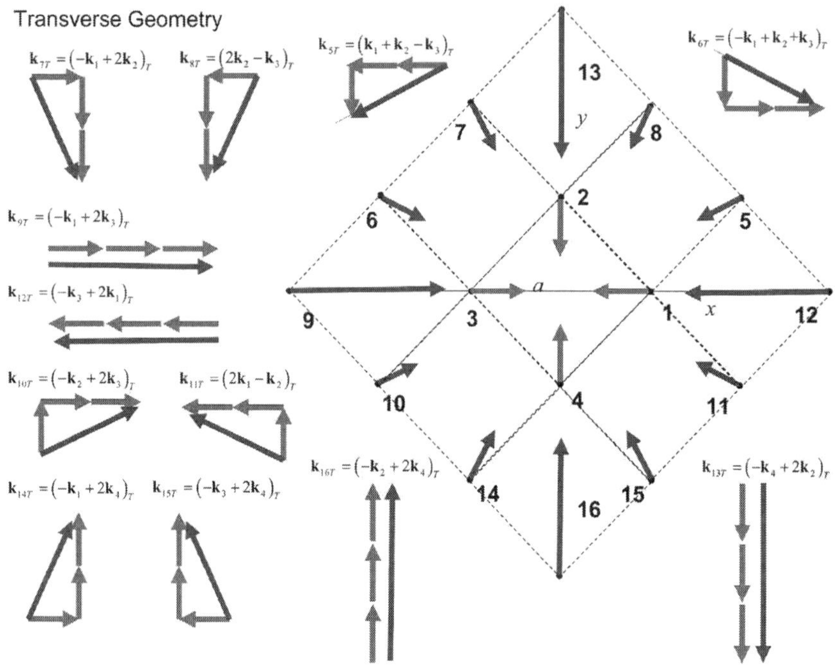

Figure 2. Relation of the transverse components of the 12 phase-mismatched beams to the four primary beams. (see color insert)

that there were no problems with wings of the beams approaching the edges. The far field was related to the near field by a two dimensional fast Fourier transform consisting of grid points separated by π/D radians.

The results of this split step numerical method were compared to the results obtained from numerical solutions of the 16 beam plane wave equations. Excellent agreement was observed between the two models for the range of intensities where the total nonlinear phase shift $2\pi n_2 I_p L/\lambda$ due to one pump beam was less than 0.24. Above that value, the self focusing and beam break up become serious and it is difficult to believe predictions of either model. Beyond this point, more than 16 beams are needed in the discrete beam model, and the near field pattern structure becomes comparable to the grid size in the split step approach. The onset of self focusing occurs at critical power $P_{cr} = 0.15\lambda^2/nn_2$ (*6*) on one of the near field beamlets produced by the interference between the pump beams. The power in one beamlet is calculated by multiplying the incident intensity by the area which is approximately $(10\lambda)^2$. For a polymer with $n_2 \approx 10^{-12}$ cm^2/Watt we obtain $P_{cr} = 8\times10^8$ Watts in four pump beams. For a beam of width ω_0, there will be a catastrophic collapse of the beamlet when the self focusing distance $z_{sf} = \left(2n\omega_0^2/\lambda^2\right)/\sqrt{P/P_{cr}}$ is equal to the sample thickness. This is expected when $P \sim 3P_{cr}$ which corresponds to approximately 400 MW/cm^2 for each beam. These estimates provide the boundaries of the regime where our models become invalid.

We also compared these two models to analytical solutions are described below that were obtained using perturbation theory up to second order and a good agreement with numerical models was observed for single beam nonlinear phase shifts up to ~0.1.

Approximate Analytical Solution

When analyzing experimental data it is helpful to have simple analytical relationships between the observable parameters even if they are only approximate. In this section we will derive some approximate analytical solutions and discuss their validity by comparing them to complete numerical solutions obtained with the numerical models described above. First, we consider interaction only between the four primary beams. Using Eqs.(6), and neglecting all higher number beams we have

$$\frac{\partial A_1}{\partial z} = i\kappa\left[A_2 A_4 A_3^* + A_1\left(P - \frac{1}{2}|A_1|^2\right)\right] \tag{8}$$

$$\frac{\partial A_2}{\partial z} = i\kappa\left[A_1 A_3 A_4^* + A_2\left(P - \frac{1}{2}|A_2|^2\right)\right] \tag{9}$$

$$\frac{\partial A_3}{\partial z} = i\kappa\left[A_2 A_4 A_1^* + A_3\left(P - \frac{1}{2}|A_3|^2\right)\right] \tag{10}$$

$$\frac{\partial A_4}{\partial z} = i\kappa\left[A_1 A_3 A_2^* + A_4\left(P - \frac{1}{2}|A_4|^2\right)\right] \tag{11}$$

where

$$P = |A_1|^2 + |A_2|^2 + |A_3|^2 + |A_4|^2 . \tag{12}$$

Approximate Analytical Solution for Magnitude of $\chi^{(3)}$

In order to simulate the three pump experiment that is commonly used to determine the magnitude of $\chi^{(3)}$, we set the initial conditions to $A_1(0) = A_2(0) = A^3(0) = A_0$, $A_4(0) = 0$. To first order in κz the solution is

$$A_1 = A_2 = A_3 \approx A_0 + i\kappa z\frac{5}{2}A_0^3, \quad A_4 \approx i\kappa z A_0^3 . \tag{13}$$

Therefore, the normalized intensity scattered into beam 4 is

$$\frac{\left|A_4(L)\right|^2}{A_0^{\ 2}} \approx \left|\kappa\right|^2 \left|A_0\right|^4 L^2 \approx 4\left(\frac{2\pi n_2 I_p L}{\lambda_0}\right)^2 . \tag{14}$$

where I_p is the input intensity of one pump beam and $n_2 = 12\pi^2 \mathrm{Re}(\chi^{(3)})/n^2 c$ (*6*) is the nonlinear index of refraction. This equation is valid for the case of real nonlinear susceptibility. The simple relationship in Eq.(14) between $|\chi^{(3)}|$ and output of beam 4 is often used to obtain $|\chi^{(3)}|$ from experimental data. Figure 3 below shows a plot of numerical solution of the differential equations and compares it with the analytical solution for 4 different values of the phase of $\chi^{(3)}$.

In general, the approximation is good only for a nonlinear coefficient at very low intensities. It is often difficult to perform measurements at such low intensities because of competition with linear scattering from the thin samples. For measurements at higher intensities, especially with significant imaginary part of the nonlinear susceptibility, considerable error may be incurred from using the simple expression.

It is interesting to note that the full calculation gives slightly higher conversion efficiency for real nonlinear susceptibility. This effect is due to the presence of phase mismatched beams which provide some additional parametric gain to the DFWM signal generated in the direction of beam 4. As expected, the result of numerical calculation for complex nonlinear susceptibility is below the approximation, because of the loss due to two photon absorption. Extending the perturbation analysis to second order should give a much better agreement with complete solution.

Approximate Analytical Solution for Phase of $\chi^{(3)}$

In order to experimentally measure the phase of $\chi^{(3)}$, we unblock beam #4 and perform measurements with the initial conditions changed to $A_j(0) = A_0 \exp(i\delta_j)$, $j = 1,2,3$, and $A_4(0) = A_0 \exp[i(\theta + \delta_4)]$. Varying the input phase θ of beam 4 results in a modulation of the output amplitudes of all four pump beams. We found that associated with the modulation of the output amplitudes is a modulation phase φ_M whose value for highly nonlinear materials can exhibit a large dependence on the phase- mismatched beams. As discussed above, there are only two distinct values of phase mismatch ΔkL for the twelve non-phasematched beams. Representative of these two types of beams are beam #5 and beam #9 and, retaining only the largest terms on the right hand side of Eq.(6), their differential equations are given by

$$\frac{\partial A_5}{\partial z} = i\frac{\kappa}{K_a}\left[\left(A_1 A_2 A_3^* + \frac{1}{2} A_1 A_1 A_4^*\right) e^{i\Delta \mathbf{k}_5 \cdot \mathbf{r}} + A_5\left(P - \frac{1}{2}\left|A_5\right|^2\right)\right], \tag{15}$$

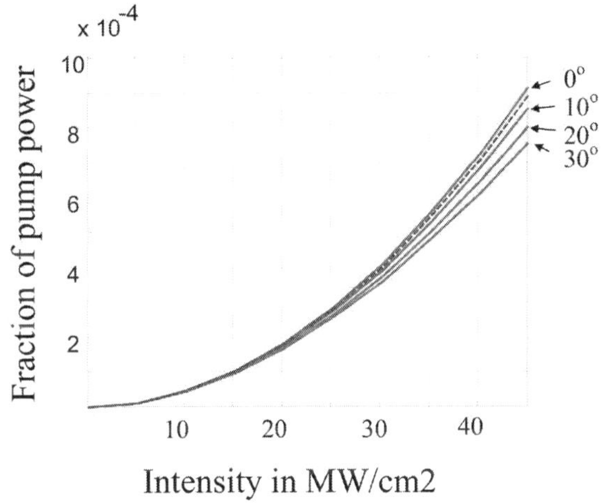

Figure 3. The blue dashed line is a plot of Eq.(14) for nonlinear material with $|\chi^{(3)}|=10^{-10}$esu, n=2,and L=100 μm, compared to the split-step method numerical calculations with four different phase values of φ_χ. (see color insert)

$$\frac{\partial A_9}{\partial z}=i\frac{\kappa}{K_b}\left[\left(\frac{1}{2}A_3A_3A_1^*\right)e^{i\Delta\mathbf{k}_9\cdot\mathbf{r}}+A_9\left(P-\frac{1}{2}|A_9|^2\right)\right], \tag{16}$$

with P given by Eq. (12).

Performing a perturbation expansion up to second order in the strength of the nonlinear interaction characterized by $X=|\kappa||A_p|^2L$, we find the following expressions for the intensities of beams #1 and #4:

$$I_1(\theta)\approx B\left[1-\Delta_M\sin(\theta+\delta+\phi_M)\right], \tag{17}$$

$$I_4(\theta)\approx B\left[1+\Delta_M\sin(\theta+\delta-\phi_M)\right], \tag{18}$$

where

$$B\approx I_p\left[1-7X\sin\phi_\chi\right], \tag{19}$$

$$\Delta_M\approx 2X, \tag{20}$$

$$\phi_M\approx\phi_\chi+X\left[\operatorname{sinc}^2\left(\frac{\Delta k_1L}{2}\right)\cos\phi_\chi-2\sin\phi_\chi\sin 2\phi_\chi\right], \tag{21}$$

and $\delta = \delta_2 + \delta_4 - \delta_1 - \delta_3$, reflecting a possible difference in initial phases of the pump beams impinging on the sample as discussed later in the Experiment section. Here, φ_χ is the sought after phase of $\chi^{(3)}$ and $\Delta k_1 L$ is the phase mismatch for beams 5-8,10,11,14, and 15. The other phase-mismatched beams 9,12,13, and 16 are found to be unaffected by modulating the input phase, θ, of beam 4. The first of the two terms in brackets in Eq.(21) is due to the generation of phase-mismatched beams and the second term results from the interaction between the 4 pump beams.

This sinusoidal variation of the output on beams #1 and #4 with θ can be observed experimentally and can be used to measure the phase. Because the phase shifts on the two beams are in the opposite directions, the relative phase shift between the two beams is equal to $2\varphi_\chi$. The value of φ_M can be readily obtained by fitting the data to sine functions, as will be discussed in the experiment section below. The important point is that

$$X \to 0 \Rightarrow \phi_M\left(X, \phi_\chi\right) \to \phi_\chi \ . \tag{22}$$

This is illustrated in Figure 4, where it can also be seen that the approximate analytical solutions give quite good estimates of the slope at low modulation depths, and that the phase-mismatched beams have a very noticeable effect on the slope of these curves. From Eqs.(20) and (21), the value of this slope is

$$\left(\frac{d\phi_M}{d\Delta_M}\right)_{\substack{Plane\\Waves}} = \frac{1}{2}\text{sinc}^2\left(\frac{\Delta k_1 L}{2}\right)\cos\phi_\chi - \sin\phi_\chi \sin 2\phi_\chi \ . \tag{23}$$

Effects of Spatial and Temporal Profiles

In the experiment, we measure pulse energies, which are intensities integrated in space and time. We assume that our profile has spatial and temporal variation that can be written in the form $I_p(t,r) = I_0 F(t,r)$. Consequently, because the nonlinear interaction parameter X is proportional to this local intensity, we also have $X(t,r) = X_0 F(t,r)$.

If we assume that the sample thickness is much smaller than the Rayleigh range and dispersion can be neglected, we can simply take a weighted average of parameters in space and time. The following approach gives us the average phase and the averge modulation phase. Using the approximation given by Eqs.(20)-(21), we obtain the slope increased from that of the plane waves by a factor that is given by the ratios of different moments of the intensity distribution in the laser beam. In particular, in place of Eq. (21), we have

$$\phi_M' = \phi_\chi + \Delta_M' \frac{\langle I_p^3\rangle\langle I_p\rangle}{\langle I_p^2\rangle^2}\left(\frac{d\phi_M}{d\Delta_M}\right)_{\text{Plane Waves}} \tag{24}$$

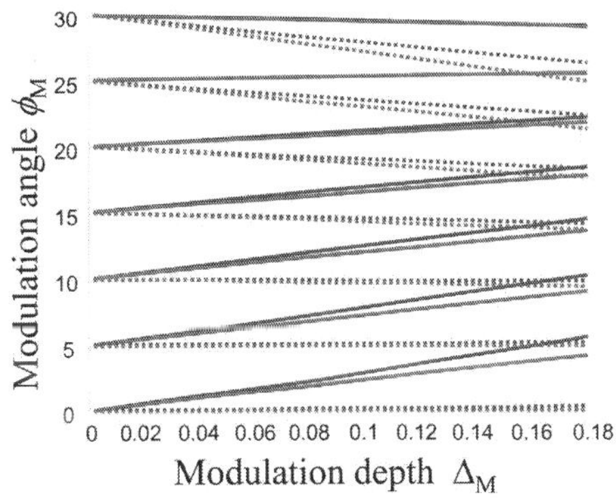

Figure 4. Simulation plots of the modulation phase vs. modulation depth for φ_χ= 0, 5°, 10°, 15°, 20°, 25°, 30°. For each φ_χ is plotted the approximate analytical expression(red) and the numerical solution of the differential equations(blue). The dotted lines are for four beams only and the solid lines include the phase mismatched beams. (see color insert)

where

$$\left\langle I_p{}^N \right\rangle \equiv \int dt \int 2\pi r dr I_p{}^N(t,r). \tag{25}$$

This modified slope $d\varphi_M'/d\Delta_M'$ depends on the spatial and temporal profiles. For example, for a Gaussian spatial and temporal distribution, we get a slope that is 1.54 times larger than calculated for the plane waves with a flat top temporal distribution. For Lorentzian shapes in time and space we get a factor of 3.1. The ratio can get much larger for more complicated profiles. For example, for Gaussian time dependence and a spatial beam consisting of two Gaussians with the radii different by a factor of 10 and the amplitudes inversely proportional to the radii, this ratio is about 4.7. We have also calculated the relationship between the modulation depth and the phase of modulation for specific temporal and spatial profiles using numerical simulations and obtained confirmation of the analytical results. Regardless of the slope however, the y-intercept is the same for any beam profile and this can used to determine the phase of $\chi^{(3)}$.

Effect of Higher Order Nonlinearities

It is also possible that higher order nonlinearities, such as those described by $\chi^{(5)}$, can also contribute to the apparent intensity dependence of the phase of $\chi^{(3)}$. For example, we found from the split step model calculations that three photon absorption could also result in larger slope $d\varphi_M/d\Delta_M$ in the phase scan plot. In principle, if we know precisely the spatial and temporal intensity variation, we can check if the experimentally observed slope is all due to $\chi^{(3)}$, or if there are some higher order effects. However, irrespective of the contributions to the slope, the calculations show that linear extrapolation of φ_M to zero pump intensity (or modulation depth) gives the proper value of the phase of $\chi^{(3)}$.

Experiment

Optical Setup

The schematic of the optical system for using femtosecond pulses to measure $\chi^{(3)}$ in polymer films is illustrated in Figure 5 below. The laser source is an OPA (Spectra-Physics OPA800-C) pumped by a regenerative Ti:sapphire amplifier (Spectra-Physics Hurricane system). The system generates 130fs output pulses at 1550 nm with a repetition rate of 1kHz. A half-wave plate and a polarizer are used to control the intensity. The beam is focused onto a phase plate shown in Figure 6a. Details of fabrication of the phase plate can be found in Appendix A. The "checker-board" pattern consists of 20 μm squares with phase contrast of π between adjacent squares. This pattern is equivalent to two crossed gratings with period of 28.3 μm as shown in the picture. This 2-D grating produces a diffraction pattern shown in Figure 6b. The calculated diffraction pattern shows that most of the energy (70%) goes into the four main beams (±1,±1 diffraction orders), with 0th order suppressed by proper design of the phase mask. Experimentally we observe a diffraction pattern closely resembled the predicted one.

Each of the four output beams is an exact replica of the original. The remaining unwanted beams were blocked. The phase of one of the beams (#4) was precisely controlled by tilting a thin 100 μm thick plate (a microscope cover slip) inserted in its path. For balance, all the other beams passed through a stationary plate of the same thickness.

The four pump beams are arranged in the folded-boxcars (*12*) geometry and are 1X imaged onto the sample with a telescope consisting of two identical spherical lenses with 10 cm focal length. After the sample, a spherical lens with 10 cm focal length is used to collimate the beams. The use of a 2-D phase grating ensures both phase matching and spatial and temporal overlap of the interacting beams. The use of a diffractive optical element beam splitter also allows the overlap of short pulses over their full aperture (*15*).

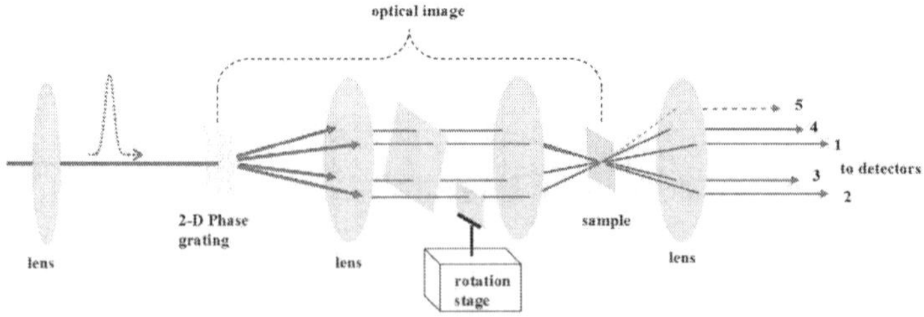

Figure 5. Optical setup for the measurement of the phase of $\chi^{(3)}$. (see color insert)

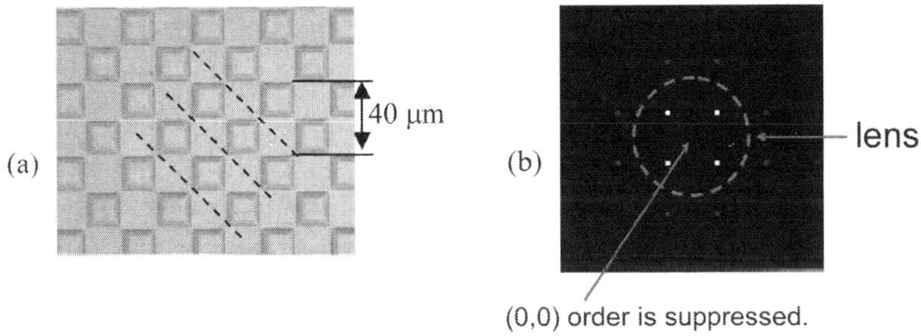

Figure 6. (a) Phase mask for multiple diffraction beam generation. (b) Far field diffraction pattern from the grating. (see color insert)

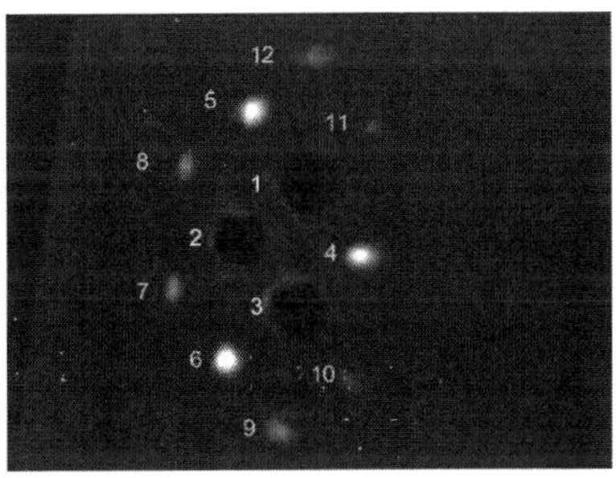

Figure 7. New beams generated by FWM of three pump beams. Beam #4 is phase matched, while the beams #5 through #12 are phase mismatched. Additionals beams are observed from mixing of four pump beams.

Figure 7 shows the beams generated by FWM in a 100 μm thick polymer sample with beam #4 input to the sample blocked. The pump beams, which were much more intense are not seen because of properly placed holes in the screen. We can see the phase-matched beam that is generated by beams #1, #2 and #3 in the direction of the blocked beam #4 as well as several other phase mismatched beams.

The magnitude of $\chi^{(3)}$ was measured by blocking beam #4 and measuring the generated FWM signal. By comparing the FWM signals from the fused silica and the nonlinear polymer we obtained $|\chi^{(3)}| \sim 0.4\times10^{-10}$ esu.

In order to measure the phase angle we used all four pump beams. The phase of beam #4 was varied in a controlled manner by propagating it through a thin tilted glass plate. A microscope coverslip (~200 μm thick) was used for this purpose as illustrated in Figure 8.

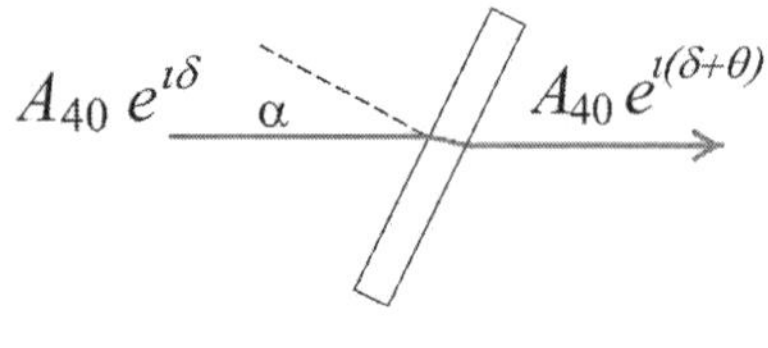

Figure 8. Depiction of the change in phase of beam #4 from a tilted plane slab. (see color insert)

The additional phase added to beam #4 on passing through the cover slip rotated at an angle α is given by

$$\theta = \frac{2\pi}{\lambda} d \left\{ n_{cs} \left(\frac{1}{\cos\beta} - 1 \right) + n_a \left[1 - \frac{\cos(\alpha - \beta)}{\cos\beta} \right] \right\}. \tag{26}$$

where $\beta = \sin^{-1}(\sin \alpha / n_{cs})$ and n_{cs} and n_a are the refractive indices of the cover slip and of air, respectively (*16*).

In the data collection, beams #1, #4, and a laser power reference signal are monitored while the cover slip is rotated from -20° to +20° in 0.2° increments. (Note that beam #2 output responds to changes in the input phase of beam #4 in the same way that beam #4 output responds; consequently, beam #2 output may be monitored instead of beam #4 output). This type of scan is then performed at different power levels. The beam #1 and #4 signals are then normalized to the reference signal and the angle of the rotation of the coverslip is converted to the phase delay of beam #4 input using Eq.(26). Typical data is shown in Figure 9. The phase of the modulation φ_M is determined from fits to the sinusoidal curves of Figure 9(b). Each of these sinusoidal curves is fit to a function of the form $U + V \sin(\theta + \psi)$. The modulation depths are obtained from the ratios V/U. The fitted phases ψ_1 and ψ_4 are related to the modulation phase φ_M introduced in Eqs.(17) and (18) by $\psi_1 = \delta + \varphi_M + \pi$ and $\psi_4 = \delta - \varphi_M$, respectively. Thus we obtain

$$\phi_M = \frac{1}{2}(\psi_1 + \pi - \psi_4). \tag{27}$$

Eliminated is the unknown constant δ, which might arise, for example, from a tilt in the stationary compensating cover slip placed between the telescope lenses shown in Figure 5. A plot of the extracted values of φ_M as a function of the depth of modulation Δ_M is shown in Figure 10, along with estimates of experimental uncertainty (*17*).

For low modulation depths $\Delta_M < 0.02$, the nonlinear FWM interaction is comparable to the noise level and experimental errors become large. (If needed, significant reduction in the noise can be achieved by using a chopper and lock-in amplifier). Extrapolating to Δ_M=0 in Fig. 10, we find the phase of $\chi^{(3)}$ to be about 6°.

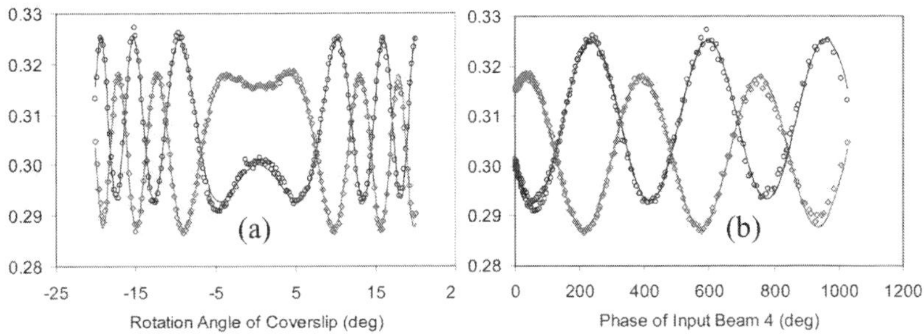

Figure 9. Power (a.u.) for beam #4 (red) and beam #1(blue) plotted vs. coverslip rotation angle (a) and vs. phase θ of beam #4 input to the sample. (b). (see color insert)

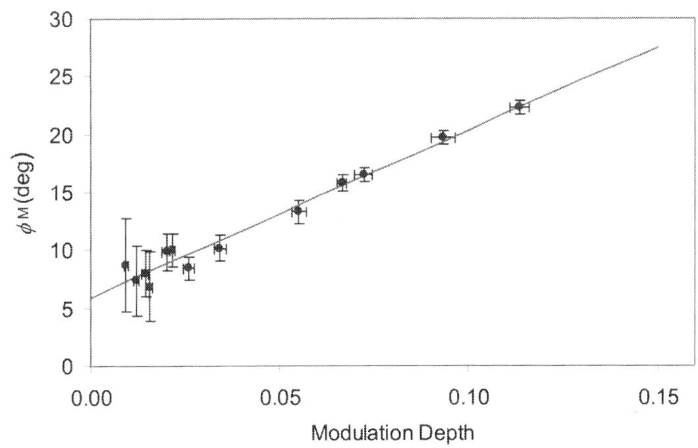

Figure 10. Plot of the phase of modulation φ_M vs. modulation depth Δ_M.

In order to provide an independent cross check of this measurement and to determine the possible contribution of higher order nonlinear absorption, we measured the transmission of a single beam through the nonlinear polymer sample with the intensity varied up to 12.6 GW/cm^2 as shown in Figure 11. A fit to this data gave a nonlinear absorption coefficient β = 2.1 cm/GW. To estimate the phase angle φ_χ of $\chi^{(3)}$ we use tan $\varphi_x = (\lambda/4\pi)(\beta/n_2)$, where $n_2 = 12\pi^2\text{Re}(\chi^{(3)})/n^2c$ (*6*) with $\chi^{(3)}$ in esu. Making the assumption that cos $\varphi_x \approx 1$ and using the measured $\left|\chi^{(3)}\right| = 3.5\times10^{-11}$ esu along with n = 2, we thus estimate the phase to be ~4 degrees. Considering the uncertainty in β, this result is consistent with the absolute measurement of phase angle in our phase scan experiment. There is no evidence of any higher order loss.

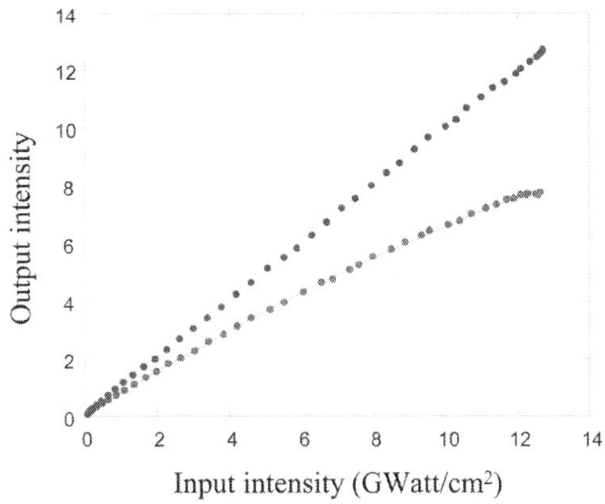

Figure 11. Plot of output intensity vs. laser input intensity. Blue dots correspond to output without the sample and red dots with the nonlinear polymer sample. The black line is the calculated transmission with 21% linear loss and a nonlinear loss coefficient of 2.1 cm/GW. (see color insert)

Discussion

In this work we described a technique for accurate measurement of the complex value of $\chi^{(3)}$. The technique involves a relatively simple apparatus and the phase angle of $\chi^{(3)}$ can be determined absolutely without a reference sample and without knowing precise spatial or temporal beam profiles. It is therefore interesting to compare our approach with the Z-scan technique. Z-scan measurements require only a focusing lens, precision translations stage an aperture and one detector. The set up for our FWM technique is a bit more complicated. It requires a special diffraction grating, two lenses, a precision rotation stage for phase delay and two detectors. In our approach, we measure the phase of $\chi^{(3)}$ by using four equal pump beams. We vary the phase of one of the pump beams (#4) at the input to the sample and observe significant variation of the output beam intensities. Let us look at the intensity distribution at the sample as a function of the phase of beam #4. Figure 12 shows a part of the near field intensity profile. We see a regular pattern of equally spaced beamlets. The series of plots below shows that as we vary the phase of beam #4, the shapes of individual beamlets vary and the intensities of the beamlets go up and down. For qualitative comparison, let us consider what happens in Z-scan. As we move the sample through the focus, the intensity and the width of the beam also varies. Plots on the figure below show how intensity of an ideal Gaussian pulse varies as the sample is moved by distance of about one Rayleigh range.

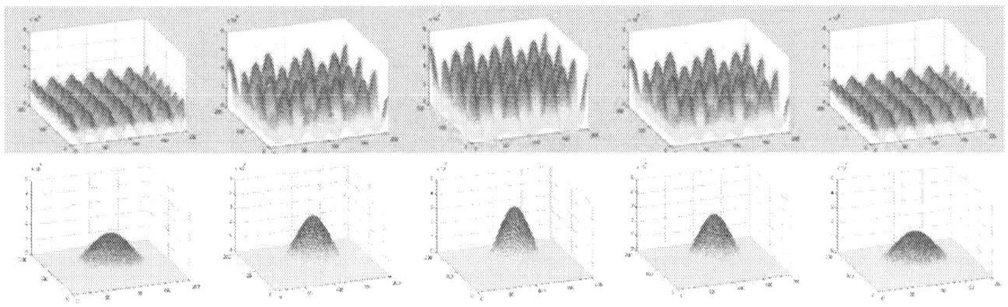

Figure 12. (see color insert)

In both experiments we can calculate the effect of the change in intensity and the beam shape and predict the effect on the output beam profile. These are basically the same type of measurement and we can learn the same amount of information from both of them. For example, if we are measuring either a nonlinear index of refraction or two photon absorption coefficient relative to some reference sample, then there is very little difference between the two approaches.

Complete knowledge of the laser beam profile is essential to both techniques for absolute measurements of $\chi^{(3)}$. There is a small difference in the level complexity of analyzing of the results. The information from discrete phase matched and phase mismatched beams in FWM provides a straight forward method of data analysis. For a reasonable range of intensities we have simple analytical expressions. In the Z-scan technique, there is a continuum of the spatial components.

The big difference is in the requirement for precise and accurate knowledge of the laser beam profile for measuring the ratio of the phase of the nonlinear susceptibility, in other words, the ratio of it's imaginary and real parts. This ratio is used to calculate the figure of merit for a nonlinear material, and it is critical for many applications (*4*). Z-scan requires complete characterization of the laser profile since measurement of each quantity depends on different parameters of the beam. In our FWM experiment, the critical information is the small scale structure of the small beamlets, and it is completely determined by the large scale far field pattern which was created by the phase mask and by a simple beam blocker.

In conclusion, we measured the intensities of all the beams generated in DFWM in a thin nonlinear polymer film. Using four pump beams and varying the phase of one of the pump beams allows determination of the phase of $\chi^{(3)}$ without the need for any reference sample, such as quartz. Using this method, a $\chi^{(3)}$ phase of 6° was obtained for a polyacetylene provided by J. Perry, S. Marder, and J. Hales at the Georgia Institute of Technology.

Appendix A: Fabrication of the Phase Mask

The phase plate is critical for this technique and was fabricated from fused silica using lithographic techniques. A contact aligner was used to transfer the two dimensional grating patterns in the photomask to photoresist on a one mm thick fused silica plate. For four micron thick photoresist OIR 908-35, the exposure time was 13 seconds. After developing it for 60 seconds and rinsing in de-ionized water for 30 seconds, the dimension of the square on four micron thick photoresist was 20 microns and the dimension of the bare fused silica square (developed) was 20 microns. The equal size of islands for photoresist and bare fused silica or 50/50 duty cycle was very important in order for us to achieve minimum power distribution in the 0th diffraction order.

A Plasma Therm 790 RIE (reactive ion etching) was used to etch fused silica plates with patterned photoresist. The RIE dry etching recipe is as follows: chamber pressure was 16 mTorr, the flow rate of CHF3 20 was set at 18 sccm (standard cubic centimeters per minute), the flow rate of O_2 20 was set at 6 sccm, and the RF power was set to 175 Watts. At these settings for RIE dry etching, the etching rates of photoresist and fused silica are 49.2 nm/min and 24.0 nm/min respectively. The phase mask was designed to have a groove depth of $\lambda/[2(n - 1]$ and to transmit minimum energy in the 0th diffraction order, where λ=1550nm is the wavelength of the OPA and n is the refractive index of fused silica window at 1550 nm. The etched island depth of the fused silica phase mask should be 1745.4 nm, which required an etching time of 72 minute and 44 seconds. After soaking in acetone for several hours to remove leftover photoresist, we obtained a fused silica phase grating mask.

References

1. Imajuku, W.; Takada, A.; Kabayashi, Y. Low-noise amplification under 3 dB noise figure in high gain phase sensitive fibre amplifier. *Electron. Lett.* **1999**, *35*, 1954.
2. Matsumoto, M. Regeneration of RZ-DPSK signals by fiber-based all-optical regenerators. *IEEE Photon. Technol. Lett.* **2005**, *17*, 1055–1057.
3. Croussore, K.; Kim, I.; Kim, C.; Han, Y.; Li, G. Phase-and-amplitude regeneration of differential phase shift keyed signals using a phase-sensitive amplifier. *Opt. Express* **2006**, *14*, 2085–2094.
4. Stegeman, G. I.; Wright, E. M.; Finlayson, N.; Zanoni, R.; Seaton, C. T. Third-order nonlinear integrated optics. *J. Lightwave Technol.* **1988**, *6*, 953–970.
5. Shen, Y. R. *The Principles of Nonlinear Optics*; John Wiley: New York, 1984.
6. Boyd, R. W. *Nonlinear Optics*; Academic Press: San Diego, CA, 1992.
7. Carter, G. M. *J. Opt. Soc. Am. B* **1987**, *4*, 1018–1024.
8. Strohkendl, F. P.; Dalton, L. R.; Hellwarth, R. W.; Sarkas, H. W.; Kafafi, Z. H. *J. Opt. Soc. Am. B.* **1997**, *14*, 92–98.
9. Sheik-Bahae, M.; Said, A. A.; Wei, T.-H.; Hagan, D. J.; Van Stryland, E. W. *IEEE J. Quantum Electron.* **1990**, *26*, 760–769.
10. Samoc, M.; Samoc, A.; Luther-Davies, B.; Bao, Z.; Yu, L.; Hsieh, B.; Scherf, U. Femtosecond Z-scan and degenerate four-wave mixing

measurements of real and imaginary parts of the third-order nonlinearity of soluble conjugated polymers. *J. Opt. Soc. Am. B* **1998**, *15*, 817–825.

11. Tseng, S.-Y.; Cao, W.; Peng, Y.-H.; Hales, J. M.; Chi, S.-H.; Perry, J. W.; Marder, S. R.; Lee, C. H.; Herman, W. N.; Goldhar, J. *Opt. Express* **2006**, *14*, 8737.
12. Shirley, J. A.; Hall, R. J.; Eckbreth, A. C. Folded BOXCARS for rotational Raman studies. *Opt. Lett.* **1980**, *5*, 380–382.
13. Maker, P. D.; Terhune, R. W. Study of optical effects due to an induced polarization third order in the electric field strength. *Phys. Rev.* **1965**, *137*, A801.
14. Jarem, J. M.; Banerjee, P. P. *Computational Methods for Electromagnetic and Optical Systems*, 2nd ed; CRC Press: Boca Raton, FL, 2000; Chapter 4.
15. Maznev, A. A.; Crimmins, T. F.; Nelson, K. A. How to make femtosecond pulses overlap. *Opt. Lett.* **1998**, *23*, 1378–1380.
16. Herman, W. N.; Roberts, M. J. The sense of chromophore orientation in films made by alternating polyelectrolyte deposition. *Adv. Mater.* **2001**, *13*, 744–746.
17. Bevington, P. R. *Data Reduction and Error Analysis for the Physical Sciences*; McGraw-Hill: New York, 1969.

Part II. Organic Optoelectronic Materials and Devices

Chapter 11

Electroluminescence with Colloidal Particles

Christopher F. Huebner, David D. Evanoff, Jr., and Stephen H. Foulger*

Center for Optical Materials Science and Engineering Technologies and School of Materials Science and Engineering, Clemson University, Clemson, SC 29634-0971, USA
***foulger@clemson.edu**

A simple route to producing electroluminescent colloidal particles that can be designed to produce a wide range of emission colors and that can be potentially converted into printable inks is presented. This approach utilizes a miniemulsion fabrication route and exploits the concept that the particles in a colloidally-based OLED can be viewed as individual "particle-devices". EL dye components were sequestered in particles that prevented any interparticle energy transfer. Red, green, and blue emitting particles were synthesized and converted into devices that exhibited emissions that spanned a large color spectrum.

Introduction

The last twenty years have seen an ongoing interest in developing all-organic light-emitting devices as alternatives to inorganic based systems. The inherent advantage in organic small molecule (*1*) and light-emitting polymer (*2*, *3*) systems, relative to inorganic systems that are fabricated through established lithography routes, is the potential that low-cost, large area electronics may be realized (*4–6*) when these materials are coupled with conventional commercial printing technologies (*7–12*).

Academic and private industry researchers are working toward efficient white light emitting devices focusing primarily on information display applications (*13*, *14*) with fewer groups showing interest in specialty applications that call for the emission of various colors within the visible spectrum (*15*) for example, automotive lighting, decorative lighting, marketing, and other novelty luminaire. To satisfy the performance criteria of these latter applications, the small molecules and π-conjugated polymers often exploited in OLEDs can require diffcult and lengthy synthetic protocols and/or complex device fabrication techniques to create devices with distinct emissions over a broad range of wavelengths (*16–21*) while inorganic quantum dot (QD) based light emitting devices (LED) are currently cost-prohibitive for widespread use (*22*). Alternative systems lies in the creation of emissive materials that utilize an organic electroluminescent (EL) dye embedded in a hole and an electron transporting polymer host (*23*). A thin film comprised of a single EL dye and hole and electron transporting host can exhibit excellent emission characteristics, but extending this device design to encompass other colors through the addition of various dyes to the host is not straightforward; a thin film containing several spatially-adjacent EL dyes can result in energy transfer processes which defeats the additivity approach (*24*, *25*). The ability to sequester EL dye components in a host with no appreciable energy transfer offers the ability to create tailored emissions over a broad range of wavelengths by mixing red, green, and blue emitters. We present a simple route to producing electroluminescent colloidal particles that can be designed to produce a wide range of emission colors and that can be potentially converted into printable inks. This approach exploits the concept that the particles in a colloidally-based OLED can be viewed as individual "particle-devices".

Results and Discussion

A schematic of the colloidally-based organic light emitting device employed in this effort is presented in Figure 1. In this construction, the anode of the device is a glass substrate which has a templated layer of indium tin oxide (ITO) on its surface. The glass/ITO surface is planarized with the addition of a ca. 150 nm layer of the transparent dielectric poly(butyl methacrylate) (PBMA) to the regions of the exposed glass surface. This dielectric layer is relatively soft at room temperature, exhibiting a glass transition temperature of ca. 35 °C, and assists in reducing field enhancements and corresponding electrical shorts at the edges of the ITO. Colloidal particles that are primarily composed of a hole-transporting polymer doped with an electron transporting small molecule and an EL dye, are spin-coated onto the anode to form the active layer of the device. The particles are formed through a mini-emulsion approach (*26*) and exhibit particle diameters which can, under varying synthetic conditions, range from 50 nm to 500 nm. Prior to their deposition on the anode, a poly(3,4-ethylenedioxythiophene):poly(styrenesulfonate)(PEDOT:PSS) dispersion is mixed into the colloids that acts as a binder for the colloids and assists in the reduction of field singularities around the particles that may result in regions of pin-hole formation. Finally, the cathode assembly is formed by the evaporation of a ca. 30 nm thick calcium layer onto the colloids, over which a thicker aluminum capping layer is evaporated.

The colloids were composed of poly(9-vinylcarbazole) (PVK), 2-(4-biphenylyl)-5-(4-tert-butylphenyl)-1,3,4-oxadiazole (tBu-PBD), and an EL dye, either Coumarin 1 (B), Coumarin 6 (G), or Nile Red (R). The formation of the colloids utilized a simple approach referred to as a "mini-emulsion" process and has been exploited by Landfester and coworkers (*26*, *27*) for the formation of a range of colloidal systems. One form of the mini-emulsion process involves the shearing of a surfactant stabilized hydrophobe in an aqueous environment. The hydrophobe can be composed of a water insoluble polymer and/or small molecule dispersed in a organic solvent. Shearing the two-phase system results in the formation of particles which can span over the range of 50 to 500 nm. A removal of the organic solvent through evaporation results in the formation of polymeric particles dispersed in an aqueous environment. A typical batch of the colloids exhibited an average weightbased composition of ca. 63 % PVK and ca. 37 % tBu-PBD, with the dyes composing less than 0.5 %. The colloids range in diameter from 20 nm up to 135 nm; the mean particle diameter was 51 nm and a standard deviation of 15 nm.

Figure 2 presents the optical images of a number of energized colloidally-based organic light emitting devices templated into a 12.7 mm wide tiger paw insignia. The devices were fabricated with varying ratios of the dye-doped colloids to produce a wide spectrum of colors. The corresponding EL spectra of the devices are presented in the right side of the figure, with an additional EL spectrum of a yellow emitting device. The excitation of the dyes in a PVK:tBu-PBD host is accomplished through at least two proposed mechanisms: (1) energy transfer or (2) carrier trapping. Energy transfer can take place through a Coulombic interaction (Förster) or electron exchange (Dexter) mechanism and occurs when an excited donor (exciton) located in either the PVK, tBu-PBD, or a PVK:tBu-PBD complex transfers energy over to a ground state dye (acceptor). To be an effective means of energy transfer, both the Förster and Dexter mechanisms have a number of conditions that must be satisfied, but two major considerations include the spatial separation between the donor and acceptor and the spectral characteristics of the donor's emission and the acceptor's absorption. A Förster-type transfer can be potentially effective up to ca. 10 nm, while a Dexter-type transfer, though highly dependent on the electronic configuration of the donor-acceptor, is roughly limited to distances under 2 nm (*28*). In addition, both modes of energy transfer require a high level of spectral overlap of the donor's emission and the acceptor's absorption. The second mechanism of exciting a dye relies on charge carrier trapping on the dye, with a subsequent recombination (*29*). Both means of exciting a dye in a PVK:tBu-PBD host are potentially operative in a EL configuration, while the reduced probability of generating free carriers in a PL study implies that a trapping mechanism will not contribute to dye emission in this configuration (*30*).

The primary RGB colors utilized in this work for creating additional colors are presented in Figure 2(1, 3, and 5). The EL spectral characteristics of the red (2-1) and green (2-3) colloidally-based OLED are very similar to the PL spectra of the individual dyes dilutely dispersed in chloroform, while the corresponding PL response of the dye-doped colloids results in the appearance of an additional peak centered at 425 nm. Recent PL studies of PVK:tBu-PBD fillms indicate that

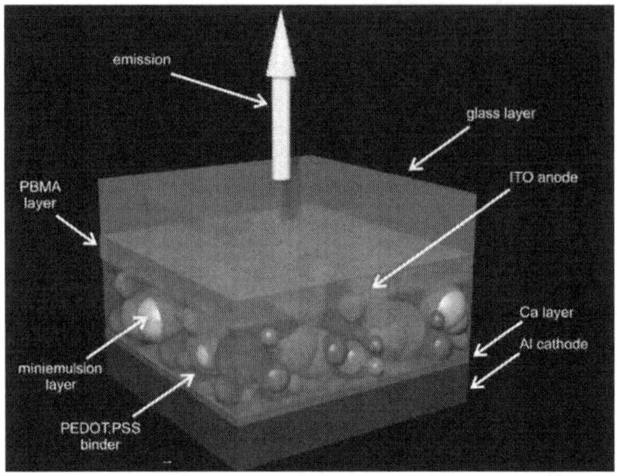

Figure 1. Schematic illustration depicting the construction of a single pixel colloidally-based organic light emitting device. A glass substrate with a templated layer of indium tin oxide serves as the anode of the device. In order to planarize the surface and prevent regions of _eld enhancement, a ca. 150 nm layer of poly(butyl methacrylate) is spin-coated onto the anode and removed from the ITO region. Emissive colloids, mixed with a binding poly(3,4-ethylenedioxythiophene):poly(styrenesulfonate) dispersion, are deposited onto this electrode assembly. The cathode is formed by the evaporation of a ca. 30 nm thick Ca layer onto the colloids, over which a thicker Al layer is evaporated. (see color insert)

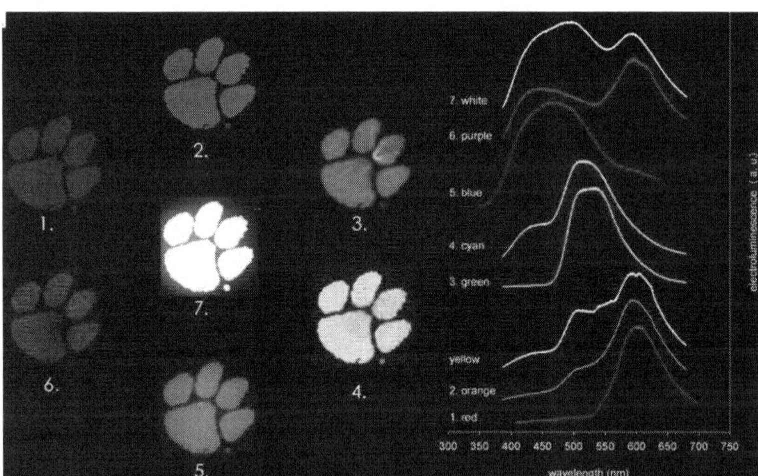

Figure 2. Optical images (1-7) of colloidally-based organic light emitting devices templated into a 12.7 mm wide tiger paw insignia with varying RGB dye ratios (left side of the figure) and the corresponding electroluminescence spectra (right side of the figure). In addition, the electroluminescence spectrum of a yellow emitting device is presented. (see color insert)

the PL emission peak centered at 425 nm will red-shift to 440 nm under an EL configuration and has been ascribed to the formation of PVK:tBu-PBD exciplex (*31*, *32*). The absence of this peak in the EL spectra of the R- and G-derived devices suggests a complete energy transfer from the PVK:tBu-PBD host to these dyes. In contrast to the R and G systems, the B-derived device exhibits an EL spectral response that is similar to the PL response of the blue-doped colloids, as indicated in Figures 2(5) and 3. Nonetheless, these latter spectra are significantly different to the PL response of the dilutely dispersed blue dye. Both the EL and PL response of the blue-doped colloids exhibit an onset to emission that is at ca. 350 nm, which is at least 40 nm less than the onset for PL of the dilutely dispersed blue dye. This can be attributed to the poor spectral overlap in dye absorption and host emission; the dilutely dispersed Coumarin 1 (B) dye exhibits an absorption peak centered at ca. 370 nm, while the emission of the PVK:tBu-PBD exiplex (cf. Figure 4) has a maximum of 425 nm. The limited energy transfer possible from the host to the dye results in the appearance of the PVK:tBu-PBD emission characteristics in the EL response of the blue dye-doped colloid.

In addition, there is a small shoulder visible within the EL response of the blue dye-doped colloid that is centered at 610 nm (cf. Figure 2(5); this weak contribution has been seen in EL studies of PVK and has been tentatively assigned to PVK electromers due to its absence in PL studies of PVK (*31*). The inclusion of the PEDOT:PSS does not appear to act as a short circuit for the devices by evidence of a lack of an Ohmic response in the current-voltage curves, suggesting that the PEDOT:PSS is unable to percolate across the electrodes at the concentration utilized. These devices exhibited luminance values ranging from under 10 cd/m^2 for the blue and red-doped colloidal devices to 30 cd/m^2 for the green doped colloidal device. Although the luminous output for these system is not optimal relative other current color-tailored OLEDS (*15*, *23*) or QD-based devices (*22*) the charge density characteristics correlate well with commercially available light emitting devices that compete in the same niche markets that call for specially tailored emissions across the visible spectrum. In addition, the calculated current density at 10 V forward bias is ca. 100 mA/cm^2; the low device turn-on voltages of the colloidally-based system, coupled with low current requirements and reasonable operating voltages, is comparable to inorganic alternatives such as QD-based devices, which are fabricated at a signifcantly higher material costs and require both a distinct hole and electron injection layer that necessitate lengthy and expensive fabrication methods.

The simple construction of these devices allowed for the rapid screening of various dye-doped colloidal ratios to explore their combined emission characteristics. Figure 2(2, yellow, 4, 6, and 7) presents a number of templated tiger paw insignias with their corresponding EL spectra. In a typical thin film device, the simple mixing of dyes at molecular lengths scales will result in the transfer of energy from shorter wavelength to longer wavelength emission, while the sequestering of the dyes in individual particles allows for their mixing with the retention of their individual emission characteristics due to a lack of energy transfer between particles. Figure 2(7) presents a device which exhibits a broad spectral emission and a "white" appearance; the ratio of RGB dye-doped colloids utilized for this device was intially estimated through a PL study.

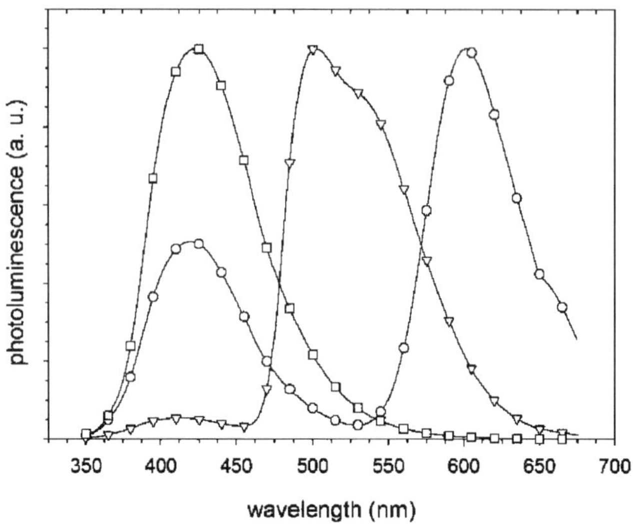

Figure 3. (a.) The photoluminescence spectra of the three RGB dye-doped colloids at an excitation wavelength of 343 nm: red (○); green (▽); and blue (□).

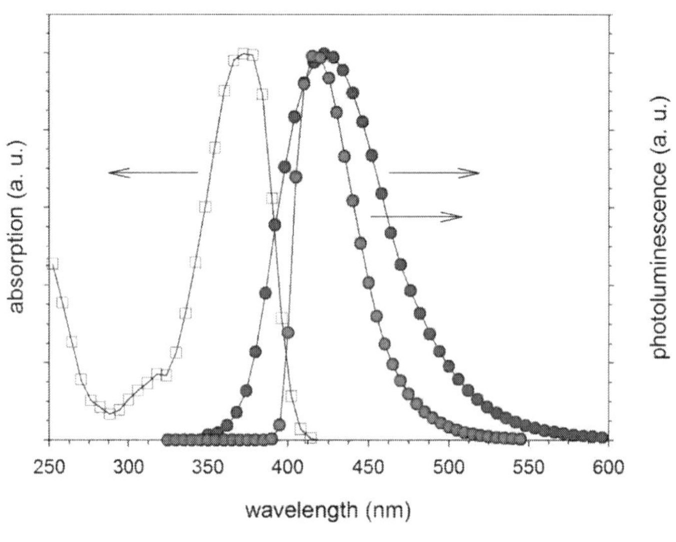

Figure 4. The absorption (2) and photoluminescence (excitation at 375 nm; red) spectra of the Coumarin 1 (B) dye in chloroform. The photoluminescence spectra (excitation at 343 nm; blue) of a colloid composed of only PVK and tBu-PBD in a ca. 63/37 weight ratio. All curves have been scaled for clarity. (see color insert)

Table 1 presents the dominant wavelength (d) and color purity (p) of the devices presented in Figure 2. The dominant wavelength for a color is denoted as the wavelength on the edge of the diagram that lies on a line joining the white point and the x,y-position of any color, while the color purity is denoted as the ratio of

Table 1. Chromaticity coordinates for colloidally-based organic light emitting devices presented in Figure 2. Dominant wavelength (d) and color purity (p) are based on white at CIE coordinates of (1/3, 1/3).

color	*CIE coordinates (x,y)*	*d (nm)*	*p*
red	0.56, 0.41	590	0.92
green	0.33, 0.57	553	0.70
blue	0.24, 0.28	484	0.35
cyan	0.28, 0.44	519	0.20
yellow	0.44, 0.44	578	0.66
orange	0.48, 0.43	583	0.73
purple	0.37, 0.33	611	0.11

distances between (1.) the color point and the white point and (2.) the dominant wavelength and the white point. The primary colors will describe the region of the chromaticity figure accessible by the additive color approach; achieving a saturated color in an OLED requires that the p values of the primary RGB colors used to generate colors within the chromaticity diagram be as saturated as possible with p values as close to 1. Monochromic or saturated colors have p values that approach 1, while a color will have a p value that approaches zero as the color becomes less vibrant or more washed-out. In the current selection of dyes, the G and R-derived dye-doped devices exhibit superior color purities with values of 0.70 and 0.92, respectively, while the B-derived device exhibits a dominant wavelength of 484 nm and a color purity of 0.35. The lower p value is in part due to the limited overlap of the PVK:tBu-PBD emission with the absorption of the Coumarin 1 (B) dye, though through a judicious choice in dye selection, it is possible to optimize the overlap to achieve high color purities in these colloidally-based organic light emitting devices.

Conclusion

The approach presented in this OLED effort describes a simple methodology by which materials designers can view their colloidal systems as individual "particle-devices". In essence, if one can effciently transport charge to an individual colloidal particle, it is not unrealistic to assume that single particle devices could be built.

Experimental

Emissive colloids and device active layer. In a typical synthetic attempt of the emissive colloids, a solution containing 1.43 ml of $CHCl_3$, 41.25 mg of poly(9 vinylcarbazole) (PVK), 33.75 mg of 2-phenyl-5-(4-biphenylyl)-1,3,4-oxadiazole (tBu-PBD), and an EL dye was formed. For the solution containing the red EL dye, 0.225 mg of Nile Red was added to the solution. Additionally, the green EL dye solution contained 0.45 mg of Coumarin 6, while the blue EL dye solution also contained 0.45 mg of Coumarin 1. These solutions were then added to an aqueous solution containing sodium dodecyl sulfate (SDS) (100 mg in 25 ml H2O). This two-phase solution was then emulsifed using a tip sonicator (VirTis Virsonic 600 Ultrasonic Cell Disruptor) for 4 minutes at a 12 W power setting. After sonication, the colloidal dispersions were placed onto a hot plate at ca. 60 °C and stirred for 12 hours to evaporate any remaining $CHCl_3$ present in the solution. The subsequent emissive colloids were cleaned through dialysis (Spectra/Por MWCO = 50000) to remove residual small molecules, dyes, or surfactant. The dialysis water was changed every ca. 12 hours with 18.2 MΩ-cm deionized water until the conductivity of the dialysis bath measured less than 0.5 μS/cm at the end of a 12 hour period.

Utilizing an additive primary color model, the red, green, and blue emitting colloids were mixed at various ratios to achieve color shades within the color space described by the RGB dyes. The ratios of RGB colloids required for a specifc color shade were estimated through the observed PL spectrum of the combined colloids. To create the active layer mixture that was initally spun-cast onto the device, the RGB colloidal mixture was combined with an electronic grade of poly(3,4-ethylenedioxythiophene):poly(styrenesulfonate) (PEDOT:PSS) (Baytron P Formulation VP CH 8000; H. C. Stark, Inc.) in a 4:1 (v/v) ratio.

Device Fabrication

Small test-pixels were fabricated from a 12.5 mm on-edge square piece of float glass with an indium-tin-oxide (ITO) coating (Delta Technologies CG-50IN-5107) that was etched with a 4 x 12.5 mm rectangular pattern and cleaned through a solvent / sonication / plasma procedure. A poly(butyl methacrylate) (PBMA):$CHCl_3$ solution (25 mg/ml) was then spun cast onto the substrate to a thickness of ca. 150 nm and then removed from the ITO region with a CHCl3 soaked cotton tipped applicator. In addition, a larger 25 mm on-edge square piece of ITO-coated oat glass was etched and patterned with a positive photoresist (HPR 504; Arch Chemicals, Inc.) to form a tiger paw template; with these structures, no PBMA layer was required. The active layer mixture (i.e. emissive colloids with PEDOT:PSS) was then spun cast onto the substrate to a thickness of ca. 100 to 150 nm. A calcium electrode was then thermally evaporated onto the active layer and then capped with a layer of evaporated aluminum.

Characterization

Thermal gravimetric analysis (TA Instruments TGA 2950) was carried out on the colloids under a nitrogen purge from 20 °C to 900 °C in a high-resolution mode, while their absorption spectra were collected with a Perkin Elmer Lambda

900 UV/Vis/NIR spectrophotometer. In addition, the photoluminescence (PL) characteristics of the colloids were collected on an Horiba Jovin-Yvon Fluorolog FL-3-22/Tau-3 Lifetime spectrouorometer over a wavelength range of 350-675 nm utilizing an excitation wavelength of 343 nm. The electroluminescence characteristics of the colloidally-based organic light emitting devices were studied in a MBraun UniLab glove box under an argon atmosphere. A computer-controlled Keithley 228A Voltage/Current Source supplied a DC voltage to the ITO anode, while a Keithley 2001 Digital Multimeter recorded the current within the device. In addition, the EL spectra were obtained with a Horiba Jovin Yvon MicroHR monochromator with a Synapse CCD detector. All data were taken at a temperature of 23 °C unless otherwise noted. All chemicals were purchased from either Sigma-Aldrich or Fisher Scientifc unless otherwise noted.

Acknowledgments

The authors thank DARPA (Grant Number: N66001-04-1-8933) and the National Science Foundation through a CAREER award (Grant No. DMR-0236692) (SHF) for financial support.

References

1. Tang, C. W.; Slyke, S. A. V. *Appl. Phys. Lett.* **1987**, *51*, 913.
2. Burroughes, J. H.; Bradley, D. D. C.; Brown, A. R.; Marks, R. N.; Mackay, K.; Friend, R. H.; Burns, P. L.; Holmes, A. B. *Nature* **1990**, *347*, 539.
3. Braun, D.; Heeger, A. *J. Appl. Phys. Lett.* **1991**, *58*, 1982.
4. Mitschke, U.; Bauerle, P. *J. Mater. Chem.* **2000**, *10*, 1471.
5. Patel, N. K.; Cinà, S.; Burroughes, J. H. *IEEE J. Sel. Top. Quantum Electron.* **2002**, *8*, 346.
6. Kukhto, A. V. *J. Appl. Spec.* **2003**, *70*, 165.
7. Hebner, T. R.; Wu, C. C.; Marcy, D.; Lu, M. H.; Sturm, J. C. *Appl. Phys. Lett.* **1998**, *72*, 519.
8. Garnier, F.; Hajlaoui, R.; Yassar, A.; Srivastava, P. *Science* **1994**, *265*, 741.
9. Pudas, M.; Halonen, N.; Granat, P.; Vahakangas *Prog. Org. Coat.* **2005**, *54*, 310.
10. Rogers, J. A.; Bao, Z.; Makhija, A.; Braun, P. *Adv. Mater.* **1999**, *11*, 741.
11. Makela, T.; Jussila, S.; Kosonen, H.; Bäcklund, T. G.; Sandberg, H. G. O.; Stubb, H. *Synth. Met.* **2005**, *153*, 285.
12. GE Demonstrates Worlds First Roll-to-Roll Manufactured Organic Light Emitting Diodes (OLEDs). http://www.ge.com/research/grc_7_1_35.html, 2008 (accessed March 12, 2008).
13. Misra, A.; Kumar, P.; Kamalasanan, M. N.; Chandra, S. *Semicond. Sci. Technol.* **2006**, *21*, R35.
14. Meerholz, K. *Nature* **2005**, *437*, 327.

15. Uchida, T.; Ichihara, M.; Tamura, T.; Ohtsuka, M.; Otomo, T.; Nagata, Y. *Jpn. J. Appl. Phys.* **2006**, *45*, 7126.
16. Liu, J.; Xie, Z.; Cheng, Y.; Geng, Y.; Wang, L.; Jing, X.; Wang, F. *Adv. Mater.* **2007**, *19*, 531.
17. Xu, Y.; Peng, J.; Mo, Y.; Hou, Q.; Cao, Y. *Appl. Phys. Lett.* **2005**, *86*, 163502.
18. Tasch, S.; List, E. J. W.; Ekstrom, O.; Graupner, W.; Leising, G.; Schlichting, P.; Rohr, U.; Geerts, Y.; Scherf, U.; Mullen, K. *Appl. Phys. Lett.* **1997**, *71*, 2883.
19. Luo, J.; Li, X.; Hou, Q.; Peng, J.; Yang, W.; Cao, Y. *Adv. Mater.* **2007**, *19*, 1113.
20. Sun, J. X.; Zhu, X. L.; Peng, H. J.; Wong, M.; Kwok, H. S. *Org. Electron.* **2007**, *8*, 305.
21. Seo, J. H.; Seo, J. H.; Park, J. H.; Kim, Y. K.; Kim, J. H.; Hyung, G. W.; Lee, K. H.; Yoon, S. S. *Appl. Phys. Lett.* **2007**, *90*, 203507.
22. Anikeeva, P. O.; Halpert, J. E.; Bawendi, M. G.; Bulovic, V. *Nano Letters* **2007**, *7*, 2196.
23. Jiang, X.; Register, R. A.; Killeen, K. A.; Thompson, M. E.; Pschenitzka, F.; Sturm, J. C. *Chem. Mater.* **2000**, *12*, 2542.
24. Forster, T. *Ann. Phys.* **1948**, *2*, 55.
25. Dexter, D. L. *J. Chem. Phys.* **1953**, *21*, 836.
26. Landfester, K. *Annu. Rev. Mater. Res.* **2006**, *36*, 231.
27. Piok, T.; Gamerith, S.; Gadermaier, C.; Plank, H.; Wenzl, F. P.; Patil, S.; Montenegro, R.; Kietzke, T.; Neher, D.; Scherf, U.; Landfester, K.; List, E. J. W. *Adv. Mater.* **2003**, *15*, 800.
28. Shoustikov, A. A.; You, Y.; Thompson, M. E. *IEEE J. Sel. Top. Quantum Electron.* **1998**, *4*, 3.
29. Pschenitzka, F.; Sturm, J. C. *Appl. Phys. Lett.* **2001**, *79*, 4354.
30. Kalinowski, J.; Giro, G.; Cocchi, M.; Fattori, V.; Marco, P. D. *Appl. Phys. Lett.* **2000**, *76*, 2352.
31. Jiang, X.; Register, R. A.; Killeen, K. A.; Thompson, M. E.; Pschenitzka, F.; Hebner, T. R.; Sturm, J. C. *J. Appl. Phys.* **2002**, *91*, 6717.
32. Kido, J.; Shionoya, H.; Nagai, K. *Appl. Phys. Lett.* **1995**, *67*, 2281.

Chapter 12

Coextruded Multilayer All-Polymer Dye Lasers

Joseph Lott,[1] Hyunmin Song,[1] Yeheng Wu,[2] Juefei Zhou,[2] Eric Baer,[1] Anne Hiltner,[1] Christoph Weder,[1] and Kenneth D. Singer*,[2]

[1]Department of Macromolecular Science and Engineering, Case Western Reserve University, Cleveland, OH 44106
[2]Department of Physics, Case Western Reserve University, Cleveland, OH 44106
***kenneth.singer@case.edu**

Coextrusion has been used to prepare multilayered polymeric films that display a photonic band gap whose position and shape can be controlled by changing layer thickness, which is achieved by adjusting processing conditions or by post processing biaxial stretching. In combination with organic fluorescent dyes, these films have been used to fabricate both distributed Bragg reflector (DFB) and distributed feedback (DFB) lasers. Both architectures displayed low lasing thresholds and high efficiencies, with DBR lasers outperforming the DFB lasers.

Since their discovery, lasers have become indispensable tools in a variety of applications such as spectroscopy, communications, information technology, and medicine (*1*). Indeed, CD and DVD players, printers, and communication systems have made lasers part of our daily lives. New materials and fabrication schemes have opened many of the opportunities for applications at various wavelengths, and will continue to do so. In addition, new concepts for low cost production, new properties, and easy systems integration will open yet more applications for lasers.

Organic materials have been used in lasers for decades as dye-doped liquid and solid-state lasers. Dye lasers dominated the market in tunable lasers for many years, while colliding pulse dye lasers were widely used in the early days of ultrashort pulsed lasers. Current research on polymer lasers is mainly aimed at fully exploiting their potential for low-cost production, ease of integration, and flexible design. Much of this recent research has been in polymer semiconductor lasers (*2*), but new schemes for dye doped polymer lasers have also been investigated as progress in polymer synthesis and processing allows for morphological control of a material's structure over multiple length scales from nano to macroscopic structures. These advances have led to the development of increasingly complex polymeric devices.

In general, the two main components required for lasing, a gain medium and a resonator for feedback, can be entirely fabricated from polymeric materials. The gain medium may consist of fluorescent dye molecules or quantum dots that one blends into a polymer host, or, alternatively, light-emitting, semiconducting polymers. As these platforms can be readily tailored, laser devices which operate at many wavelengths can be fabricated. A polymer resonator cavity can be created using periodic structures acting as multilayer interference reflectors or photonic crystals. Current methods for creating such periodic structures include embossing (*3*), (nano)imprinting (*4*), conventional or soft lithography (*5, 6*), and repetitive spin-coating (*7–9*). However in general, all of these methods are somewhat complex making them less attractive in commercial applications.

We have been investigating the use of coextruded multilayer polymer films as components in all-polymer lasers. This melt processing technique allows rapid, large-volume low-cost production of reflecting multilayer structures and is amenable to roll-to-roll mass production. The resulting polymer film lasers are low-cost elements that can be laminated onto diode pump lasers or optical circuits for a variety of sensing and optical data processing applications. We have applied the layer-multiplying coextrusion technique to fabricate all polymer distributed Bragg reflector (DBR) (*10*) as well as distributed feedback (DFB) lasers. In the DBR scheme, two multilayer polymer films act as tuned mirrors on both surfaces of a gain medium, i.e., a monolithic thin polymer film doped with a laser dye. In a DFB laser, a single multilayer film, where one of the polymer components is doped with the laser dye, serves a dual role of both the gain medium and feedback structures. This DFB laser can be fabricated in a single coextrusion step. Both structures are shown in Figure 1.

In both types of lasers, the feedback is provided by a stack of A-B quarter wave layers possessing a relatively narrow-band reflection due to interference of the optical waves in the periodic structure. In the case of DBR lasers, the multilayer films act as Bragg reflectors for the monolithic gain medium in between. These DBR films provide for low cost mirrors that would otherwise be produced by vacuum deposition of metals, a more complicated and expensive process. For DFB lasers, the photonic crystal concept provides for the lasing action. Photonic crystals (PhC) are periodic dielectrics with, for example, quarter-wave thickness layers. In this concept, the reflection band constitutes the photonic bandgap in a one-dimensional PhC. The enhanced density of states near the band edge provides the necessary feedback to achieve lasing action. In this chapter, we will describe the fabrication and properties of both DBR and DFB all-polymer lasers made using our layer-multiplying coextrusion process.

Polymer Coextrusion

Layer multiplying coextrusion is a useful technique for building layered polymeric systems and was developed in the early 1970s by Alfrey and Shrenk (*11*). The continuous nature of melt processing by coextrusion allows for large quantities of material to be fabricated, bridging the gap between applied research and commercialized product. Layered polymeric systems have been used to study a wide variety of phenomena, such as: morphology (*12*), polymer interfaces (*13*), gas barrier (*14*), and dielectric response (*15*). For optical applications, layered polymeric systems have been used to produce one dimensional photonic crystals (*16*) and to tailor the refractive index of multilayered films (*17*).

As shown in Figure 2, the layer multiplying coextrusion process begins by combining the melt of two thermoplastic polymers using a special feedblock. Melt pumps control the flow of polymer, thereby determining the volumetric composition of the melt. Once the polymers are combined in the first feedblock into a bilayer scheme, they enter the first multiplying element. The polymer melt is first sliced vertically to form two melt streams, which are subsequently spread horizontally to conform to the original geometry. Finally, the melt streams are stacked one on top of another, forming a four layered system. Each subsequent multiplier acts in a similar manner, doubling the number of layers, from 4 to 8 to 16 to 32 and so on. Films comprising hundreds to thousands of layers can be easily produced using this process. Peelable skin layers can be added to the top and bottom surfaces of the melt using a skin layer feedblock prior to the flow entering the exit die. The skin layers smooth the film and protect the core layers during handling. A number of different spreaders can be used as the exit die to form tapes, films, or large sheets. Once the melt is spread into the desired dimension, a chill roll is used to quench the melt and impart a smooth surface finish.

The melt viscosity between two polymer materials during layer multiplying coextrusion is an important parameter in obtaining uniform layers: a simple plug flow rheometer or melt flow indexer is typically used to predict the melt viscosity of the polymer yielding a plot of the log of melt viscosity against processing temperature. Uniform layers are obtained by processing two polymers having similar melt viscosities at a given temperature.

Controlling the ratio between the melt pumps of three extruders is important to achieve the targeted geometry of the overall structure. The skin layers used in layer multiplying coextrusion typically occupy 50 to 90% by volume of the total structure. Skin layers are vital for photonic applications since visible light reflection in photonic films requires layer thicknesses to range between 50 and 130 nm, which result in films 3 to 12 µm thick. Films that are less than 10 µm thick can be unstable during the spreading process at the exit die. To counteract this instability, thick skin layers are used so that the overall film thickness exceeds 50 µm for easy processing and handling while simotaneously protecting the core multilayers. In photonic films, the volume composition of polymer A to polymer B is typically 50:50. Uniform distributions in layer thickness result in narrow reflection bands whose wavelength and shape depend on the number of layers, refractive indices of the two polymers, and layer thickness.

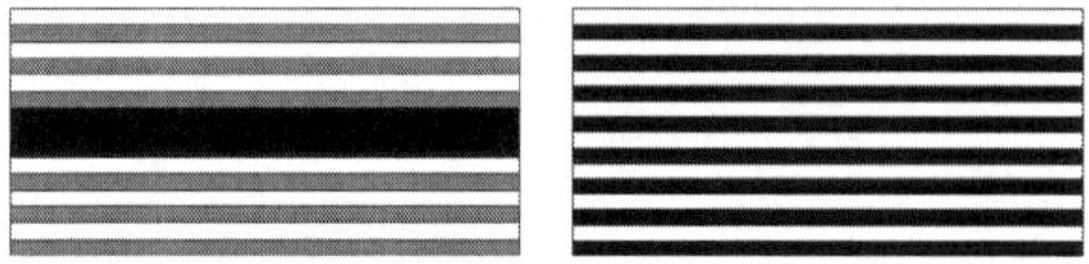

Figure 1. Structures of DBR (left) and DFB (right) lasers.

Distributed Bragg Reflector Lasers

In this section we describe the fabrication and characterization of distributed Bragg reflection lasers, and compare and optimize their performance with an end-pumped laser model. A DBR laser in micro scale can be considered as a micro-resonator, which has a total thickness of less than 100 μm. Differerent DBR resonators were fabricated by a layer-multiplying co-extrusion process and consisted of 128 alternating layers of poly(methyl methacrylate) (PMMA) and polystyrene (PS).

The difference in the refractive indices *n* of PS (n=1.585±0.002) and PMMA (*n*=1.489±0.002) caused the multilayer films to display a sharp reflection band. The centers of the resonator reflection bands were matched with the emission spectra of two series of gain media by controlling the thickness of the polymer layers during the extrusion process, and fine-tuning the properties by reducing the film thickness by subsequent biaxial stretching (*16*). During the stretching, the films were heated above the glass transition temperatures of their components and stretched evenly in two directions. The layers thicknesses were designed so that the reflection band center coincided with the emission maximum of the laser dye as seen in Figure 3.

The gain media consisted of fluorescent monolithic polymer films, which were eventually sandwiched between the two Bragg mirrors. Standard melt processing protocols were used to blend suitable fluorescent dyes into the polymer host of interest and press thin films to produce the gain media (*18*). We have focused on using small organic fluorescent chromophores for the gain media. As in any traditional dye laser, candidate chromophores must be highly fluorescent with a high quantum yield, exhibit a satisfactory Stokes shift, and should be stable to photodegradation. In addition, these dyes must be thermally stable so as to survive the coextrusion process and be readily soluble in the polymer hosts of interest. Rhodamine 6G (R6G), a commercial chromophore and 1,4-bis-(α-cyano-4-methoxystyryl)-2,5-dimethoxybenzene (C1-RG) (*19*) were used. Cyano OPVs were included as a chromophore platform because their emission color and solubility can be tailored by changing the makeup of the R groups as shown in Figure 4 and Table I (*20*).

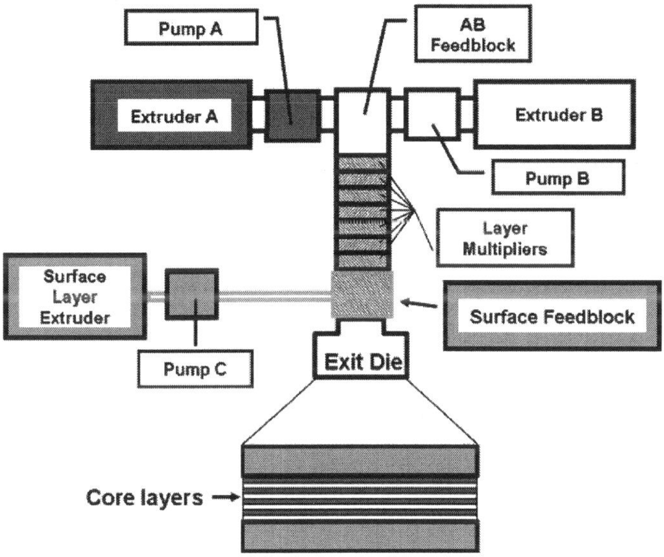

Figure 2. Schematic of the layer multiplying coextrusion process.

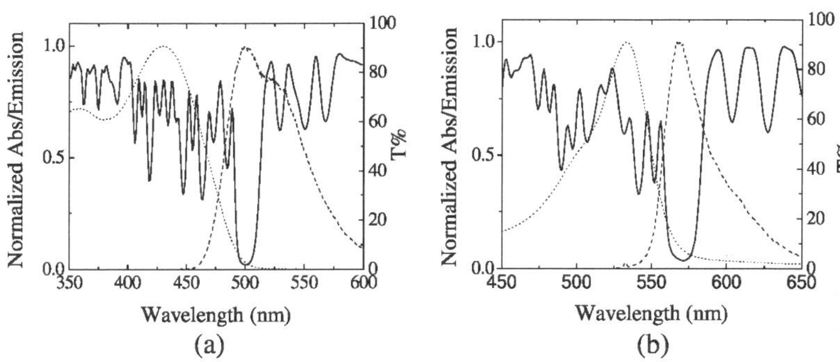

Figure 3. (a) Reflection band (solid), absorption (dotted) and emission (dashed) of C1-RG. (b) Reflection band (solid), absorption (dotted) and emission (dashed) of R6G.

The DBR dye lasers investigated here can be quantitatively described by a 4-level end-pumped laser model including re-absorption and scattering of the laser output, as well as the case of non-uniform pumping in an absorbing gain medium. For steady-state lasing, the threshold gain, g_{th}, can be expressed by (*21*):

$$g_{th} = -\frac{1}{2l}\ln(r_1 r_2) + a \tag{1}$$

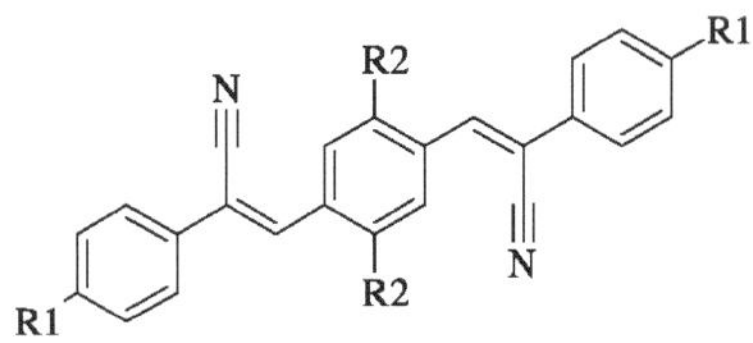

Figure 4. Structure of cyano OPVs.

Table I. Emission color of various cyano OPVs

Dye	*R1*	*R2*	*Abs. λ_{max} Solution, (nm)*	*PL. λ_{max} Solution, (nm)*	*PL. λ_{max} Solid, (nm)*	*PL. $\lambda_{monomer}$ LLDPE, (nm)*	*PL. $\lambda_{excimer}$ LLDPE, (nm)*
C1-YB	OCH_3	H	389	461,486	550	448,478	573
C1-RG	OCH_3	OCH_3	365,436	506,538	644	497,535	644
C18-RG	$OC_{18}H_{37}$	OCH_3	434	502,537	650	447	650

where l is the length of the gain medium, r_1, r_2 are the reflectivities of the two mirrors and a is the general loss at the lasing wavelength (arising from absorption, collision, scattering, etc.) in the cavity. This expression indicates that a device with thinner gain medium should have higher threshold gain requiring higher pump intensity. In a four-level system, the population inversion is proportional to pumping density. Further, the required minimum population inversion ΔN_{th} is determined by g_{th} and the stimulated emission cross section $\sigma(\omega)$.

$$\Delta N_{th} = g_{th} / \sigma(\omega) \tag{2}$$

Considering the case where the pumping is nonuniform, i.e. when the absorption is sufficiently high so that the pump intensity attenuates over the thickness of the active core, the threshold power, P_{th}, can then be expressed as,

$$\frac{P_{th}(1-e^{-\alpha l})}{V} \approx \hbar\omega\Delta N_{th}\Gamma_{e'g'} \tag{3}$$

where α is the absorption coefficient at the pump wavelength, V the volume of the gain medium illuminated by the pump, which, in the low attenuation limit, equals the product of the spot area S and the length of the gain medium l. $\Gamma_{e'g'}$ is the transition rate from e' to g', the lasing transitions. In the left hand side of eq 3, the average pump density was used. Therefore, with eqs 13, we can express the pump threshold P_{th} by,

$$P_{th} = \frac{K(-\ln(r_1 r_2) + 2(\alpha l)a / \alpha)}{1-e^{-\alpha l}} \tag{4}$$

The quantity K is a constant determined only by the type of molecule and the focusing of the pump beam. Equation 4 further implies that, for a specific dye concentration, there is an optimum absorption at which the device exhibits a minimum lasing threshold. The optimized absorption is determined by the output coupling and the ratio of loss to pump efficiency, described by a/α. For example, considering the systems we studied, a typical value of r_1r_2 would be around 0.9 and $a/\alpha \sim 0.005$, if we only consider the absorption loss from the dye molecule itself. Then the optimized absorption would be 1.13 OD. From the modeling above, one can estimate the order of pump threshold in the micro-resonator system. Taking $\sigma \sim 10^{-16}$ cm^2, $\Gamma_{e'g'} \sim 1/(3\text{ns})$, the pump threshold should be approximately 10^5 W/cm^2. Thus, a pulsed laser is required to achieve such high pump intensity without generating extra heat to burn the sample.

Our samples were pumped at oblique incidence with a tunable optical parametric oscillator (OPO), which itself was pumped by a frequency-tripled Nd:YAG laser (λ = 355 nm, pulse length 7 ns, frequency 10 Hz). A rotatable half-wave plate in combination with a linear polarizer was used to control the power of the incident beam in a continuous manner (Figure 5a). The p-polarized pump beam, focused by a lens with a 7.5 cm focal length, was incident at an angle of 10° from normal, allowing the pump light to fully penetrate the DBR reflectors by avoiding the reflection band, as the reflection band location and shape depends on the incident angle.

An objective lens was employed to collect the emission at the normal direction. Color filters were used to block the scattered pump light. An ICCD camera spectrometer was used to measure the emitted spectrum, and calibrated photodiodes were employed to measure the input and output power.

Intense laser emission was observed from all devices as high quality transverse Gaussian shaped modes. The minimum threshold fluence observed in C1-RG lasers was 435 μJ/cm^2 in a laser having a 135 μm thick gain layer comprising the dye in a concentration of 1.6 x 10^{-3} M. In the case of R6G lasers, the minimum threshold fluence was much lower at 90 μJ/cm^2 in a device having a 40 μm thick gain layer at 5.4 x 10^{-3} M.

Typical emission spectra above and below the lasing threshold for an R6G laser are shown in Figure 5c. Multimode emission above threshold was observed with the number of modes and spacing, of course, depending on the film thickness. Emission near the band center is as expected. The spectral width (full width half maximum) of a single lasing mode is about 0.4 nm, near the instrumental resolution. The emission spectra of the photoluminescent dyes employed are rather broad with distinct phonon modes, as shown in Figure 3. By contrast, the emission spectra of the DBR devices pumped below the lasing threshold are more complex and include contributions that arise from the reflection of the multilayer film and also periodic oscillations as shown in Figure 5c (dot-dash-dot curve). The oscillations are due to interference effects associated with DBR cavity reflection and are indicative of the cavity modes.

The energy conversion efficiency was measured in the linear region above pump threshold. A typical plot of output power in the forward direction versus input power is shown in Figure 5b. The slope efficiency for this sample is 14.3%, the highest efficiency observed for C1-RG lasers. The highest efficiency observed

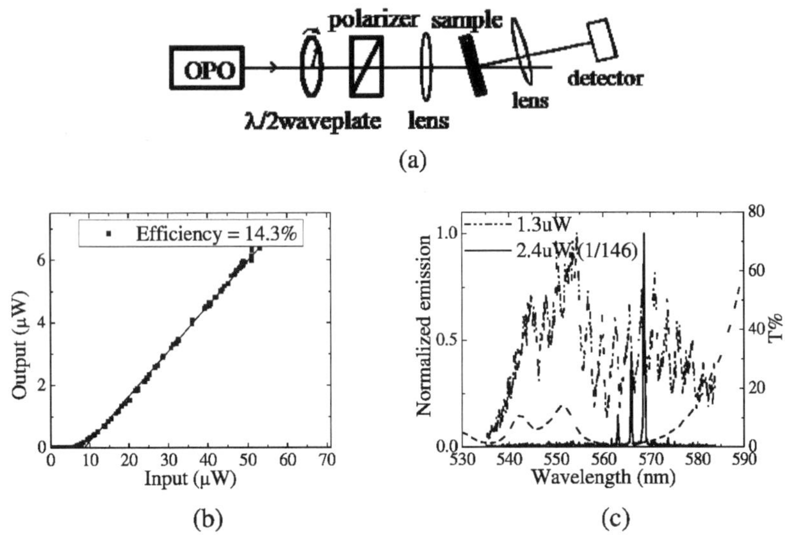

Figure 5. (a) Schematic of the experiment. (b) Slope efficiency (14.3%) of a C1-RG laser with 120 mm thick gain layer at 1.3 x 10^{-2} M in PMMA. (c) Emission spectrum of a R6G laser above (solid) and below(dot-dash-dot) threshold. Transmission curve of the whole assemnly is shown as the dashed curve.

for R6G lasers was 19.3%. These measured efficiencies take into account that the laser emits in both the forward and backward directions.

The experiment measured lasing thresholds as a function of thickness for R6G films having a gain medium doped at 5.4 x 10^{-3} M are shown in Figure 6. The relatively large uncertainties are due to sample to sample variations, as well as variations across the sample surface of the DBR films that are correspondingly due to variations in the layer thicknesses across the surface as noted above. The trend of the data shown in Figure 6 is fully consistent with eq 4 with K as an adjustable fit parameter. Despite the uncertainties, it is obvious from Figure 6 that R6G lasers comprising a gain medium doped with 5.4 x 10^{-3} M dye exhibit a threshold minimum when the core thickness is about 25 μm. This corresponds to an optical density (O.D.) of about 1.1 in agreement with our theoretical estimate. For lasers containing either R6G or C1-RG doped gain media of various dye concentrations, the minimum threshold was observed to be in the 1.1-1.4 O.D. indicating this is a generally favorable design criterion.

We have also studied the effect of layer nonuniformity on the laser emission. We measured the layer thicknesses in the distributed Bragg mirrors using atomic force microscopy (AFM), with a typical cross section shown in Figure 7. From these images, the histogram indicates a variation in layer thickness of approximately 22% for a 128-layer film. These fluctuations occur due to the coextrusion process.

Penetration of the optical mode into Bragg reflectors has been previously described (*22, 23*). The penetration depth, L_P, at wavelength λ is related to the reflection delay and to the phase dispersion. The two are related by (*24*):

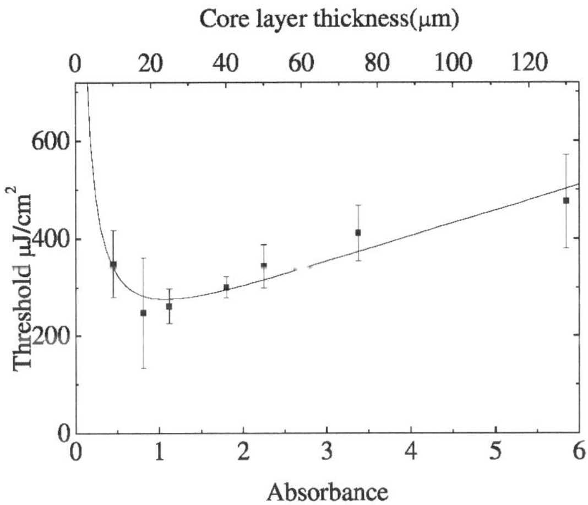

Figure 6. Threshold trend observed on R6G samples, which have different optical densities. Data points were fitted by a 4-level lasing model (eq 4), solid curve.

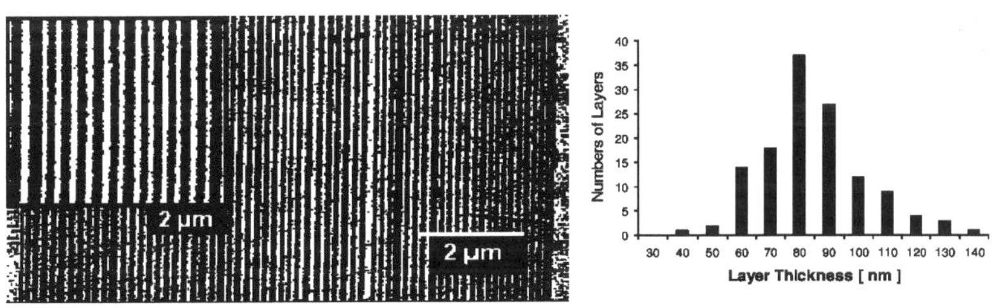

Figure 7. (a) AFM image of a 128 layer PMMA/PS multilayer cross-section. (b) Dimensional analysis of the AFM image shown in (a).

$$L_p = \frac{\partial \Delta\varphi}{2\partial k} = -\frac{\lambda^2}{4\pi n}\frac{\partial \Delta\varphi}{\partial \lambda} \tag{5}$$

where $\Delta\varphi = 2kL_p$. In a periodic structure, the phase shift of the reflected wave is based on the superposition of the waves reflected from all the interfaces. The phase dispersion can be calculated by considering a complex transfer matrix calculation (*25*). Transfer matrix calculations were carried out both by assuming a perfectly periodic structure as well as using the actual layer thicknesses as determined from the image in Figure 7. Calculations for a uniform film yields a penetration depth of less than 1 μm, while those for the film in Figure 7 yields a penetration depth in the range of 3 μm in closer agreement with our determination from the laser modes, which were calculated from the difference between the effective cavity length given by the mode spacing and the gain layer thickness.

Distributed Feedback Lasers

Unlike the DBR laser, where the gain medium was sandwiched by two polymer mirrors that constitute the laser cavity, a distributed feedback (DFB) laser is typically defined as a laser where the gain medium is an integral part of the cavity. The optical feedback is achieved by optical interference in the periodic structure containing the gain medium. This periodic structure forms a one-dimensional PhC. In this case, emission is not allowed within the reflection band, also known as the PhC bandgap. However, near the band edge of the PhC, the group velocity of photons approaches zero, equivalent to an enhanced density of photon states leading to enhanced gain and thus providing the necessary optical feedback (*26*). Low thresholds can be attained when the band edge coincides with the gain medium emission peak. Polymeric one dimensional PhC can be fabricated by spin coating (*7*), self-assembly of block copolymers (*27*) and coextrusion (*10*, *28*).

The lasers fabricated here and shown in Figure 8, are optically-pumped surface-emitting lasers. In this section, we demonstrate lasing from a distributed feedback (DFB) structure fabricated by the single roll-to-roll multilayer processing technique.

The output power, emission spectrum, slope efficiency and threshold for various designs were measured. Lasing efficiency and laser threshold were found to be pump-angle dependent. In the case of DFB lasers, more careful consideration of the dye is required. The emitting species such as fluorescent dyes or quantum dots are doped directly into the multilayered PhC, and are subject to the melt extrusion process. For example, at the temperatures involved in the coextrusion process, small molecules can easily diffuse throughout the multilayered film. This hinders control over the placement of emitting species within layers of a specific type. There are available strategies to minimize and prevent this diffusion. Covalent attachment of small molecules to larger species making them less prone to diffusion, is one approach. In this scheme, fluorescent chromophores can be covalently attached to polymers as part of the chain backbone or as pendant groups, bound to the surface of a nanoparticle, or even trapped within the core of a core-shell nanoparticle. In addition, one can exploit non-bonding interactions to limit diffusion by selecting polymers for which the chromophore has a strong affinity for one of the host layers, and a strong aversion to the other.

DFB film lasers were fabricated with alternating layers of SAN25 (a styrene acrylonitrile – random copolymer with 25 mol% acrylonitrile, $n = 1.56$), and THV 220G (a fluoroelastomer terpolymer of vinylidiene fluoride, hexafluoropropylene and trifluoroethylene, $n = 1.37$). The SAN25 layers were doped with 10.8×10^{-3} M R6G. To minimize diffusion of the chromophore from the SAN25 layers, THV was chosen as the undoped buffer layer because of its low affinity for R6G. The multilayered films with 64 and 128 layers were fabricated using the continuous layer-multiplying coextrusion process. The film with 128 layers was approximately 10 μm thick with each layer having a thickness of about 85 nm. We found that a minimum of 128 layers was required to provide sufficient optical feedback for lasing. Films with more layers are readily made, however in our coextrusion process the layer uniformity can be compromised as the number of layers increases. Films with 64 layers are expected to have better layer uniformity than films with a higher number of layers. Stacking several 64-layer films is an attractive alternative to coextruding a single film with more layers. Stacked film lasers exhibit better energy conversion efficiency and lower threshold than single films of the same thickness because of their improved distributed feedback. We demonstrate lasing from both single 128-layer films and from stacked 64-layer films.

The experimental layout is similar to the one used in the DBR laser measurement (Figure 5a) except that the sample and detector (photodiode and fiber detector head from the ICCD camera spectrometer) were mounted on a common rotation stage, allowing for convenient variation of the pump angle. DFB lasers with R6G dye are pumped at a wavelength of 532 nm. The DFB lasers emit at about 570 nm as controlled by the thickness of the layers.

The highest efficiency measured for a 128 single film laser was 2.6% with a threshold of 1.9 mJ/cm^2 (see Figure 9a) as measured by extrapolating the slope line to the x axis (input power) and taking the intercept. Conversion efficiency was improved by stacking a few thin films. Figure 9b shows that the slope efficiency in a stack of five 64-layer films is about 8% with a reduced threshold of $0.8 mJ/cm^2$. We note that single 64-layer films did not have sufficient gain to lase.

All multilayer film based lasers exhibit pump angle dependent output because the reflection band shape and location varies with the angle of incidence. The absorbance spectrum blue shifts with increasing pump angle. Lasing occurs where the transmission window at normal incidence and spontaneous emission peak area overlap. Strong lasing in a single 128 layer film occured near 15 and 55 degrees corresponding to the high effective absorption of pump energy as the reflection band of the multilayer film shifts across pump wavelength with incident angle. We have found that DFB lasers usually show better conversion efficiency at larger incident angles as the pump wavelength moves entirely out of the reflection band into a region of high transmission. This is also consistent with the fact that p-polarized beam transmission increases with incident angles due to higher reflection. We have confirmed this with transfer matrix calculations.

In a DFB laser, emission should occur at the band edge due to the enhanced density of photon states and concomitant low group velocity. Our DFB lasers are found to emit either at the band edge or somewhere inside of band gap. Figure 3 indicates a complicated structure of the reflection band due to the nonuniformity of the layers. Such structures were reproduced in transfer matrix calculations assuming variations in layer thicknesses. Such defect-like structures can act to produce emission within the nominal gap. In addition, the film nonuniformity acts to increase the threshold and decrease the efficiency. Process improvements are underway to improve performance with more uniform layer thicknesses.

Figure 8. Lasing R6G-DFB laser pumped by a 530 nm pulse generated from a YAG-third harmonic-operated OPO. The pump was set at 18 degrees from the sample surface normal. The small green spot on the right is the pump residue. (see color insert)

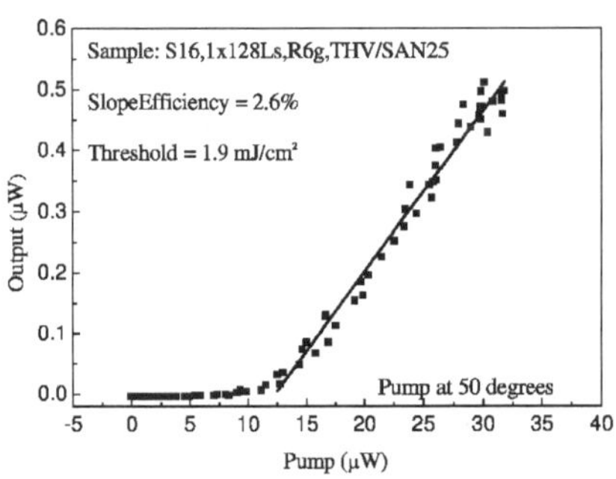

(a)

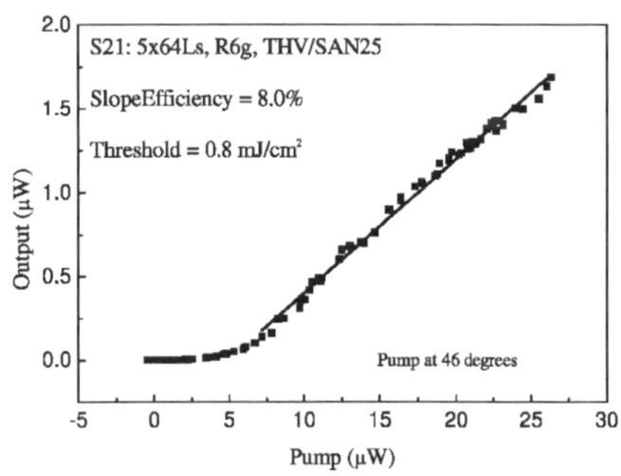

(b)

Figure 9. (a) Slope efficiency and threshold in a single run 1x128 layers THV/SAN25:R6G laser. (b) Slope efficiency and threshold in a 5x64 layers THV/SAN25:R6G stacked laser.

Conclusions

The use of layer-multiplying coextrusion to fabricate polymeric dye lasers has been shown to be extremely versatile. The photonic bandgaps can be controlled across the visible spectrum while the emission properties of organic chromophores can also be tailored to match the wavelenth of the bandgap. This allows for the fabrication of both DBR and DFB lasers capable of emitting throughout the visible spectrum. The fabrication process is continuous, fast, and inexpensive, lending itself to large scale mass production. As these laser dyes are unstable to singlet oxygen, the use of more stable gain media could lead to low-cost, commercial lasers.

References

1. Hecht, J. *Understanding Lasers: An Entry-Level Guide*, 3rd ed.; IEEE Press, John Wiley & Sons: Piscataway, Hoboken, N.J., 2008; pp xiii, 478.
2. Samuel, I. D. W.; Turnbull, G. A. *Chem. Rev.* **2007**, *107* (4), 1272–1295.
3. Kallinger, C.; Hilmer, M.; Haugeneder, A.; Perner, M.; Spirkl, W.; Lemmer, U.; Feldmann, J.; Scherf, U.; Mullen, K.; Gombert, A.; Wittwer, V. *Adv. Mater.* **1998**, *10* (12), 920–923.
4. Chou, S. Y.; Krauss, P. R.; Renstrom, P. J. *Appl. Phys. Lett.* **1995**, *67* (21), 3114–3116.
5. Berggren, M.; Dodabalapur, A.; Slusher, R. E.; Timko, A.; Nalamasu, O. *Appl. Phys. Lett.* **1998**, *72* (4), 410–411.
6. Rogers, J. A.; Meier, M.; Dodabalapur, A.; Laskowski, E. J.; Cappuzzo, M. A. *App. Phys. Lett.* **1999**, *74* (22), 3257–3259.
7. Komikado, T.; Yoshida, S.; Umegaki, S. *Appl. Phys. Lett.* **2006**, *89* (6), 061123/1–061123/3.
8. Takeuchi, H.; Natsume, K.; Suzuki, S.; Sakata, H. *Electron. Lett.* **2007**, *43* (1), 30–32.
9. Yoon, J.; Lee, W.; Caruge, J. M.; Bawendi, M.; Thomas, E. L.; Kooi, S.; Prasad, P. N. *Appl. Phys. Lett.* **2006**, *88* (9), 091102/1–091102/3.
10. Singer, K. D.; Kazmierczak, T.; Lott, J.; Song, H.; Wu, Y. H.; Andrews, J.; Baer, E.; Hiltner, A.; Weder, C. *Opt. Express* **2008**, *16* (14), 10358–10363.
11. Alfrey, T.; Gurnee, E. F.; Schrenk, W. J. *Polym. Eng. Sci.* **1969**, *9*, 400–404.
12. Bernal-Lara, T. E.; Masirek, R.; Hiltner, A.; Baer, E.; Piorkowska, E.; Galeski, A. *J. Appl. Polym. Sci.* **2006**, *99* (2), 597–612.
13. Liu, R. Y. F.; Bernal-Lara, T. E.; Hitner, A.; Baer, E. *Macromolecules* **2005**, *38* (11), 4819–4827.
14. Wang, H. P.; Keum, J. K.; Hiltner, A.; Baer, E.; Freeman, B.; Rozanski, A.; Galeski, A. *Science* **2009**, *323* (5915), 757–760.
15. Wolak, M. A.; Pan, M. J.; Wan, A.; Shirk, J. S.; Mackey, M.; Hiltner, A.; Baer, E.; Flandin, L. *App. Phys. Lett.* **2008**, *92* (11), 113301/1–113301/3.
16. Kazmierczak, T.; Song, H. M.; Hiltner, A.; Baer, E. *Macromol. Rapid Commun.* **2007**, *28* (23), 2210–2216.
17. Beadie, G.; Shirk, J. S.; Rosenberg, A.; Lane, P. A.; Fleet, E.; Kamdar, A. R.; Jin, Y.; Ponting, M.; Kazmierczak, T.; Yang, Y.; Hiltner, A.; Baer, E. *Opt. Express* **2008**, *16* (15), 11540–11547.
18. Crenshaw, B. R.; Weder, C. *Macromolecules* **2006**, *39* (26), 9581–9589.
19. Lowe, C.; Weder, C. *Synthesis (Stuttgart)* **2002** (9), 1185–1190.
20. Crenshaw, B. R.; Weder, C. *Adv. Mater.* **2005**, *17* (12), 1471–1476.
21. Milonni, P. W.; Eberly, J. H. *Lasers*; Wiley: New York, 1988; pp xvi, 731.
22. Koyama, F.; Suematsu, Y.; Arai, S.; Tawee, T. E. *IEEE J. Quantum Electron.* **1983**, *19* (6), 1042–1051.
23. Jewell, J. L.; Lee, Y. H.; Mccall, S. L.; Harbison, J. P.; Florez, L. T. *Appl. Phys. Lett.* **1988**, *53* (8), 640–642.
24. Babic, D. I.; Corzine, S. W. *IEEE J. Quantum Electron.* **1992**, *28* (2), 514–524.
25. Yeh, P. *Optical Waves in Layered Media*; Wiley: New York, 1988.

26. Dowling, J. P.; Scalora, M.; Bloemer, M. J.; Bowden, C. M. *J. Appl. Phys.* **1994**, *75* (4), 1896–1899.
27. Urbas, A.; Sharp, R.; Fink, Y.; Thomas, E. L.; Xenidou, M.; Fetters, L. J. *Adv. Mater.* **2000**, *12* (11), 812–814.
28. Sandrock, M.; Wiggins, M.; Shirk, J. S.; Tai, H. W.; Ranade, A.; Baer, E.; Hiltner, A. *Appl. Phys. Lett.* **2004**, *84* (18), 3621–3623.

Chapter 13

Solid State Organic Solar Cells

Paul A. Lane*

**U.S. Naval Research Laboratory, Optical Sciences Division,
4555 Overlook Ave. SW, Washington, DC 20375-5611
*paul.lane@nrl.navy.mil**

Organic semiconductors are a remarkable class of materials with various applications in opto-electronic devices. Starting with the earliest devices and their limitations, the various approaches to improve device performance will be reviewed. The operation of an organic solar cell is governed by the excitonic nature of these materials. Light absorption in an organic semiconductor generates a neutral excited state akin to a Frenkel exciton. The photocurrent in an organic solar cell is generated by a photoinduced charge transfer reaction between a donor and an acceptor molecule. A difference in the electron affinity or ionization potential or both drives the charge transfer reaction rather than the built-in electric field as in a traditional solar cell. Interfaces and morphology are therefore key to the properties of an organic solar cell. Charge separation occurs at the interface between the electron donor and acceptor—the organic heterojunction—and then charge transport through the respective layers. Through optimizing materials properties and film morphology, the performance of solid-state organic solar cells has improved dramatically in the last several years, with independently verified power conversion efficiencies over 5%.

Introduction

Organic semiconductors are a class of materials that have shown significant promise for applications in displays, lighting, printed circuits, and sensors. Most recently, there has been remarkable progress in organic photovoltaics, including multiple reports of cells with efficiencies over 5%. This chapter reviews progress in organic photovoltaics, concluding with prospects for high efficiency devices. While the focus of this work is on solar cells made from organic components, the considerable work on hybrid structures incorporating organic and inorganic materials must be noted. Dye-sensitized cells are particularly notable in that their efficiencies have yet to be matched by their organic analogs.

The design of an organic solar cell is governed by the excitonic nature of organic semiconductors. The primary photoexcitation is a bound state requiring a signficant amount of energy to separate into charge carriers, at least several tenths of an eV. Consequently, photogeneration of charge carriers requires an electron donor and and electron acceptor. The difference in electron affinity or ionization potential or both drives the charge transfer reaction that in turn generates a photocurrent. It is also important for there to be a relatively slow back-transfer rate so that photocarriers escape their mutual Coulomb attraction. Figure 1 illustrates the two primary classes of organic solar cells. Charge transfer in multi-layer structures takes place at an interface between layers of the electron donor and the electron acceptor. Charge transfer in bulk heterojunction devices takes place in a composite layer that contains both the electron donor and acceptor.

The processing methods generally used for molecular and polymer semiconductors also governs the structure of devices using these materials. Devices made purely from molcular semiconductors are generally fabricated by sublimation under vacuum of successive layers of electron- and hole-transporting materials. In contrast, polymer cells are generally prepared by solution processing. The most efficient devices consist of blends of a conjugated polymer and a molecular sensitizer or blends of two different conjugated polymers. Bulk heterojunction devices have been prepared by co-evaporation of molecular conductors, but have limited performance compared to solution-processed devices using polymers. Likewise, multi-layer structures have been prepared by solution processing at least one layer, but the performance of these devices lags that of their vacuum-evaporated counterparts.

Solar cells are generally characterized by the short-circuit current (J_{sc}), the open-circuit voltage (V_{oc}), and the fill factor (*ff*). The fill factor of a device, is defined as the ratio between the maximum power delivered to an external circuit and the product of V_{oc} and J_{sc}. The power conversion efficiency of a device is defined as the ratio between the maximum electrical power generated and the incident optical power P_0. The spectral response is also an important way in which to characterize OPVs and optimize their performance. It also serves as an important check against possible errors in characterization under simulated solar conditions. The external quantum efficiency (EQE), also known as the incident photon conversion efficiency, is given by the number of electrons generated per incident photon

The electronic behavior of a solar cell is often modeled in terms of discrete components. The ideal solar cell can be treated as a current source in parallel with a diode. The real world performance of a solar cell is limited and so a series resistance and shunt (parallel) resistance are added to the model. An ideal cell would have zero series resistance and infinite shunt resistance. The modified Schottky equation is given below:

$$J(V) = J_0\{\exp[q(V - JR_S)/nkT]-1\} + (V - JR_S)/R_P + J_{SC} \quad (1)$$

where R_S is the series resistance, R_P is the shunt resistance, and n is the diode ideality factor. Figure 2 shows the effect of (a) increasing series resistance and (b) decreasing shunt resistance on the J-V characteristics of a cell. In both cases, deviations from ideality can affect the fill factor dramatically and both V_{oc} and J_{sc} decreasing to a lesser extent. Sources of series resistance include the intrinsic mobility of the organic semiconductor, contact resistance at interfaces, and electrode resistance. A primary source of shunt resistance can be partial electric shorts as well as photoenhanced conductivity. It is worth noting that significant deviations from Schottky-like behavior have been observed for organic solar cells as the device parameters can be dependent upon the illumination intensity and electrical bias.

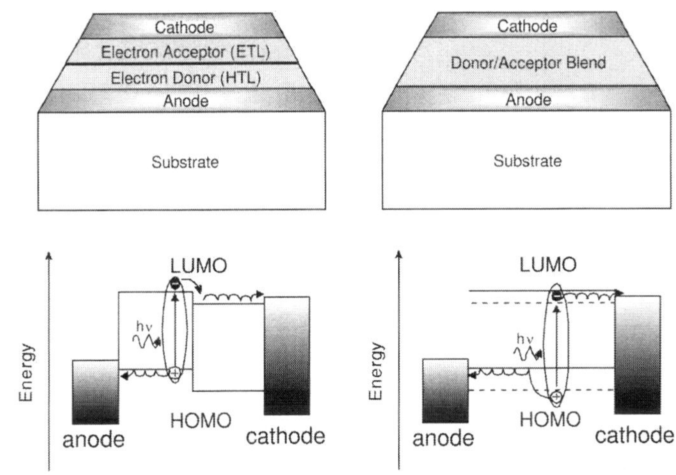

Figure 1. Schematic diagrams of multi-layer (top-left) and bulk heterojunction (top right) solar cells. The respective energy levels are shown below each type of cell.

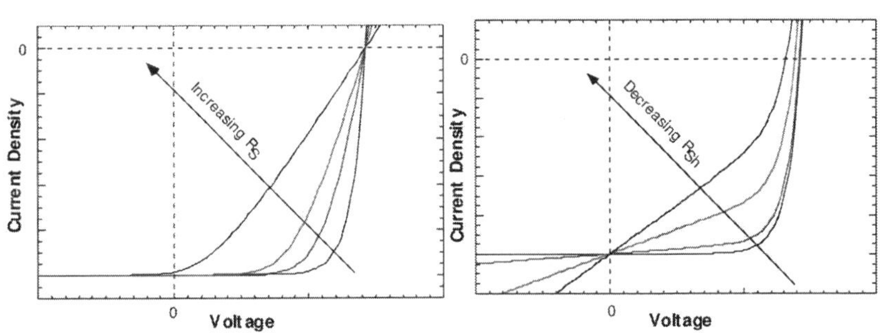

Figure 2. The effect of (a) increasing series resistance and (b) decreasing shunt resistance on the J-V characteristics of a solar cell.

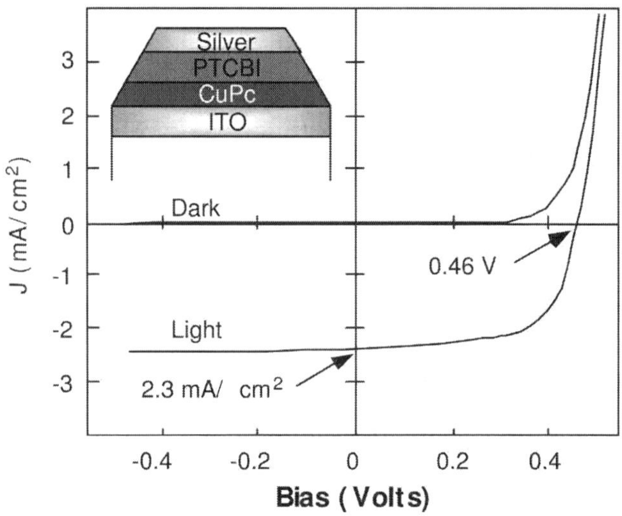

Figure 3. J-V characteristics of the a bi-layer organic solar cell (2). The device structure is inset. Copyright 1986 American Institute of Physics.

Organic Solar Cells Based on Molecular Semiconductors

The earliest organic solar cells were Schottky devices in which a rectifying contact is formed at one interface in a metal-organic-metal sandwich (*1*). Such devices are inherently inefficient, because charge photogeneration takes place in a thin layer near the metal/organic interface. A breakthrough in organic solar cell performance was reported by C. W. Tang in 1985, who used a bi-layer structure and obtained a power efficiency of 0.95% under AM2 simulated solar illumination (*2*). The device was fabricated by evaporating a 25 nm thick layer copper phthalocyanine (CuPc) onto a glass substrate coated by indium tin oxide (ITO). A 45 nm layer of 3,4,9,10-perylene tetracarboxylic-bis-benzimidazole (PTCBI) was evaporated onto the CuPc layer and the entire structure was capped by a silver cathode. Figure 3 shows the current-voltage characteristics of the cell in the dark and under illumination. A short-circuit current density J_{SC} = 2.3 ± 0.1 mA/cm^2, an open-circuit voltage V_{OC} = 0.24 ± 0.02 V, and fill factor ff = 0.65 ± 0.03 were reported. The power conversion efficiency was an order of magnitude better than previous efforts. The photovoltaic spectral response of the CuPc/PTCBI solar cell followed the absorption spectrum of the bi-layer, reaching a peak collection efficiency of about 15% at λ = 620 nm. The optical density of the CuPc/PTCBI film at this wavelength is approximately 0.4 and the corresponding internal quantum efficiency is roughly 25%.

Many research groups continued to study molecular solar cells, though device performance remained limited to ~1%. A key innovation by the Forrest group at Princeton University led to a breakthrough in OPV perfromance (*3*, *4*). Higher performance was achieved by inserting a layer between the electron acceptor and the cathode to prevent photogenerated excitons from being quenched at the cathode. Peumans *et al.* used copper phthalocyanine (CuPc) as the electron donor/hole transport layer (HTL), 3,4,9,10-perylene tetracarboxylic-bis-benzimidazole (PTCBI) as the electron acceptor/electron transport layer (ETL), and bathocuproine (BCP) as an exciton blocking layer (EBL). Excitons are excited in PTCBI and CuPc, dissociate at the organic heterojunction, and then holes are transported through CuPc and electrons though PTCBI and BCP. Although the energy of the lowest unoccupied molecular orbital (LUMO) of BCP is well above the LUMO of PTCBI, comparison of the devices with and without BCP shows that this material does not interfere with electron transport. The organic layers consisted of 15 nm of CuPc, 6 nm of PTCBI, and 15 nm of BCP: with 10% PTCBI to prevent recrystallization. Even though the absorbing layer is relatively thin, devices show a power conversion efficiency of ~1%. A light-trapping structure was fabricated in order to determine the potential collection efficiency of an OPV with an exciton blocking layer. The power conversion efficiency reached 2.4% for a cell with 6 nm thick layers of CuPc and PTCBI.

Even better results were obtained when PTCBI was replaced by C_{60}. The unit triplet yield of C_{60} is a consequence of rapid intersystem crossing (ISC) and resulting in excited state lifetimes >10 μs (*5*) and long diffusion lengths (7.7 ± 1.0 nm) (*6*). Furthermore, C^{60} is a strong electron acceptor with good electron mobility. Transient fluorescence measurements have shown that charge transfer to C_{60} proceeds on a sub-picosecond timescale (*7*). The second criterion for use in solar cells is a slow back transfer rate. Photoinduced absorption spectroscopy was used to measure the lifetime of photoinduced charge was on the order of milliseconds at 80K (*8*). Hence, the back transfer rate was many orders of magnitude slower than the charge transfer rate. The attractive photophysics of C_{60} led Peumans and Forrest to use C_{60} in OPVs with exciton blocking layers (*4*). Such devices reach power conversion efficiencies of 3.6% under AM1.5 spectral illumination at 150 mW/cm^2. Figure 4 shows the current-voltage characteristics of the tri-layer device; the energy levels of CuPc, C_{60} and BCP are inset.

The foregoing illustrates the value of exciton confinement, but the need to optimize the heterojunction itself remains. The performance of a multi-layer organic solar cell is limited by the efficiency of charge photogeneration. As charge transfer takes place at the organic heterojunction, absorption must take place at the interface or within the exciton diffusion length in the respective materials. Given that the extinction coefficient of an organic semioconductor is limited to ca. 100,000 cm^{-1} and even assuming diffusion lengths as long as 20 nm, the absorbance within the charge generation region will be limited to less than half the incident light intensity. Two approaches have been undertaken to improve charge photogeneration in molecular solar cells and are illustrated in Figure 5.

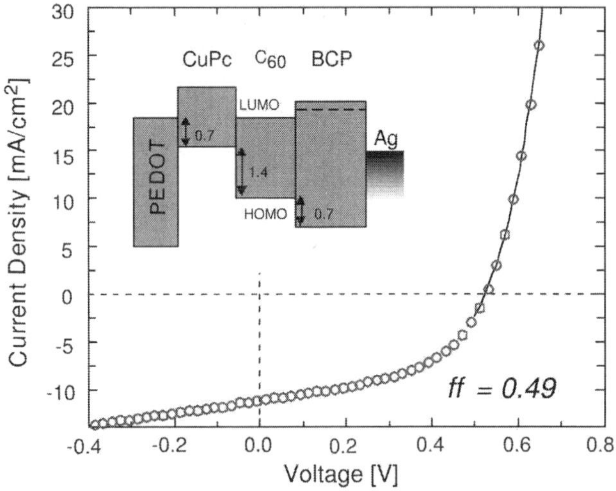

Figure 4. J-V characteristics of an organic solar cell with a CuPc electron donor, C_{60} electron acceptor, and BCP as an exciton blocking layer. The energy levels of the constituent materials are inset. Adapted from Ref. (4). Copyright 2001 American Institute of Physics.

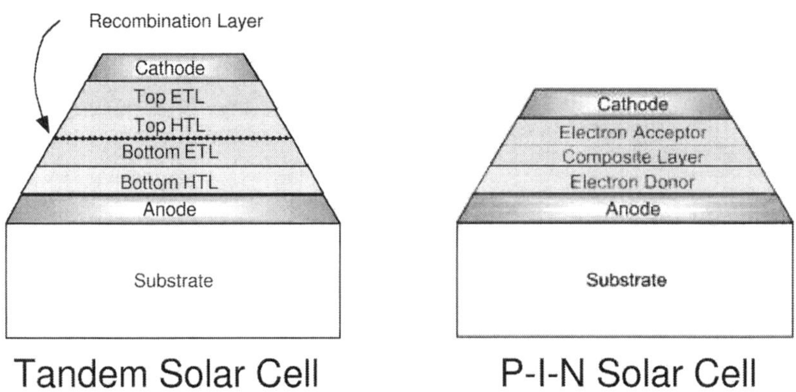

Figure 5. Schematic diagrams of (left) tandem and (right) p-i-n solar cells.

One approach is to increase the width of the charge generation region, typically by co-depositing the donor and acceptor to create a bulk heterojunction. These cells have also been referred to as p-i-n structures, with the composite layer performing the function of the intrinsic layer in an inorganic solar cell. The first tri-layer cells were reported by Hiramoto *et al.* (*9*), who compared devices with and without a composite layer. The cells used N,N′-dimethyl-3,4:9,10-perylene tetracarboxylic acid diimide (Me-PTCDI) as the electron acceptor and either a free base phthalocyanine (H_2Pc) or CuPc as the electron donor. The quantum efficiency of charge photogeneration was significantly improved for both materials combinations; J_{sc} improved by 60% for the PTCBI/CuPc device and by 125% for the PTCDI/H_2Pc device. Device efficiencies were not as high as those reported by Tang, but the improved photocurrent reflected the promise of this approach. Recently, several groups have reported organic solar cells with p-i-n structures having efficiencies of 2% or better. Männig *et al.* reported an efficiency of 1.9% from a cell using a mixture of ZnPc and C_{60} as the charge generation layer (*10*) and Suemori obtained 2.5% from a structure using a co-deposited film of C_{60} and H_2Pc (*11*). Careful control of the deposition conditions to encourage crystallinity was key to obtaining such a high efficiency. Several groups have explored the self-organization properties of liquid crystals to create separate comains for charge transport within a molecular bulk heterojunction (*12*). Another route is to deposit alternating ultrathin layers of the electron donor and acceptor, inducing phase segregation not seen in thicker layers (*13*).

An alternate way to increase charge photogeneration is to stack multiple cells separated by an internal electrode that serves as a recombination layer for charge generated in individual sub-cells. This type of structure has generally been referred to as a tandem cell. In contrast to inorganic tandem cells, the structure of the individual sub-cells may be identical. Tandem OPVs were first proposed by Hiramoto *et al.* (*14*) Tandem OPVs were constructed by successively depositing a free-base phthalocyanine (H_2Pc) and Me-PTCDI onto an ITO-coated glass substrate. The individual sub-cells were separated by an intermediate gold film (< 3 nm). The open-circuit voltage (0.78 V) of the tandem structure was nearly double that of the single cell (0.44), but the short-circuit photocurrent of the tandem cell was only one-third that of the bi-layer cell. This points out the need to very carefully match the photocurrent generated by individual sub-cells.

Power efficiencies of 2.5% were obtained by Yakimov and Forrest from stacked solar cells with using CuPc and PTCBI (*15*). Much thinner layers were used and the tandem cell contained up to five phthalocyanine/perylene bi-layers. Silver interlayers were used for the internal electrode and the optimum layer thickness was found to be 0.5 nm. Transmission electron microescopy indicated that the silver layer consisted of clusters of 1 – 5 nm in diameter with an average separation distance of 10 nm. The maximum photovoltage as a function of the number of layers. V_{oc} increases linearly up to three bi-layers and saturates; the five-heterojunction device also requires much higher light intensity to reach the maximum V_{oc}. Most recently, the Leo group at Dresden has combined these two concepts (*16*). Figure 6 shows the structure of the tandem p-i-n cell and the J-V characteristics of optimized individual and tandem cells. The electron transport layer (ETL) is C_{60} doped by a cationic electron donor; the charge generation layer is a blend of C_{60} and zinc phthalocyanine (ZnPc); the hole transport layer (HTL) is a phenyl amine doped by the strong electron acceptor tetrafluorotetracyano-quinodimethane. Typical doping levels are 1% - 4%. The efficiency of the tandem structure (3.8%) is nearly double that of the single cell (2.1%). It is notable that the fill factor of the tandem cell is higher than the single cell.

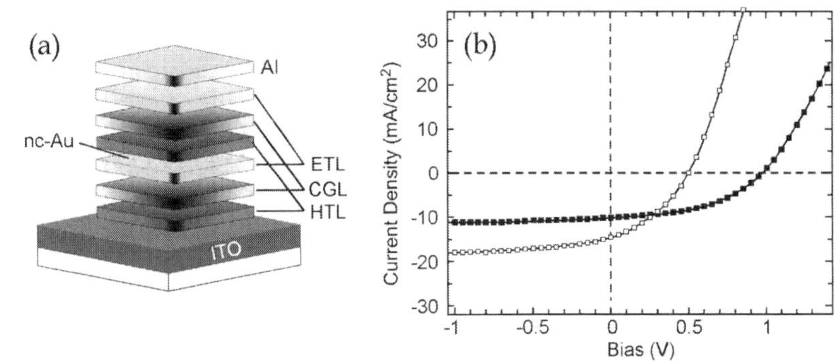

Figure 6. (a) Structure of the tandem p-i-n cell. (b) J-V characteristics of a single (open symbols) and the tandem structure (filled symbols) (16). Copyright 2005 American Institute of Physics.

Organic Solar Cells Based on π-Conjugated Polymers

Reports of photovoltaic effects in as semiconducting (or π-conjugated) polymers date back to 1981 (*17*). The earliest polymer cells were based upon trans-poly(acetylene) (t-CH_x), the absorption spectrum of which overlaps the most intense portion of the solar spectrum. While thin film sandwich structure devices with active layers of t-CH_x exhibited photovoltaic effects, both the photocurrent and fill factor were quite low. For example, Schottky-diode structures with an active layer of t-CH_x prepared by Kanicki and Fedorko (*18*) had a power efficiency of 0.1% under white light illumination of 50 mW/cm_2 from a xenon lamp. As with molecular semiconductors, it is necessary to sensitize charge photogeneration by the addition of an electron acceptor. Bi-layer polymer cells based on the organic heterojunction concept were initially tried. A hole-transporting polymer film is spin-coated onto the anode and then a sensitizer layer was either evaporated or spin-coated onto the underlayer. For example, one of the earliest such devices was prepared by spin-coating poly(2-methoxy-5-ethyl(2′-hexyloxy) *para*-phenylene vinylene) (MEH-PPV) onto the substrate, followed by vacuum evaporation of C_{60} to form a *p-n* heterojunction (*19*). Such devices suffer from the same limitations of the comparable molecular structures. Furthermore, the lower purity of polymer vis-à-vis purified molecular semiconductors leads to lower carrier mobility and, hence, worse device performance. There have been significant efforts to increase the width of the charge generation region, but such devices prepared to date are not competitive with either multi-layer molecular cells.

Polymer solar cells based on the bulk heterojunction concept, however, have been remarkably successful. The highest efficiencies have been achieved with a polymer used as the electron donor and of C_{60} derivatives as the acceptor, though other dyes as well as polymers have been used as the acceptor. Because the donor and acceptor are in intimate contact with one another, nearly all absorbed photons result in charge photogeneration. The main challenge is to minimize carrier trapping or recombination. An ideal heterojunction consists of two interpenetrating networks wherein charge generation occurs at interfaces between the two domains and charge transport within domains. Obtaining the right film morphology to obtain this structure has been critical to the success of bulk heterojunction cells.

Bulk heterojunction devices made from a blend of MEH-PPV and C_{60} showed improved performance with respect to bilayer devices (*20*). The bulk heterojunction layer consisted of a 10:1 ratio of MEH-PPV to C_{60}. Under monochromatic illumination (2.8 mW/cm^2 at 500 nm), the open circuit voltage was 0.8 V, double that of the bi-layer structure. This is also much higher than the V_{oc} of the typical molecular heterojuncdtion cell. As efficient charge generation occurs at a C_{60} concentration of only a few percent, the fundamental problem lie in obtaining efficient charge transport through the active layer. The concentration of both components of the blend must be high enough so that two percolation networks are formed in the active layer. This is problematic for C_{60} as it crystallizes during film formation and its solubility is relatively low in organic solvents used for spin-casting polymer films (*21*). The solution to this problem lie in the field of fullerene chemistry. A variety of fullerene derivatives have been prepared through modification, including functionalized fullerenes suitable for solution processing (*22*). Dyes other than fullerenes have also been used as dopants in polymer solar cells, though none have approached the performance levels of fullerene-based cells.

Yu *et al.* used the methanofullerene derivative 1-(3-methoxycarbonyl)-propyl-1-phenyl-(6,6)C_{61} or [6,6]-PCBM in bulk heterojunction devices (*23*). Films with up to 80 wt% PCBM could be fabricated, roughly one fullerene per polymer repeat unit. A device based on neat MEH-PPV had a quantum efficiency of roughly 0.1-0.2% and devices with 50% PCBM or 25% C_{60} have quantum efficiencies of about 15% at low intensities, falling to 5-8% at 20 mW/cm^2. The device containing 80wt% [6,6]PCBM, in contrast, reaches an EQE of 45% at low intensities, falling to 29% at 20 mW/cm^2. The open-circuit voltage of the device was 0.82 V for Ca/ITO electrodes and 0.68 V for Al/ITO electrodes. Due to rise in the photovoltage with intensity, the power efficiency remained relatively constant; 3.2% at 10 μW/cm^2, falling to 2.9% at 20 mW/cm^2.

Sariciftci and co-workers studied the effect of the solvent on the performance of polymer:fullerene bulk heterojunction devices using an active layer containing 80% by weight [6,6]-PCBM and 20% poly[2-methyl,5-(3′,7′-dimethyloctyloxy)]-1,4-phenylene vinylene) (MDMO-PPV). The first study compared films spin-cast from toluene or chlorobenzene solutions, finding that chlorobenzene suppresses aggregation and that devices using such films are more efficient (*24*). Figure 7 shows atomic force microscopy images of the surfaces of the polymer/fullerene film. The toluene-cast film has features on the order of 0.5 μm wide and with surface roughness of up to 10 nm. Such features were not observed in neat films of MDMO-PPV cast from toluene. These features were assigned to phase-segregated domains with different fullerene concentrations. In contrast, the film cast from chlorobenzene is much smoother and no large-scale features can be discerned. The latter study concentrated on effects of the lithium fluoride layer between the aluminum cathode and the polymer:PCBM film (*25*). By optimizing the thickness of the LiF layer, devices with a power conversion efficiency of 3.3% under simulated AM1.5 illumination were demonstrated.

A number of polymers other than PPV derivatives have been tried in organic solar cells, most notably polyfluorene co-polymers and polythiophenes. For example, Svensson *et al.* (*26*) used an alternating fluorene copolymer (PFDTBT) and a power conversion efficiency of 2.2% under solar AM1.5 illumination. Although the power conversion efficiency is lower than the PPV-based devices discussed above, the open-circuit voltage is significantly higher, 1.04 V. This opens the prospect for higher power conversion efficiencies from optimized devices. Even higher efficiencies have been reported from devices using polythiophene derivatives. Polythiophene has red-shifted absorption relative to MDMO-PPV, permitting better harvesting of incident photons, and has a relatively high hole mobility for a π-conjugated polymer. Multiple groups have reported polymer cells with efficiencies of 4% or higher. One particularly notable report came from Plextronics, who reported a 5.4% cell efficiency, certified by the National Renewable Energy Laboratory (NREL) (*27*). We note a report in December 2008 that an NREL-certified polymer solar cell with 6% efficiency was fabricated by Konarka in collaboration with Profs. Mario Leclerc and Alan Heeger (*28*).

Finally, we close with a brief discussion of tandem solar cells based on polymer:fullerene bulk heterojunctions. Molecular semiconductors are particularly well suited to tandem cells due to the sequential deposition method. This is more challenging with polymers as second layer would partially dissolve the underlying film. It is possible to avoid dissolving the first layer by thermal converting the first film to an insoluble form (*29*) or using a solvent for the second layer in which the already spun polymer is insolvent. However, this approach has not yielded efficient cells. Recently, a group at the University of California, Santa Barbara demonstrated highly efficient polymer tandem solar cells that used two different polymers with complementary absorption spectra (*30*). Figure 11(a) shows the structure of the tandem polymer cell. The charge-separation layer for the front cell is a blend of of poly[2,6-(4,4-bis-(2-ethylhexyl)-4H-cyclopenta[2,1-b;3,4-b′] dithiophene)-alt-4,7-(2,1,3-benzothiadiazole)] and [6,6]PCBM. The charge-separation layer for the back cell is a bulk heterojunction composite of poly(3-hexylthiophene) (P3HT) and [6,6]-phenyl-C71 butyric acid methyl ester ($PC_{70}BM$). The two layers are separated by a titanium dioxide layer that was coated by conductive polymer film prior to deposition of the second sub-cell. Figure 8 compares the J-V characteristics of the cells based on the individual blends and those of the tandem cell. The current density of the tandem cell is nearly that of the individual cells and the open circuit voltage has doubled. The calculated efficiency of this cell is 6.5%, the highest to date for an organic solar cell.

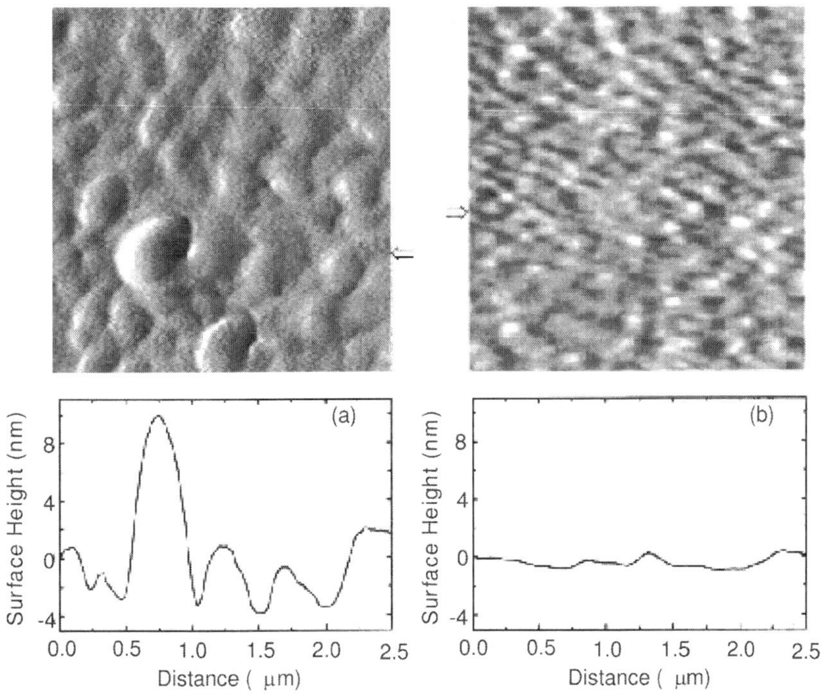

Figure 7. Atomic force microscopy images of MEH-PPV:PCBM films cast from toluene (a) or chlorobenzene (b) solutions (24). Surface profiles are shown below the images of each film. Copyright 2001 American Institute of Physics.

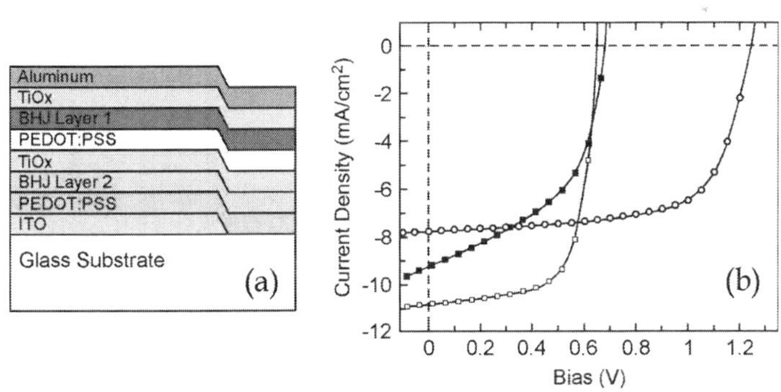

Figure 8. (a) Schematic diagram of the tandem bulk heterojunction cell. See above text for the composition of the two bulk heterojunction layers. (b) J-V characteristics of individual sub-cells and the tandem cell (30). Copyright 2008 AAAS.

Conclusions

Organic photovoltaics are an exciting and rapidly advancing field of research. Recent advances have brought performance levels to the threshold of commercialization. The development of tandem cells has increased the ultimate limits of both molecular and polymer solar cells. Using reasonable device fill factors and a known acceptor (PCBM), Dennler *et al.* have shown that device efficiencies as high as 15% are feasible from tandem cells using polymers with optimized energy levels as electron donors (*31*). These efficiencies combined with the unique processing advantages of organic semiconductors will lead to a promising future for organic solar cells.

References

1. Chamberlain, G. A.; Cooney, P. J.; Dennison, S. Photovoltaic properties of merocyanine solid-state photocells. *Nature* **1981**, *289* (5793), 45–47.
2. Tang, C. W. Two-layer organic photovoltaic cell. *Appl. Phys. Lett.* **1986**, *48* (2), 183–185.
3. Peumans, P.; Bulovic, V.; Forrest, S. R. Efficient photon harvesting at high optical intensities in ultrathin organic double-heterostructure photovoltaic diodes. *Appl. Phys. Lett.* **2000**, *76* (19), 2650–2652.
4. Peumans, P.; Forrest, S. R. Very-high-efficiency double-heterostructure copper phthalocyanine/C_{60} photovoltaic cells. *Appl. Phys. Lett.* **2001**, *79* (1), 126–128.
5. Hung, R. R.; Grabowski, J. J. A precise determination of the triplet energy of C_{60} by photoacoustlc calorimetry. *J. Phys. Chem.* **1991**, *95* (16), 6073–6075.
6. Pettersson, L. A. A.; Roman, L. S.; Inganäs, O. *J. Appl. Phys.* **1999**, *86*, 487.
7. Kraabel, B.; Lee, C. H.; McBranch, D.; Moses, D.; Sariciftci, N. S.; Heeger, A. J. Ultrafast photoinduced electron-transfer in conducting polymer buckminsterfullerene composites. *Chem. Phys. Lett.* **1993**, *213* (3-4), 389–394.
8. Smilowitz, L.; Sariciftci, N. S.; Wu, R.; Gettinger, C.; Heeger, A. J.; Wudl, F. Photoexcitation spectroscopy of conducting-polymer-C(60) composites: photoinduced electron-transfer. *Phys. Rev. B* **1993**, *47* (20), 13835–13842.
9. Hiramoto, M.; Fujiwara, H.; Yokoyama, M. *p-i-n* like behavior in three-layered organic solar cells having a co-deposited interlayer of pigments. *J. Appl. Phys.* **1992**, *72* (8), 3781–3787.
10. Männig, B.; Drechsel, J.; Gebeyehu, D.; Simon, P.; Kozlowski, F.; Werner, A.; Li, F.; Grundmann, S.; Sonntag, S.; Koch, M.; Leo, K.; Pfeiffer, M.; Hoppe, H.; Meissner, D.; Sariciftci, N.S.; Riedel, I.; Dyakonov, V.; Parisi, J. Organic *p-i-n* solar cells. *Appl. Phys. A: Mater. Sci Process.* **2004**, *79*, 1.
11. Suemori, K.; Miyata, T.; Yokoyama, M.; Hiramoto, M. Three-layered organic solar cells incorporating a nanostructure-optimized phthalocyanine:fullerene codeposited interlayer. *Appl. Phys. Lett.* **2005**, *86*, 063509.

12. Gregg, B. A.; Fox, M. A.; Bard, A. J. Photovoltaic effect in symmetrical cells of a liquid crystal porphyrin. *J. Phys. Chem.* **1990**, *94* (4), 1586–1598.
13. Hong, Z. R.; Maennig, B.; Lessman, R.; Pfeiffer, M.; Leo, K.; Simon, P. Improved efficiency of zinc phthalocyanine/C_{60} based photovoltaic cells via nanoscale interface modification. *Appl. Phys. Lett.* **2007**, *90* (20), 203505.
14. Hiramoto, M.; Suezaki, M.; Yokoyama, M. Effect of thin gold interstitial-layer on the photovoltaic properties of tandem organic solar-cell. *Chem. Lett.* **1990** (3), 327–330.
15. Yakimov, A.; Forrest, S. R. High photovoltage multiple-heterojunction organic solar cells incorporating interfacial metallic nanoclusters. *Appl. Phys. Lett.* **2002**, *80* (9), 1667–1669.
16. Drechsel, J.; Männig, B.; Kozlowski, F.; Pfeiffer, M.; Leo, K.; Hoppe, H. Efficient organic solar cells based on a double *p-i-n* architecture using doped wide-gap transport layers. *Appl. Phys. Lett.* **2005**, *86* (3), 244102.
17. Tsukamoto, J.; Ohigashi, H.; Matsumura, K.; Takahashi, A. A Schottky-Barrier type solar-cell using polyacetylene. *Jpn. J. Appl. Phys.* **1981**, *20* (2), L127–L129.
18. Kanicki, J.; Fedorko, P. Electrical and photovoltaic properties of trans-polyacetylene. *J. Phys. D: Appl. Phys.* **1984**, *17*, 805–817.
19. Sariciftci, N. S.; Braun, D.; Zhang, C.; Srdanov, V. I.; Heeger, A. J.; Stucky, G.; Wudl, F. Semiconducting polymer-buckminsterfullerene heterojunctions: Diodes, photodiodes, and photovoltaic cells. *Appl. Phys. Lett.* **1993**, *62* (6), 585–587.
20. Yu, G.; Pakbaz, K.; Heeger, A. J. Semiconducting polymer diodes: Large size, low cost photodetectors with excellent visible-ultraviolet sensitivity. *Appl. Phys. Lett.* **1994**, *64* (25), 3422–3424.
21. Ruoff, R. S.; Tse, D. S.; Malhotra, R.; Lorents, D. C. Solubility of C_{60} in a variety of solvents. *J. Phys. Chem.* **1993**, *97* (13), 3379–3383.
22. Taylor, R.; Walton, D. R. M. The chemistry of fullerenes. *Nature* **1993**, *363* (6431), 685–693.
23. Yu, G.; Gao, J.; Hummelen, J. C.; Wudl, F.; Heeger, A. J. Polymer Photovoltaic cells: Enhanced efficiencies via a network of internal donor-acceptor heterojunctions. *Science* **1995**, *270* (5243), 1789–1791.
24. Shaheen, S. E.; Brabec, C. J.; Sariciftci, N. S.; Padinger, F.; Fromherz, T.; Hummelen, J. C. 2.5% efficient organic plastic solar cells. *Appl. Phys. Lett.* **2001**, *78* (6), 841–843.
25. Brabec, C. J.; Shaheen, S. E.; Winder, C.; Sariciftci, N. S.; Denk, P. Effect of LiF metal electrodes on the performance of plastic solar cells. *Appl. Phys. Lett.* **2002**, *80* (7), 1288–1290.
26. Svensson, M.; Zhang, F. L.; Veenstra, S. C.; Verhees, W. J. H.; Hummelen, J. C.; Kroon, J. M.; Inganas, O.; Andersson, M. R. High-performance polymer solar cells of an alternating polyfluorene copolymer and a fullerene derivative. *Adv. Mater.* **2003**, *15* (12), 988.
27. Laird, D. W.; Vaidya, S.; Li, S.; Mathai, M.; Woodworth, B.; Sheina, E.; Williams, S.; Hammond, T. *SPIE Proc.* **2007**, 6656.
28. See http://www.pv-tech.org/news or www.konarka.com.

29. Burn, P. L.; Grice, A. W.; Tajbakhsh, A.; Bradley, D. D. C.; Thomas, A. C. Insoluble poly[2-(2′-ethylhexyloxy)-5-methoxy-1,4-phenylenevinylene] for use in multilayer light-emitting diodes. *Adv. Mater.* **1997**, *9* (15), 1171.
30. Kim, J.Y.; Lee, K.H.; Coates, N.E.; Moses, D.; Nguyen, T.-Q.; Dante, M.; Heeger, A.J. Efficient tandem polymer solar cells fabricated by all-solution processing. *Science* **2007**, *317*, 222–225.
31. Dennler, G.; Scharber, M. C.; Ameri, T.; Denk, P.; Forberich, K.; Waldauf, C.; Brabec, C. J. Design rules for donors in bulk-heterojunction tandem solar cells-towards 15% energy-conversion efficiency. *Adv. Funct. Mater.* **2008**, *20* (3), 579.

Chapter 14

High Fill-Factor Organic Bulk Heterojunction Photovoltaic Devices Using a Highly Conducting Hole-Doped Polymer Transport Layer

M. Ballarotto,* W. N. Herman, and D. B. Romero

Laboratory for Physical Sciences, University of Maryland, 8050 Greenmead Drive, College Park, Maryland 20740, USA
***mihaela@lps.umd.edu**

This work investigates the effects of the conductivity of the conducting hole-doped polymer transport layer on the fill-factor of organic bulk heterojunction photovoltaic devices. Devices with a *p-i-m* (hole-doped layer / intrinsic layer / metal) structure were fabricated using poly- [3,4-ethylenedioxythiophene] doped with poly [styrenesulfonate] (PEDOT:PSS) as the *p*-type transport layer and a blend of regio-regular poly(3-hexylthiophene-2,5-diyl) (P3HT) and {6}-1-(3-(methoxycarbonyl)propyl)-{5}-1-phenyl-[5,6]-C61 (PCBM-C_{60}) as the intrinsic active layer. The dependence of the device photovoltaic characteristics on the conductivity of the *p*-layer was determined by employing different formulations of the PEDOT:PSS while keeping the P3HT:PCBM-C_{60} blend ratio fixed. High fill-factor was achieved in devices that employed the high conductivity formulation of PEDOT:PSS. We attribute this enhancement to the contribution of the *p*-type polymeric transport layer to the voltage-dependent charge-collection efficiency.

Introduction

While conjugated conducting polymers have been widely investigated as electrode material in the fabrication of organic devices (*1–14*), their role in defining the device characteristics remains to be elucidated. Owing to their high processibility (*1–8*), excellent film-forming properties (*1, 3–6*), and favorable band alignment (*1–4, 14*) with the organic semiconductors most commonly used for optoelectronic applications, conducting polymers have been employed either as interfacial layers (*9–14*) or stand-alone polymeric anodes in devices (*1–8*). As such, their role has been singularly attributed to band-offset adjustment at the anode-active layer interface (*1–14*). On the other hand, because of their low-mobility and propensity to form interface states that can serve as charge trapping and recombination centers, ICPs can also be expected to have a strong influence on the nonlinear current versus voltage characteristics of organic devices.

For organic photovoltaic devices, the electronic properties of the conducting polymeric hole-transport layer can significantly affect the voltage dependence of the photocurrent, $J_{ph}(V)$, generated in the device. It is important to establish the materials properties that determine $J_{ph}(V)$ since it is intimately connected to the device fill-factor (*FF*), a figure of merit that parameterizes the maximum electrical power that can be extracted from the photovoltaic device under loaded conditions. In bulk heterojunction organic photovoltaic devices, previous works have attributed the observed $J_{ph}(V)$ entirely to the effects of the active layer morphology (*4, 9–11*) and blend composition (*12, 13*). Theoretical modeling of $J_{ph}(V)$ has considered a metal / intrinsic semiconductor / metal device structure and has focused mainly on space-charge formation (*15*) as well as the voltage dependent geminate-pair dissociation and charge-carrier recombination (*16, 17*) associated with optoelectronic processes in the bulk of the device active layer. However, the *p*-type conducting polymer layer (typically, a few hundred angstrom thick) that is employed in bulk heterojunction photovoltaic devices is also expected to play an essential role in shaping $J_{ph}(V)$. Aside from its well-known effect of pinning the Fermi level of the high workfunction anode (*1–4, 14*), its behavior as a hole-transport layer and the interface it forms with the active layer must be included as well in a quantitative description of $J_{ph}(V)$. We note that in an earlier work (*18*) on a device based on low molecular weight organic molecules with a multilayer structure akin to that of the bulk heterojunction photovoltaic devices, an analysis of the device characteristics based on a *p-i-m* device structure was proposed. It was suggested that incorporation of *n*- and *p*- type doped charge-transport layers results in reduced ohmic losses as well as provide exciton blocking layers to avoid quenching at the electrodes in organic photovoltaic devices (*18*).

In this work, we shall present evidence of the effects of the electronic properties of the conducting polymer charge-transport layers on $J_{ph}(V)$. We find a systematic enhancement of the fill-factor in P3HT:PCBM-C_{60} based bulk heterojunction photovoltaic devices upon increasing the conductivity of the PEDOT:PSS hole-transport layer. We attribute the observed enhancement to improved charge-collection efficiency in devices employing high conductivity PEDOT:PSS layer.

Experimental Details

Bulk heterojunction organic photovoltaic devices were fabricated as a multilayer structure sandwiched between a transparent anode and a cathode. The anode consists of a photolithographically patterned indium-tin-oxide (ITO) on glass substrate (~ 150 nm thick with $R_{\square}$ = 4-8 Ω/□ from Delta Technologies). The patterned ITO substrate was cleaned by ultrasonic treatment sequentially in detergent, deionized water and acetone followed by rinsing in isopropyl alcohol. Subsequently, thin layers of PEDOT:PSS (~ 500 Å) were spun-cast at 5000 rpm for 60 seconds onto the patterned ITO/glass substrates, and then annealed at 140 °C for 10 minutes. The PEDOT:PSS films that were deposited on the patterned ITO/glass substrates were subsequently patterned such that only the ITO layer was encapsulated by the PEDOT:PSS. The samples were then transferred into a nitrogen-filled glove box to complete the device fabrication. The active layer of the devices was a blend of regioregular P3HT received from Rieke Metals and PCBM-C_{60} purchased from Nano-C with a 1:1 ratio, mixed in a 3% solution by weight in dichlorobenzene. Thin P3HT:PCBM-C_{60} films (~ 250 nm) were spun-cast onto the PEDOT:PSS coated ITO/glass substrate at 400 rpm for 30 seconds. The films were allowed to dry slowly in a saturated dichlorobenzene atmosphere before depositing the cathode consisting of 10 Å of LiF followed by 1000 Å aluminum through a shadow mask. The completed devices were annealed at 90°C for 10 minutes before characterization. The active area of the device, defined by the overlap between the ITO and Al electrodes, was estimated to be A_d = 0.1 x 0.1 inch square.

To highlight the effect of the PEDOT:PSS layer on the device photovoltaic characteristics, the same active P3HT:PCBM-C_{60} blend composition was used in the devices. Three different formulations of aqueous PEDOT:PSS solutions (purchased from H.C. Starck) were employed in this work: Baytron P VP Al4083 (used as received) and Baytron P and Baytron PH500 mixed with various percentages (0 to 30% by volume) of dimethyl sulfoxide (DMSO). Thin films of these various PEDOT:PSS formulations were also deposited on glass substrates without ITO for measurement of the film conductivity using a four-point probe technique.

The device photovoltaic characteristics were extracted from the current versus voltage curves measured using a simulated AM 1.5 Global solar illumination (Oriel 300 W) with ~ 95 mW/cm^2 integrated intensity as determined by a crystalline silicon photovoltaic cell calibrated at the National Renewable Energy Laboratories (NREL). The spectral dependence of the device external quantum efficiency, as well as the spectral irradiance of the simulated solar source, was obtained using a set of 10 nm-width narrow bandpass filters covering the wavelength range of 400 nm to 1600 nm. The spectral mismatch factor (*19*) was estimated following the standard reporting conditions recommended by NREL to account for the differences between the simulated source and the standard AM 1.5 Global solar spectrum as well as the spectral responsivities of our photovoltaic devices and the Si photovoltaic reference cell.

From these characteristics, the fill-factor was extracted as,

$$FF = \frac{V_M J_M}{V_{OC} J_{SC}}$$

where V_M and J_M are the voltage and current density, respectively, at the maximum power point, V_{OC} is the open-circuit voltage and J_{SC} is the short-circuit current density. The power conversion efficiency is calculated as,

$$\eta_P = \frac{V_{OC}\left(J_{SC}/MMF\right)}{P_{in}} FF$$

with P_{in} as the total incident illumination intensity. The values of η_P reported in this work have been corrected using an estimated $MMF = 0.84$.

Results and Discussion

Tuning the PEDOT:PSS Conductivity

Because of its commercial availability, PEDOT:PSS is the intrinsically conducting polymer that is widely used in the fabrication of organic optoelectronic devices. At the molecular level, PEDOT oligomers are electrostatically attached to the PSS polymeric chain to produce hole-rich segments with the sulfonic acid sidegroups providing the background counterion that maintain overall charge neutrality along the PSS chain. The composite PEDOT:PSS molecule forms a nanosize gel particle in the aqueous solution with stacked hole-doped PEDOT oligomers surrounded by negatively-charged PSS chains (*20*). Various formulations of aqueous PEDOT:PSS solutions are available that allow the fabrication of polymeric anodes with different conductivities. For our work, we used Baytron P VP Al4083 with PEDOT:PSS ratio of 1:6, and Baytron P and PH500 containing a higher PEDOT to PSS content of 1:2.5. Earlier studies have demonstrated that the conductivity of the PEDOT:PSS films can be tuned by the addition of polar solvents to the aqueous solution (*3*, *5*, *8*, *14*, *21*, *22*). In this work, we systematically adjusted the PEDOT:PSS conductivity using dimethyl sulfoxide (DMSO) as the polar additive. Figure 1 displays the conductivity (σ_{PEDOT}) from four-point probe measurement of the PEDOT:PSS films as a function of DMSO content. For films prepared from the non-DMSO treated aqueous solution of all three PEDOT:PSS formulations, $\sigma_{PEDOT} \approx 1$ S/cm. Higher σ_{PEDOT} values were obtained with increasing DMSO content. The PH500 formulation yielded the film with the best saturation conductivity at $\sigma_{PEDOT} \approx 300$ S/cm while that from Baytron P has a maximum at $\sigma_{PEDOT} \approx 60$ S/cm. These observations can be understood in terms of the morphological changes in the thin-film structure induced by the presence of DMSO in the aqueous solution (*14*, *21*, *22*). For the PEDOT:PSS films from the non-DMSO treated aqueous solutions, there are less conducting percolative paths connecting the PEDOT chains due to the preponderance of PSS barriers between chains. This accounts for the low value of σ_{PEDOT} seen in Figure 1. An atomic-force microscopy study (*14*) has shown that the addition of the polar solvent in the aqueous solution leads to the swelling of the PEDOT:PSS gel nanoparticle which subsequently drives their aggregation to form larger clusters of overlapping PEDOT chains in the films. As a consequence, percolative connectivity between PEDOT chains is enhanced resulting in the observed higher conductivity. The difference in the saturation value of σ_{PEDOT} in Baytron P and PH500 seen in Figure 1 is due to the broader dispersion in the size of the gel nanoparticle in the former than in the latter.

Effect of the PEDOT:PSS Conductivity on Fill-Factor

A systematic improvement in the fill-factor is obtained in devices employing PEDOT:PSS transport layer with high conductivity. One practical issue in the characterization of these organic photovoltaic devices is the uncertainty in determining the device area introduced by the highly conductive PEDOT:PSS layer. To avoid such uncertainty, an A_L = 0.05 x 0.05 inch square aperture was used to define an exposed area for illumination. Figure 2 compares the normalized illuminated current density (J_L) versus voltage (V) characteristics of representative devices from such measurements. It clearly shows that FF increases upon using a hole-transport layer with high conductivity. The inset to Figure 2 demonstrates that the fill-factor of our best P3HT:PCBM-C_{60} photovoltaic device (FF = 0.73) approaches that of our reference silicon-based solar cell (FF = 0.78). To ensure that the trend illustrated in Figure 2 is repeatable, we performed the same measurements on several devices. Table 1 summarizes the statistical average of the photovoltaic device characteristics from these measurements. All devices portray very similar open-circuit voltage (V_{oc}) and short-circuit current density (J_{sc}). Increasing σ_{PEDOT} leads to an improvement in FF from FF = 0.61±0.02 in device #4 that used the least conducting Baytron P VP AI 4083 while FF = 0.71±0.01 in device #1 in which the most conducting PH500 mixed with 5% DMSO is the hole-transport layer. Such an increase in the device fill-factor results in significant improvement in the power conversion efficiency from 3.19±0.15% to 3.59±0.20%.

To eliminate the possibility that the observed change in the device fill-factor as a function of the formulation of the PEDOT:PSS layer is not an artifact introduced by the use of an aperture during the device characterization, the above-described measurements were repeated without the aperture on similar devices. The results of these measurements are summarized in Figure 3 and Table 2. For clarity, only the data for devices with the highly conducting PH500 + 5% DMSO and the insulating Baytron P VP AI 4083 layers are included in Figure 3. The comparison of these two sets of devices, as well as the values displayed in Table 2, clearly shows that a higher fill-factor is obtained in the device with highly conducting PEDOT:PSS hole-transport layer. The inset in Figure 3 illustrates that the same trend is seen as the device active area is increased. A useful implication of this latter observation is that the use of highly conducting PEDOT:PSS as hole-transport layer can lead to the fabrication of large-area organic photovoltaic devices with high fill-factor.

Table 1. Comparison of the device photovoltaic characteristics obtained using a $A_L = 0.05$ x 0.05 square inch aperture for illumination. Device active area is $A_d = 0.1$ x 0.1 square inch.

PEDOT:PSS layer	σ_{PEDOT} *[S/cm]*	J_{SC} *[mA/cm²]*	V_{oc} *[V]*	*FF*	η_P *[%]*
PH500 + 5% DMSO	294 ± 94.4	7.87 ± 0.50	0.51 ± 0.01	0.71 ± 0.01	3.59 ± 0.20
Baytron P + 20% DMSO	54.3 ± 14.4	7.92 ± 0.54	0.51 ± 0.01	0.67 ± 0.02	3.44 ± 0.24
PH500	0.94 ± 0.18	7.93 ± 0.29	0.51 ± 0.01	0.60 ± 0.05	3.06 ± 0.31
Baytron P VP AI 4083	0.75 ± 0.07	8.22 ± 0.18	0.51 ± 0.01	0.61 ± 0.02	3.19 ± 0.15

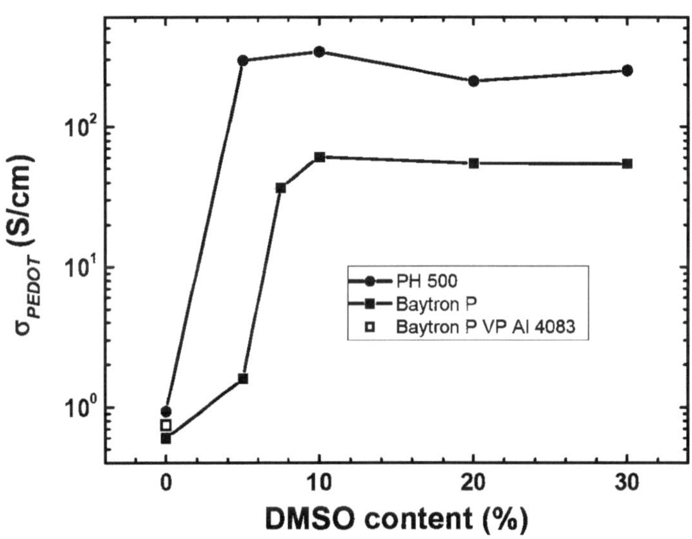

Figure 1. Conductivity of PEDOT:PSS thin films versus the DMSO volumetric content of the aqueous solution.

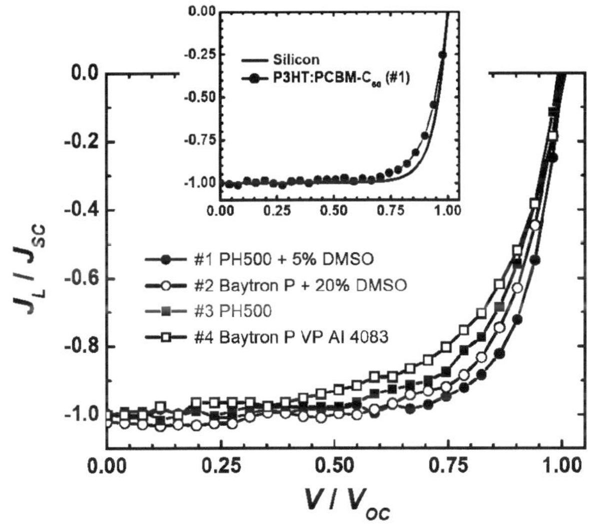

Figure 2. Normalized illuminated J_L/J_{sc} versus V/V_{oc} characteristics of P3HT:PCBM-C_{60} bulk heterojunction devices fabricated with different types of conducting PEDOT:PSS anodes, obtained using a 0.05 x 0.05 inch square aperture (see text). Inset: Comparison of P3HT:PCBM-C_{60} and a commercial crystalline silicon photovoltaic devices. (P3HT:PCBM-C_{60} device: FF = 0.73, V_{oc} =0.51 V, J_{sc}= 8 mA/cm^2; Silicon device: FF = 0.78, V_{oc} =0.54 V, J_{sc}= 29 mA/cm^2)

An interesting correlation is observed in the values of the photovoltaic device characteristics presented in Tables 1 and 2. As noted above, the values of V_{oc} and J_{sc} reported in Table 1 for measurements using an illumination aperture are nearly the same; it is only *FF* that displays a systematic variation between devices. However, without the aperture, the values of V_{oc} and J_{sc} reported in Table 2 are generally higher while *FF* is lower than those given in Table 1 for the corresponding device. The difference in V_{oc} can be explained by larger dark current contribution to the J_L versus V characteristics in the measurement with the aperture due to an illuminated area that is smaller than the active geometric area of our devices. The values of J_{sc} and *FF* are almost identical in both measurements in devices utilizing the least conducting Baytron P VP AI 4083 and PH500 formulations of PEDOT:PSS. On the other hand, for the more conducting PEDOT:PSS formulations (PH500 + 5% DMSO and Baytron P + 20% DMSO), J_{sc} is higher and *FF* is lower in the measurement without the aperture. It should be emphasized that an uncertainty in defining the device active area is introduced by contributions from PEDOT:PSS just outside the perimeter defined by the ITO electrode. This contribution can be viewed as a second diode in parallel with the one determined by the common geometric area defined by the ITO and Al electrodes. In this scenario, the J_{sc} and *FF* behaviors just described can be understood. For the insulating PEDOT:PSS formulations, the extraneous diode contribution can be ignored leading to identical J_{sc} and *FF* values in both measurements. However, when a conducting PEDOT:PSS formulation is used, non-negligible extra photocurrent is generated outside the device active area accounting for the higher J_{sc} that is determined in the measurement without the aperture. Furthermore, the extra diode provides a low parallel resistance load to the photovoltaic device explaining the lower *FF* values. Such effects is the main justification for using the illumination aperture in our measurements.

Table 2. Comparison of the device photovoltaic characteristics obtained without using an illumination aperture. Device active area is A_d = 0.1 x 0.1 square inch.

PEDOT:PSS layer	σ_{PEDOT} *[S/cm]*	J_{SC} *[mA/cm²]*	V_{oc} *[V]*	*FF*	η_P *[%]*
PH500 + 5% DMSO	294 ± 94.4	10.09 ± 0.51	0.56 ± 0.01	0.62 ± 0.01	4.34 ± 0.27
Baytron P + 20% DMSO	54.3 ± 14.4	9.09 ± 0.47	0.55 ± 0.01	0.60 ± 0.02	3.76 ± 0.35
PH500	0.94 ± 0.18	9.11 ± 0.95	0.56 ± 0.01	0.54 ± 0.01	3.43 ± 0.30
Baytron P VP AI 4083	0.75 ± 0.07	8.32 ± 0.11	0.57 ± 0.01	0.57 ± 0.01	3.36 ± 0.08

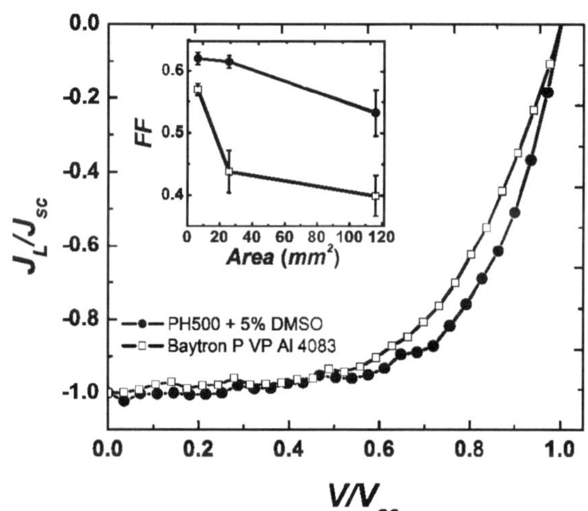

Figure 3. Normalized illuminated J_L/J_{SC} versus V/V_{OC} characteristics of P3HT:PCBM-C_{60} bulk heterojunction devices fabricated with the conducting PH500 + 5% DMSO and insulating Baytron P VP AI 4083 as the hole-transport layer, obtained without the illumination aperture. The device area is 0.1 x 0.1 square inch. Inset: Dependence of the fill-factor as a function of the area of the device active layer.

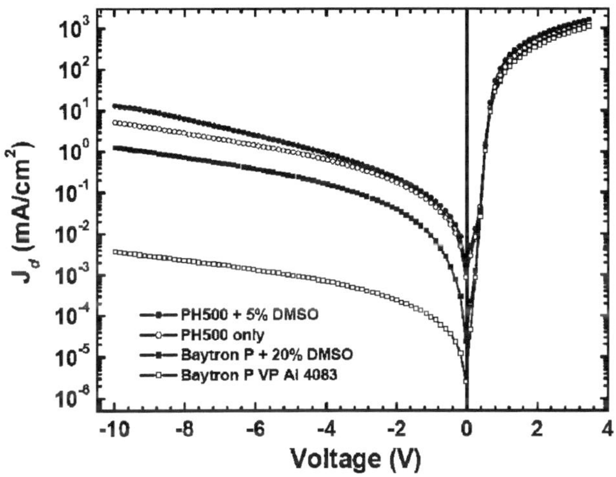

Figure 4. Dark current (J_d) versus voltage (V) characteristics of P3HT:PCBM-C_{60} bulk heterojunction devices fabricated with different types of conducting PEDOT:PSS anodes.

1. Origin of the Enhancement of the Device Fill-Factor

One possible explanation for the observed variation in the device fill-factor is the change in the series and parallel resistance of the diodes associated with using different conducting PEDOT:PSS layer. Figure 4 displays the dark current density (J_d) versus voltage (V) characteristics of our devices. At forward-bias ($V >$ 0 V), charge-transport is characterized by Ohmic conduction below 0.5 V followed by a nearly exponential increase up to 1 V and then a linear saturation region thereafter. We observe small differences in the J_d versus V characteristics of the devices suggesting a bulk dominated charge-carrier transport for $V >$ 0.5 V. On the other hand, the devices display very different J_d under reverse bias ($V <$ 0 V) that depends on the type of conducting PEDOT:PSS used as hole-transport layer. At $V = -10$ V, $J_d \approx 10^{-3}$ mA/cm^2 for the least conducting Baytron P VP AI 4083 layer while $J_d \approx 10$ mA/cm^2 for the most conducting transport layer of PH 500 mixed with 5% DMSO.

The correction in the device fill-factor introduced by the series and parallel resistances of the diode is given by (*23*),

$$FF\left(R_S, R_P\right) \approx FF\left(0, \infty\right)\left(1 - \frac{J_{SC}R_S}{V_{OC}} - \frac{V_{OC}}{J_{SC}R_P}\right)$$

where R_S and R_P are the area-corrected series and parallel resistance, respectively, extracted from the J_d versus V characteristics, FF (R_S, R_P) is the experimentally measured fill-factor, and FF (0, ∞) represents the intrinsic fill factor for the ideal diode with $R_S = 0$ and $R_P = \infty$. Based on the diode equivalent circuit (*23*), a model independent way of estimating R_S and R_P is to consider the asymptotic behavior of the J_d versus V characteristic. At sufficiently high forward-bias voltage for which the voltage drop across the diode can be neglected, J_d varies linearly with V with a slope that yields a value for R_S. Similarly, under reverse-bias voltage around $V = 0$ V, the current flowing through the diode can be ignored. The resistance corresponding to the observed linear J_d versus V behavior in this voltage range gives $R_S + R_P$ from which we calculate R_P. The results of this analysis are summarized in Table 3. Since the values of FF (R_S, R_P) are close to the corresponding FF (0, ∞) obtained for each device, we conclude that the changes in R_S and R_P introduced by using different conducting PEDOT:PSS layer are not responsible for the observed enhancement in the device fill-factor.

In general, $J_{ph}(V)$, which determines the value of the device fill-factor, is related to the voltage-dependent charge-collection efficiency, $\eta_{CC}(V)$. Based on our present findings, this efficiency can now be expressed as,

$$\eta_{CC}(V) = \eta_{CC}^{bulk}(V)\,\eta_{CC}^{PEDOT}(V)$$

where the bulk contribution, η_{CC}^{bulk}, involves the transport of the photogenerated charges in the device active layer while η_{CC}^{PEDOT} is associated with the fraction of the photogenerated holes collected through the PEDOT:PSS charge-transport layer. Since we employed the same P3HT:PCBM-C_{60} composition, η_{CC}^{bulk} is the same in our devices suggesting that the improvement in fill-factor that we observe is associated with a voltage dependent η_{CC}^{PEDOT}. This voltage dependence can come from tunneling and emission as well as surface recombination processes at the anode-active layer interface and hole-transport in the conducting PEDOT:PSS layer.

Table 3. Comparison of the device intrinsic fill-factor

PEDOT:PSS layer	σ_{PEDOT} *[S/cm]*	R_s *[Ω-cm²]*	R_p *[MΩ-cm²]*	$FF(R_s,R_p)$	$FF(0,\infty)$
PH500 + 5% DMSO	294	1.95	0.25	0.73	0.75
Baytron P + 20% DMSO	54.3	2.50	0.06	0.69	0.72
PH500	0.94	2.60	2.52	0.65	0.68
Baytron P VP AI 4083	0.75	2.68	5.87	0.59	0.62

Conclusion

We identified a contribution to the voltage-dependent photocurrent in organic bulk heterojunction photovoltaic devices associated with hole-transport in the *p*-doped conducting polymer layer. In P3HT:PCBM-C_{60} based devices, this contribution lead to high fill-factor when highly conducting PEDOT:PSS is used as a hole-transport layer. Our results suggest that the influence of the conducting charge-transport layers is an important consideration in the design of high-efficiency organic photovoltaic devices.

References

1. Carter, S. A.; Angelopoulos, M.; Karg, S.; Brock, P. J.; Scott, J. C. *Appl. Phys. Lett.* **1997**, *70*, 2067–2069.
2. Lee, T.-W.; Kwon, O.; Kim, M.-G.; Park, S. H.; Chung, J.; Kim, S. Y.; Chung, Y.; Park, J.-Y.; Han, E.; Huh, D. H.; Park, J.-J.; Pu, L. *Appl. Phys. Lett.* **2005**, *87*, 231106.
3. Fehse, K.; Walzer, K.; Leo, K.; Lövenich, W.; Elschner, A. *Adv. Mater.* **2007**, *19*, 441–444.
4. Levermore, P. A.; Chen, L.; Wang, X.; Das, R.; Bradley, D. D. C. *Adv. Mater.* **2007**, *19*, 2379–2385.
5. Zhang, F.; Johansson, M.; Andersson, M. R.; Hummelen, J. C.; Inganäs, O. *Adv. Mater.* **2002**, *14*, 662–665.
6. Kushto, G. P.; Kim, W.; Kafafi, Z. H. *Appl. Phys. Lett.* **2005**, *86*, 093502.
7. Huang, J.; Wang, X.; Kim, Y.; de Mello, A. J.; Bradley, D. D. C.; de Mello, J. C. *Phys. Chem. Chem. Phys.* **2006**, *8*, 3904–3908.
8. Ahlswede, E.; Mühleisen, W.; bin Moh Wahi, M. W.; Hanisch, J.; Powalla, M. *Appl. Phys. Lett.* **2008**, *92*, 143307.
9. Li, G.; Shrotriya, V.; Huang, J.; Yao, Y.; Moriarty, T.; Emery, K.; Yang, Y. *Nat. Mater.* **2005**, *4*, 864–868.
10. Ma, W.; Yang, C.; Gong, X.; Lee, K.; Heeger, A. J. *Adv. Func. Mater.* **2005**, *16*, 1617–1622.
11. Reyes-Reyes, M.; Kim, K.; Carroll, D. L. *Appl. Phys. Lett.* **2005**, *87*, 083506.
12. Barber, R. P.; Gomez, R. D.; Herman, W. N.; Romero, D. B. *Org. Electron.* **2006**, *7*, 508–513.
13. Mihailetchi, V. D.; Koster, L. J. A.; Blom, P. W.; Melzer, M. C.; de Boer, B.; van Duren, J. K. J.; Janssen, R. A. J. *Adv. Func. Mater.* **2005**, *15*, 795–801.
14. Snaith, H. J.; Kenrick, H.; Chiesa, M.; Friend, R. H. *Polymer* **2005**, *46*, 2573–2578.
15. Mihailetchi, V. D.; Wildeman, J.; Blom, P. W. M. *Phys. Rev. Lett.* **2005**, *94*, 126602.
16. Koster, L. J. A.; Smits, E.C.P.; Mihailetchi, V. D.; Blom, P. W. M. *Phys. Rev. B* **2005**, *72*, 085205.

17. Koster, L. J. A.; Mihailetchi, V. D.; Blom, P. W. M. *Appl. Phys. Lett.* **2006**, *88*, 052104.
18. Drechsel, J.; Männig, B.; Gebeyehu, D.; Pfeiffer, M.; Leo, K.; Hoppe, H. *Org. Electron.* **2004**, *5*, 175–186.
19. Seaman, C. H. *Sol. Energy* **1982**, *29*, 291–298.
20. Simpson, J.; Kirchmeyer, S.; Reuter, K. *Proceedings of the 2005 AIMCAL Fall Techical Conference and 19th International Vacuum Web Coating Conference.*
21. Timpanaro, S.; Kemerink, M.; Touwslager, F. J.; de Kok, M. M.; Schrader, S. *Chem. Phys. Lett.* **2004**, *394*, 339–343.
22. Ouyang, J.; Chu, C.-W.; Chen, F.-C.; Xu, Q.; Yang, Y. *Adv. Funct. Mater.* **2005**, *15*, 203–208.
23. Bube, R. H.; Fahrebruch, A. L. *Advances in Electronics and Electron Physics* Academic Press: New York 1981, pp 163−217.

Chapter 15

Photophysical and Photoconductive Properties of Novel Organic Semiconductors

A. D. Platt, J. Day, W. E. B. Shepherd, and O. Ostroverkhova*

Department of Physics, Oregon State University, Corvallis, OR 97331
*** oksana@science.oregonstate.edu**

Optical, fluorescent, and photoconductive properties of novel functionalized anthradithiophene (ADT), pentacene, and dicyanomethylenedihydrofuran (DCDHF) derivatives are presented. Fast charge carrier photogeneration and high photoconductive gains were observed in solution-deposited films. By creating mixtures of various derivatives, different charge carrier dynamics could be achieved at ps-μs time-scales after 100-fs pulsed photoexcitation. In most derivatives, high fluorescent quantum yields (QYs) of 70-85% were observed in solutions. In fluorinated ADT films, QYs were at least 40-50% at room temperature and decreased by a factor of 3-5 as the temperature increased from 20 °C to 80 °C.

Introduction

Organic semiconductors have been investigated as an alternative to conventional inorganic semiconductors due to their low cost, ease of fabrication, and tunable properties (*1*). Applications envisioned for organic semiconductors include xerography, thin-film transistors, light-emitting diodes, lasers, solar cells and photorefractive devices. Low-cost, solution processable, high-purity materials that can be cast into high-performance (photo)conductive thin films are of particular technological interest. This motivated the selection for our studies of small-molecular-weight novel photoconductive materials that form thin films with good coverage by solution deposition, namely functionalized anthradithiophene (ADT), functionalized pentacene, and dicyanomethylenedihydrofuran (DCDHF) derivatives. Functionalized ADT and pentacene polycrystalline films have recently been extensively explored in thin-film-transistor (TFT) applications, and TFT mobilities of over 1-1.5 $cm^2/(Vs)$ have been recently reported in solution-deposited pentacene and ADT thin films (*2*, *3*). DCDHF derivatives are glass-forming molecules originally developed for photorefractive applications, and DCDHF-based monolithic amorphous glasses are among the best-performing organic photorefractives (*4*, *5*). In addition, a number of functionalized ADT and DCDHF derivatives are highly fluorescent, which broadens the range of their possible applications. For example, DCDHF derivatives have been utilized as fluorophores in single-molecule spectroscopy (*6*, *7*).

Solution-processable materials also simplify tailoring materials for specific applications by creating mixtures of various derivatives, thus tuning optical and electronic properties of the resulting composites. Polymer composites have been extensively studied and designed to improve charge carrier mobility, photogeneration efficiency, luminescence, photorefractive properties, etc. (*8–10*). Composites based on small molecular weight materials have not been explored to the same extent due to reduced solubility of many high-performance compounds, thus requiring, for example, thermal co-evaporation in vacuum, which limited practical use of such composites.

In this chapter, we present an overview of optical, fluorescent, and photoconductive properties of novel functionalized ADT, pentacene, and DCDHF derivatives and their composites.

Experimental

Materials

The molecules explored in this study are (Fig.1): novel high performance fluorinated ADT derivatives functionalized with triisoproplysilylethynyl (TIPS) or triethylsilylenthynyl (TES) side-groups (ADT-TIPS-F and ADT-TES-F, respectively) or CN-containing ADT-TIPS (ADT-TIPS-CN) (*11*, *12*), a pentacene derivative functionalized with TIPS groups (TIPS pentacene) (*13*, *14*), and a 2,6-naphthalene functionalized with dihexylamino donor and dicyanomethylenedihydrofuran acceptor (DCDHF) (*7*). TIPS pentacene and fluorinated ADT derivatives exhibit brick-wall-type crystal packing with significant π-overlap, favoring high TFT mobility (*2*, *3*). These materials form polycrystalline films when deposited from solution. Under similar conditions, ADT-TIPS-CN forms an amorphous film, and DCDHF forms an amorphous glass (*11*) with a glass transition temperature (T_g) of 32 °C.

Sample Preparation

Stock solutions of functionalized ADT and pentacene derivatives were prepared at ~1 wt% in toluene. For quantum yield (QY) measurements of solutions, stock solutions were diluted to < 10^{-5} M. For studies of solid-state

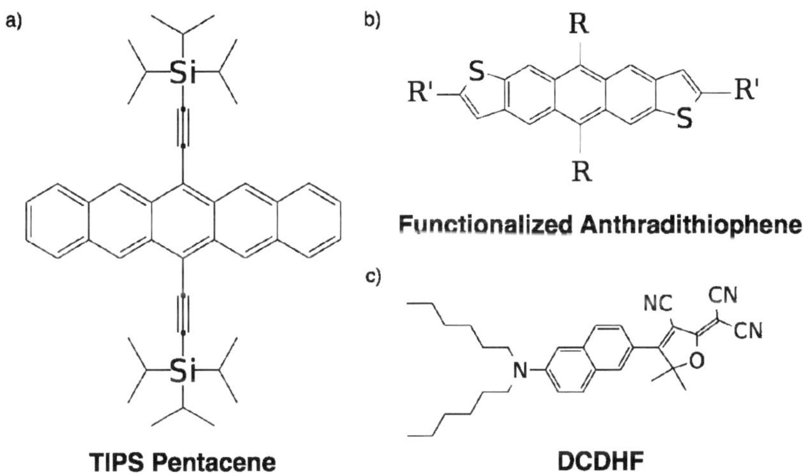

Figure 1. Molecular structures of the compounds used in our study: (a) pentacene functionalized with TIPS side-groups, (b) anthradithiophene (ADT) functionalized with side- and end- groups, and (c) dicyanomethylenedihydrofuran (DCDHF) derivative. In (b), R is either triisopropylsilyl (TIPS) or triethylsilylethynyl (TES) and R' is F or CN.

fluorescent and (photo)conductive properties, films with thickness of 0.5-2 μm were prepared by drop-casting stock solutions onto glass substrates at 60 °C.

For measurements of dark currents and photoresponse, glass substrates were prepared by photolithographic deposition of 5 nm/50 nm thick Cr/Au or 100 nm thick aluminum (Al) electrode pairs. Each pair consisted of 10 interdigitated finger pairs, with 1 mm finger length, 25 μm finger width and 25 μm gaps between the fingers of opposite electrodes. Coplanar electrodes, with 25 or 50 μm gap, were also prepared. Films were drop-cast onto the interdigitated or gap (in the case of coplanar geometry) regions.

For studies of properties of organic semiconductor composites, each of the compounds was dissolved in toluene. Various mixtures of these solutions were then prepared to yield the following compositions: (i) ADT-TES-F/C_{60}, with C_{60} added at the concentration of 2, 5, or 10 wt%, (ii) ADT-TES-F(90 wt%)/TIPS pentacene (10 wt%), and (iii) ADT-TES-F(90 wt%)/ADT-TIPS-CN (10 wt%). The mixtures were drop-cast onto the substrates.

For fluorescence imaging of the individual molecules, the stock solutions were diluted to ~10^{-10} M in a 1% wt. poly(methyl methacrylate) (PMMA) in toluene solution and spun coat onto a clean cover slip at 2000 rpm for 55 s. We also prepared PMMA films with significantly higher concentration of fluorescent molecules (~10^{-3} M).

Optical Measurements

Optical absorption spectra were measured using halogen lamp and a fiber-coupled Ocean Optics USB2000 spectrometer. Absorption of solutions was measured using a standard 1 cm path length quartz cuvette with a halogen light source fiber-optically delivered to the sample holder and spectrometer. Solution and film emission spectra were acquired in a custom fluorescence measurement setup with laser excitation at wavelengths of either 400 nm (frequency-doubled mode-locked Ti:Sapphire laser from KM Labs), 532 nm (Nd:YVO_4 laser from Coherent, Inc.), 355 nm (Nd:YAG laser), or 633 nm (HeNe laser). Emitted photons were collected using a parabolic mirror and detected with a fiber coupled spectrometer. Fluorescence quantum yields (QYs) in solution were referenced against standards with known quantum yields and corrected for differences in optical density and solvent refractive index. The ADT derivatives were measured against rhodamine 6G in ethanol (Φ_f=0.95) and verified against DCDHF derivative in Fig.1 in toluene (Φ_f=0.85) (*7*). The QY of TIPS pentacene solution was measured against rhodamine 6G in ethanol and Alexa Fluor 647 in a phosphate buffer solution (pH 7.2, Φ_f = 0.33) (*28*). The QYs in films were estimated using DCDHF-N-6 in PMMA (Φ_f = 0.98) (*7*) as a reference and assuming a value of 1.7 for the index of refraction. The detection limit of the setup was estimated to be at QY≈0.5%.

Fluorescence lifetime measurements were performed using a frequency-doubled mode-locked Ti:Sapphire laser with a repetition rate of 93 MHz picked at 9.3 MHz using a home-built pulse picker (based on a TeO_2 acousto-optic modulator from NEOS) and 80 fs pulses as the excitation source. Single-photon avalanche photodiode (SPAD -- Molecular Photonic Devices) was used in conjunction with a time-correlated single-photon counter (TCSPC) data analysis board (PicoQuant TimeHarp 200) for detection. The instrument response function (IRF) was recorded using scattered light from an etched microscope slide.

For measurements of temperature dependence of film spectra, samples on microscope cover slips were mounted on a custom electrically heated and water cooled stage (range: 280 – 360 K) for temperature control in the fluorescence setup for spectra and relative QY measurements at the excitation wavelength of 532 nm. Absorption measurements were taken *in situ* over the entire temperature range. Most measurements were performed in ambient air. Experiments were also performed under N_2 atmosphere and showed no discernable difference.

For single-molecule fluorescence imaging, the samples were studied using an Olympus IX-71 inverted microscope. The samples were excited with 700-μW cw 532-nm light from Nd:YVO_4 laser, and the fluorescence was collected in a wide-field configuration with an NA of 1.4, 100x oil objective and an Andor iXon DU897E EMCCD camera.

Dark and Photocurrent Measurements

For transient photoconductivity measurements, an amplified Ti:Sapphire laser (800 nm, 100 fs, 1 kHz) was used in conjunction with a frequency-doubling beta-barium borate (BBO) crystal to excite the samples. Voltage was supplied by a Keithley 237 source-measure unit, and light pulse-induced transient photocurrent was measured with a 50 Ω load by a 50 GHz CSA8200 digital sampling oscilloscope (*11*). Average electric field E was calculated as $E=V/L$, where V is the applied voltage, and L is the gap between the electrodes.

For dark current and continuous wave (cw) photocurrent measurements, the samples were embedded in a fixture incorporating a thermoelectric unit for temperature control (range: 285 – 350 K). The Keithley 237 source-measure unit was used to measure current through the sample in the absence and in the presence of 532-nm cw photoexcitation with a Nd:YVO_4. The photocurrent was calculated as the difference between the two.

Results and Discussion

Optical and Fluorescent Properties

In Solution

Figure 2 shows optical absorption and fluorescence spectra of several functionalized ADT derivatives in toluene solution. Optical properties were determined primarily by the molecular core and were not affected by TIPS or TES side-groups, which resulted in identical spectra of, for example, ADT-TES-F and ADT-TIPS-F in Fig.2. Small Stokes shifts of <10 nm were observed in all solutions and are due to rigidity of their molecular core. Fluorescence lifetime decay of solutions was well described by a single-exponential function (~exp[-t/τ],where τ is the fluorescence lifetime). ADT-TIPS-F and ADT-TES-F derivatives exhibited similar lifetimes (τ) of ~9 ns (Fig.5) and high QYs (Φ_f) of ~70% in toluene. The QYs of ADT-TIPS-CN and TIPS pentacene in toluene solutions were comparable, ~75%, and their lifetimes were ~12-12.5 ns (Fig.5).

In Film

Figure 3 shows absorption and fluorescence spectra measured in several ADT films. Optical absorption spectra of films (Fig.3(a)) exhibited a redshift, or displacement Δ, with respect to those in solutions, due to enhanced Coulomb interaction of the molecule with its surrounding and exchange interaction between translationally equivalent molecules (*15*). In contrast to identical absorption and fluorescence spectra of ADT-TIPS-F and ADT-TES-F in solution (Fig.2), those of corresponding films were considerably different, which we attribute to differences in packing of these molecules in the crystal and film crystallinity (*16*). Although Δ exhibited slight sample-to-sample variation, it was always larger in ADT-TES-F films compared to ADT-TIPS-F films, indicative of a higher degree of exciton delocalization in ADT-TES-F films. Also redshifted were fluorescence spectra of ADT-TES-F with respect to those of ADT-TIPS-F films (Fig.3(b)). Figure 4

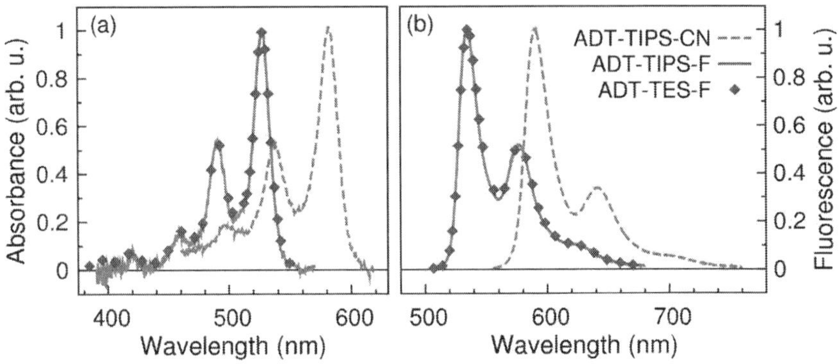

Figure 2. Normalized optical absorption (a) and fluorescence (b) spectra of ADT molecules in toluene solution.

illustrates a change in the emission spectra of ADT-TIPS-F films with changing thickness or crystallinity. When low concentration of the ADT-TIPS-F molecules is dispersed in PMMA, the emission spectrum measured in ADT-TIPS-F/PMMA is similar to that of ADT-TIPS-F in toluene solution shown in Fig.2. In contrast, pristine ADT-TIPS-F films show considerable variation with thickness and crystallinity due to both self-absorption and formation of physical dimers and aggregates (Fig.4) (*15*). In all materials studied, the fluorescence decay dynamics in films were faster than those in solution (Fig.5) and could be described by a bi-exponential function ($a_1\exp[-t/\tau_1]+a_2\exp[-t/\tau_2]$, where $\tau_{1(2)}$ and $a_{1(2)}$ are shorter (longer) lifetimes and their relative amplitudes, respectively), indicating several relaxation pathways. Both τ_1 and τ_2 were shorter than lifetimes τ measured in solutions of the same molecules. The weighted average lifetimes in films,

$$\tau_{av} = a_1\,\tau_1 + a_2\,\tau_2,$$

were typically on the order of 0.4-3 ns at room temperature, depending on the material, and varied somewhat with film quality (e.g. from 1.1 to 2.5 ns in ADT-TIPS-F films) (*12*).

Although all ADT derivatives exhibited considerable solid-state quenching of the fluorescence due to intermolecular interactions, thin films of ADT-TIPS-F and ADT-TES-F were still highly fluorescent, with QYs ($\Phi_{f,film}$) of at least 40-50% depending on the film thickness and morphology. These values represent a lower limit (since effects of self-absorption are significant in even the thinnest of our films); they are considerably higher than, for example, $\Phi_f = 0.008$ % obtained in tetracene single crystals (*17*).

Fluorescense response in films was strongly temperature dependent, in contrast to that of solutions of the same molecules, which suggests temperature dependent intermolecular coupling. Fluorescence spectra obtained upon cw 532 nm excitation of an ADT-TIPS-F film at 25°C, 45°C, and 80°C are shown in Fig.6(a). In order to quantify temperature dependence of the fluorescence observed in films, we fitted the QY data (e.g. ADT-TIPS-CN in Fig.6(b)) with

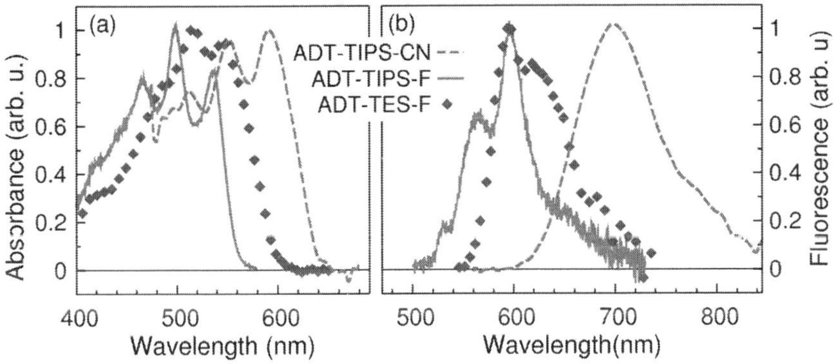

Figure 3. Normalized optical absorption (a) and fluorescence (b) spectra of ADT films.

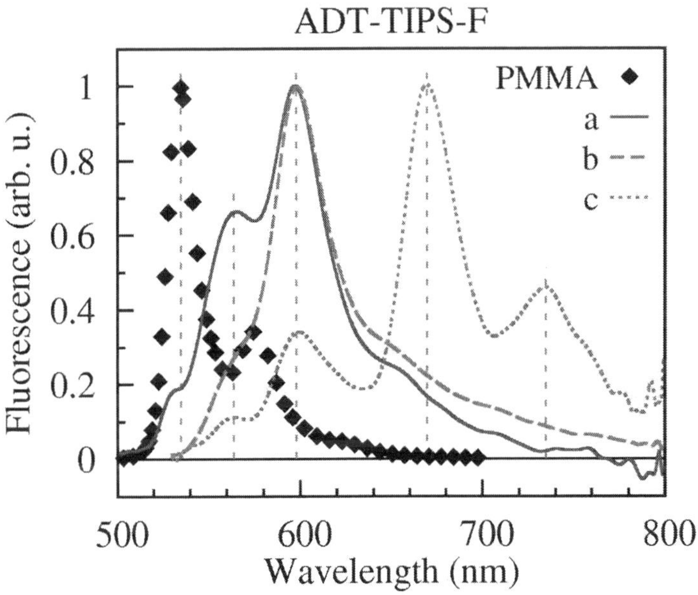

Figure 4. Fluorescence spectra of ADT-TIPS-F dispersed in PMMA in low concentration, and of ADT-TIPS-F films of different thicknesses and crystallinity. The samples a and b are morphologically distinctive from sample c. The samples a, b, and c have optical densities of 2.2, 2.1, and 3.4, respectively, at the wavelength of maximal absorption.

$$1/\Phi_{f,film} \sim 1+a \exp[-\Delta_{fl}/k_B T],$$

where Δ_{fl} is the fluorescence quenching activation energy, a is a fitting parameter related to the ratio of the radiative rate and temperature-independent non-radiative rate prefactor, k_B is the Boltzmann constant, and T is the temperature. Fluorescence quenching activation energies Δ_{fl}, obtained from fits of data to the equation above yielded 0.13±0.01 eV, 0.14±0.01 eV, and 0.17±0.01 eV in the case of ADT-TES-F, ADT-TIPS-CN, and ADT-TIPS-F films, respectively (Fig.6 (b)).

Single-Molecule Fluorescence Imaging

Several derivatives under study exhibited high enough fluorescence and stability to enable fluorescence imaging on the single-molecule level. Inset of Figure 7 shows fluorescence image of individual DCDHF molecules (Fig.1(c)) (white spots) (*7*). Blinking behavior characteristic of a single-molecule emitter is apparent from Fig. 7. Single-molecule-fluorescence data and comparative analysis of the ADT derivatives as single-molecule probes will be reported elsewhere.

Transient Photocurrent

Pristine Materials

Transient photocurrents obtained in ADT-TIPS-F, TIPS pentacene and DCDHF films upon excitation with 400-nm pulses at the fluence of 30 μJ/cm^2 and average electric field of 1.2x10^4 V/cm are shown in Fig.8. The rise-time of the transient photoresponse observed in all samples was about 30-70 ps (e.g. ~40 ps in the inset of Fig.8), depending on the fixture, limited by the time resolution of our setup. The data suggest that charge carriers in our materials can be generated very fast, in agreement with similar studies of tetracene single crystals (*18*) and with optical pump-THz probe data for similar materials (*14*).

Transient photocurrent dynamics observed in ADT-TIPS-F, ADT-TES-F, and TIPS pentacene films exhibit behavior previously observed in several organic crystals (*19*, *20*). The initial decay is fast, most likely due to deep trapping and initial recombination. This is followed by a slow component that can be fitted with a power-law function ($I_{ph} \sim t^{-\beta}$) with β~0.2-0.6 over at least three orders of magnitude (Fig.9), depending on the sample and on the material, but independent of the temperature in the range studied for a given sample (*11*). In agreement with previous studies (*20*), TIPS pentacene films of different quality showed some variation in the relative contribution of the fast and slow components to decay dynamics, particularly initial decay, which reflects film-to-film deep trap density variation. In any ADT-TIPS-F sample, the initial decay was less pronounced than that in the best TIPS pentacene films and even TIPS pentacene single crystals (*21*). In ADT-TES-F films, the initial decay was even less pronounced than in ADT-TIPS-F, and the power-law exponent β of 0.2-0.3, describing dynamics at time scales from ~100 ps to at least 100 μs after photoexcitation, was lower than that in ADT-TIPS-F (e.g. β ~0.45 in Fig.9) and in TIPS pentacene films. The long decay in ADT-TES-F films could be made even slower by introducing additives (such as C_{60} or ADT-TIPS-CN in the inset of Fig.9), as discussed below. In contrast, in the DCDHF amorphous glass, fast decay, attributed to charge trapping

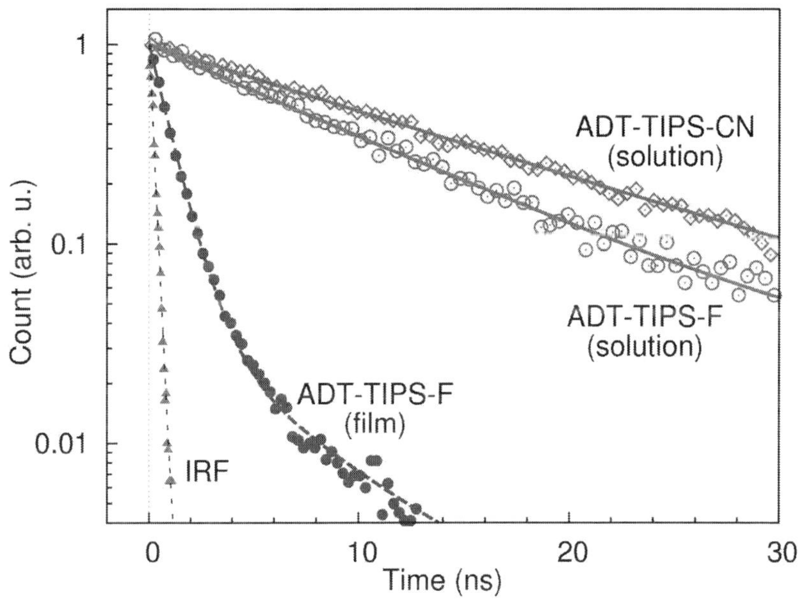

Figure 5. Fluorescence decay obtained in response to 400-nm 80-fs pulsed excitation in ADT-TIPS-F and ADT-TIPS-CN solutions in toluene and in an ADT-TIPS-F film. IRF is also shown.

at deep traps, was observed (Fig.8). This prevented observation of mobile carriers in DCDHF at times longer than approximately 200 ps after photoexcitation.

Temperature dependence of the amplitude of transient photocurrent in ADT-TIPS-F and TIPS pentacene samples is shown in Fig.10 (left axis). The amplitude decreased as the temperature increased from 285 K to 350 K, a trend qualitatively similar to that observed in GaAs (*11*). A possible explanation for such behavior is that charge carrier mobility decreases as the temperature increases, which has been attributed to bandlike charge carrier transport in crystalline materials (*14*). In the same samples, the dc photocurrent under cw photoexcitation (I_{cw}) was thermally activated (right axis), as discussed below.

Composites

Inset of Figure 11 shows peak amplitude of the transient photocurrent as a function of concentration of C_{60} in the ADT-TES-F/C_{60} composite, at 40, 70, and 100 V. Addition of C_{60} at a concentration of 2 and 5% increased the photocurrent amplitude by a factor of ~3 and 10, respectively, at 40 V. This increase is most likely due to fast photoinduced electron transfer (a mechanism similar to that leading to sensitization of C_{60}-containing polymer composites and organic glasses (*10*)), based on our observation of significant fluorescence quenching in ADT-TES-F/C_{60} composite films compared to pristine ADT-TES-F films (*22*). Further addition of C_{60} lowered the measured photocurrent amplitude (e.g. by a factor of

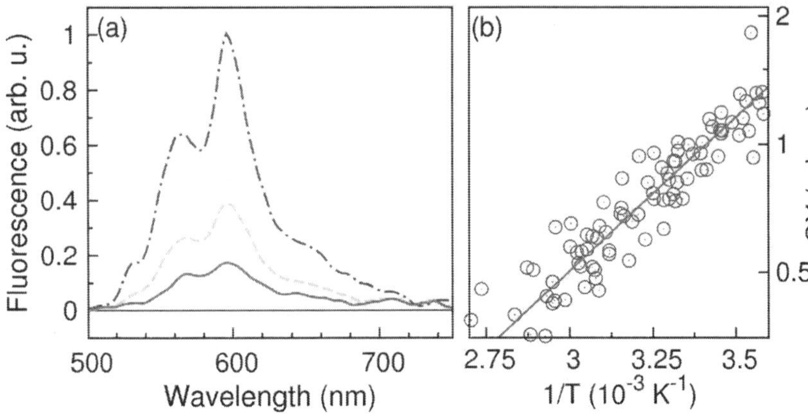

Figure 6. (a) Fluorescence spectra measured in an ADT-TIPS-F film at temperatures of 22°C (dot-dashed), 51°C (dashed), and 85°C (solid). (b) Fluorescence QY (normalized at its value at room temperature of 20°C) measured in an ADT-TIPS-CN film as a function of inverse temperature (1/T). The fit to the equation described in the text is also shown.

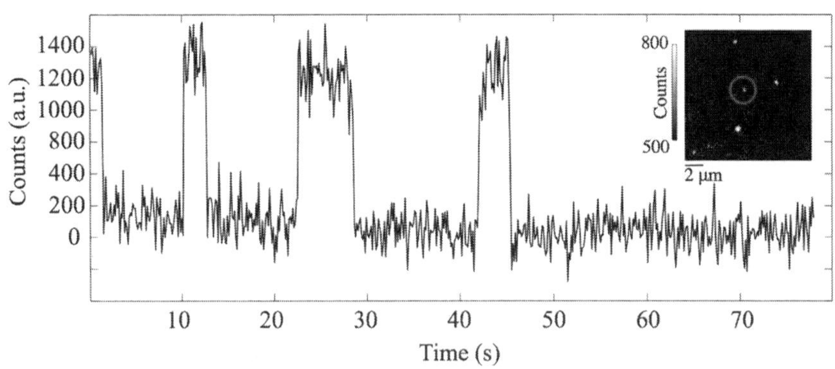

Figure 7. Blinking behavior exhibited by a DCDHF molecule labeled on the image (inset) in PMMA matrix under wide-field 700-μW 532-nm cw illumination. The data was taken with the integration time of 100 ms per frame.

~2 at 10% of C_{60}, at 100 V), most likely due to enhanced initial recombination, occurring at times below ~100 ps, not resolved in our experiments.

Addition of either TIPS pentacene or ADT-TIPS-CN to ADT-TES-F lowered the amplitude of the transient photocurrent by a factor of ~3-4 at all applied voltages in the range studied. The most interesting effect of adding these molecules to the ADT-TES-F host was, however, the change in the initial photocurrent decay dynamics (Fig. 11). In particular, upon addition of TIPS pentacene, the initial decay became faster, the effect similar to that achieved by the addition of C_{60}. For example, only ~32% of "initially" photogenerated carriers remained mobile at 5 ns after photoexcitation in the ADT-TES-F/TIPS pentacene composite. We attribute this to fast hole transfer from ADT-TES-F to

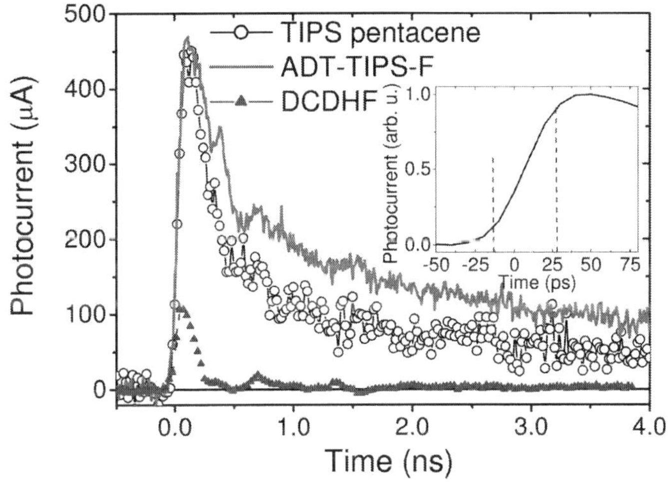

Figure 8. Transient photocurrents obtained upon 400-nm 100-fs excitation of ADT-TIPS-F, TIPS pentacene and DCDHF films. Inset illustrates sub-40 ps rise dynamics of the photocurrent, limited by the time resolution of our setup. Adapted from Ref. (11), with permission. Copyright American Institute of Physics (2008).

TIPS pentacene followed by trapping at the TIPS pentacene molecules (*22*). In contrast, addition of ADT-TIPS-CN to ADT-TES-F introduced a slow component into the photocurrent *rise* dynamics and completely removed the fast initial decay (Fig. 11). In particular, the fast rise of the photocurrent, limited by the time resolution of our setup, accounted only for ~70% of all photogenerated carriers, whereas the other 30% were generated over ~0.1-20 ns. As a result, the peak of the photocurrent in the ADT-TES-F/ADT-TIPS-CN composite was achieved at about 20 ns after photoexcitation, after which a slow decay, characterized by a power-law function with $\beta < 0.1$ (e.g. $\beta = 0.059 \pm 0.002$ in the inset of Fig. 9), persisted to at least 1 ms after photoexcitation. Our observations of complete quenching of the fluorescence of ADT-TES-F, while magnifying that of the ADT-TIPS-CN in the ADT-TES-F/ADT-TIPS-CN composite (*22*) suggest efficient energy transfer from ADT-TES-F to ADT-TIPS-CN. Therefore, it is possible that the slow component in the rise dynamics of the transient photocurrent in Fig. 11 is due to a multistep process that involves excitation of ADT-TES-F, followed by energy transfer to ADT-TIPS-CN, which then leads to energetically favorable hole transfer back to ADT-TES-F, while the electron remains trapped on the ADT-TIPS-CN molecules.

Dark Current and CW Photocurrent

Dark currents measured in the ADT-TIPS-F, ADT-TES-F, and TIPS pentacene samples with Au electrodes are shown in Fig.12. Au forms a hole-injecting contact to these materials, leading to space-charge-limited currents (SCLCs) in these samples. Effective charge carrier mobilities (μ_{eff}) were

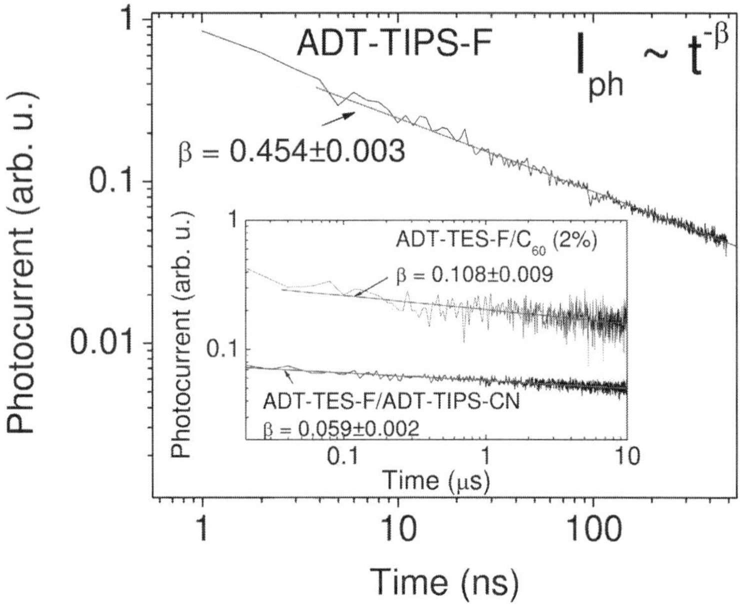

Figure 9. Long-time scale dynamics of the transient photocurrent (I_{ph}) decay in an ADT-TIPS-F film (main figure) and in ADT-TES-F/C_{60} (2%) and ADT-TES-F/ADT-TIPS-CN composites (inset). Fits with the power-law function(I_{ph}~$t^{-\beta}$) are also shown. Adapted from Ref. (22), with permission. Copyright American Institute of Physics (2009).

calculated from the slope of the fits of the dark current as a function of applied voltage squared (*23*). SCLC effective mobilities, which represent a lower bound of charge carrier mobilities in these films (*24*), showed sample-to-sample variation, especially significant in the case of ADT-TIPS-F and TIPS pentacene films. On average, however, μ_{eff} in ADT-TES-F (0.033-0.092 $cm^2/(Vs)$) was at least a factor of ~3 higher than that in ADT-TIPS-F, and a factor of ~7 higher than in TIPS pentacene films.

Figure 12 also shows cw photocurrent obtained at 532-nm excitation of ADT-TIPS-F, ADT-TES-F, and TIPS pentacene films on Au electrodes. In all samples, cw photocurrent measured at light intensity of 0.58 mW/cm^2 was higher than the dark current. Especially strong cw photoresponse was observed in best ADT-TES-F films, with photocurrents of over 200 μA at the average electric field of $4x10^4$ V/cm at low light intensities. This corresponds to linear photocurrent densities ($j_{cw} = I_{cw}/d$, where *d* is the channel width) of over 0.1 mA/cm, which, under assumption that photons absorbed throughout the entire thickness of a 1 μm thick film may contribute to the photocurrent, yields (area) current densities (J_{cw}) of over 1 A/cm^2. If photoconductivity (σ_{ph}) is calculated using $\sigma_{ph} = J_{cw}/E$, then values of σ_{ph} ~$2.5x10^{-5}$ S/cm are obtained at $4x10^4$ V/cm at 0.58 mW/cm^2 in ADT-TES-F, which are considerably higher than those obtained in conductive polymers such as PPV (*25*) and in tetracene crystals (*18*).

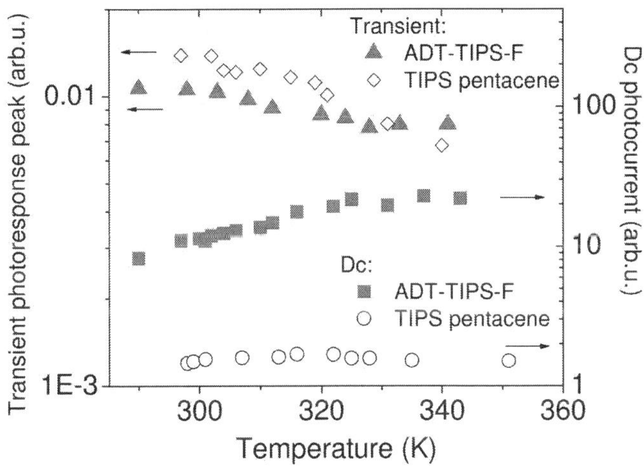

Figure 10. Temperature dependence of the amplitude of the transient photocurrent (left axis) and of the dc photocurrent (right axis) obtained in ADT-TIPS-F and TIPS pentacene films.

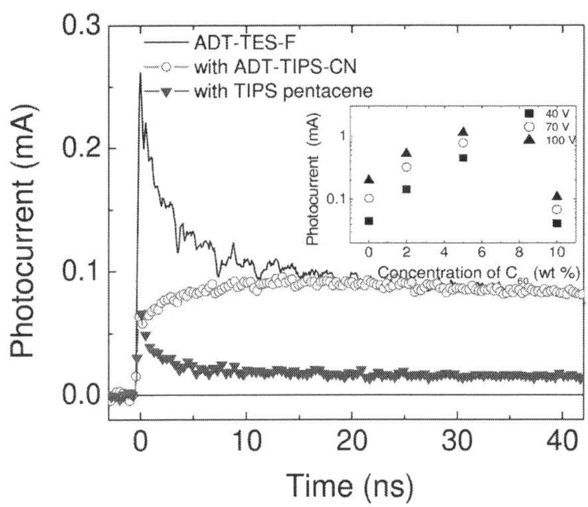

Figure 11. Transient photocurrents measured in a pristine ADT-TES-F film and in ADT-TES-F/ADT-TIPS-CN and ADT-TES-F/TIPS pentacene composites. Inset shows dependence of the transient photocurrent amplitude on the concentration of C_{60} in the ADT-TES-F/C_{60} composites. Adapted from Ref. (22), with permission. Copyright American Institute of Physics (2009).

Photoconductive gain *G* was calculated from the cw photocurrents, absorption coefficients, and light intensity as the ratio between the number of carriers flowing in the film and the number of absorbed photons (*26*). In the case of hole-transporting materials and hole-injecting electrodes (such as Au in the case of ADT-TIPS(TES)-F and TIPS pentacene), bulk photoconductive gain *G* is

$$G \approx \eta_0\, \tau_c/t_{tr},$$

where τ_c is the carrier lifetime, and t_{tr} is the time for the hole to transit through the film. At long carrier lifetimes, it can be much larger than the initial photogeneration efficiency η_0 ($\eta_0<1$). Gain G values of much larger than unity were indeed observed in all our films (*12*). The highest photoconductive gains, up to 130 at 4×10^4 V/cm at 0.58 mW/cm^2, were achieved in ADT-TES-F films, which is consistent with highest effective mobility μ_{eff} (shorter transit time t_{tr}) and longest carrier lifetimes τ_c (observed in SCLC and in the transient photocurrent measurements, respectively) in ADT-TES-F films, as compared to other materials under study. The values of G measured in ADT-TES-F films were similar to those in GaN photodetectors at similar light intensity levels (*27*) and at least an order of magnitude higher than those in unsubstituted pentacene and in most functionalized pentacene films (*28*, *29*).

The temperature dependence of the cw photocurrent (I_{cw}) was described by $I_{cw} \sim \exp[-\Delta_{cw}/k_BT]$, with activation energies Δ_{cw} ranging between 0.05 and 0.24 eV depending on the sample and on the material (*11*). For example, values of ~0.13 and 0.17 eV were obtained in ADT-TES-F and ADT-TIPS-F polycrystalline films, respectively, while higher activation energy of ~0.24 eV were measured in DCDHF amorphous glass samples.

It is interesting to note that in ADT-TES(TIPS)-F films the temperature dependencies of the fluorescence QY and of the bulk photoconductive gain G, are similar. In particular, Δ_{fl} (which characterizes a decrease in QY with temperature) is approximately equal to Δ_{cw} (which characterizes an increase in cw photocurrent with temperature) measured in a similar sample. It has been shown theoretically (*30*) that in materials with π–stacking, such as functionalized pentacene and ADT derivatives, thermal fluctuations lead to dramatic changes in intermolecular coupling, which could be important for both fluorescence and photoconductivity in films. In particular, it appears that thermally activated nuclear motions (*30*) can both enhance nonradiative recombination, leading to a decrease in fluorescence, and facilitate charge carrier detrapping from shallow traps and transport of *localized* carriers, which would improve cw photoconductivity. It is also possible that while improving transport of localized carriers, the same thermal motion reduces intrinsic bandlike charge carrier mobility observed in the time-resolved photocurrent and in ultrafast optical pump-terahertz (THz) probe spectroscopy experiments on ps time-scales after photoexcitation (*11*, *14*, *18*).

Conclusions

In summary, we presented a detailed study of fluorescent and photoconductive properties of several organic semiconductive derivatives.

Single exponential fluorescence decays with lifetimes τ between 9 and 13 ns and QYs of ~70-75%, depending on the molecule, were observed in toluene solutions of ADT-TES(TIPS)-F, ADT-TIPS-CN, and TIPS-pentacene. In films, bi-exponential decays were obtained, with an average lifetime τ_{av} ~0.4-3 ns

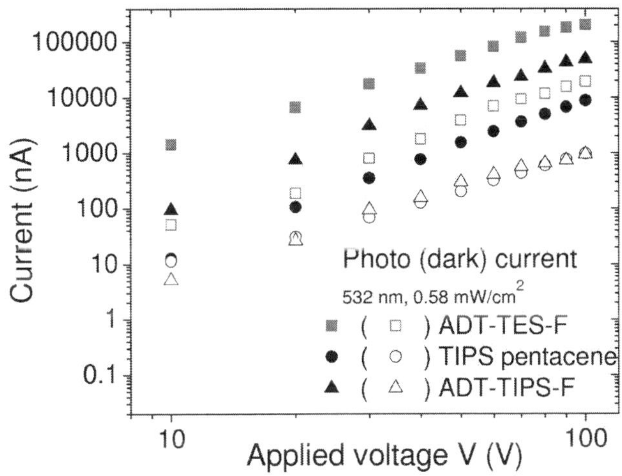

Figure 12. Dependence of the dark current and cw photocurrent under 532 nm excitation on applied voltage measured in ADT-TES-F, ADT-TIPS-F, and TIPS pentacene films on interdigitated Au electrodes.

and fluorescence QYs of ~40-50% in ADT-TES(TIPS)-F. Strong temperature dependence of the QYs, with a QY reduction by a factor of 3-5 as the temperature increased from 20°C to 80°C, was observed in films, but not in solutions.

Effective SCLC mobilities μ_{eff} of ~0.03-0.09 $cm^2/(Vs)$ were obtained in ADT-TES-F films. On average, μ_{eff} in ADT-TIPS-F and TIPS pentacene films were at least a factor of ~3 and ~7 lower, respectively, as compared to that in ADT-TES-F films.

In all films, fast, sub-30 ps charge photogeneration, limited by the time resolution of our setup, and power-law decay of the transient photocurrent (I_{ph}~ $t^{-\beta}$) with β~0.2-0.6, depending on the material, were observed. Both amplitude and decay dynamics of the transient photocurrent could be manipulated by adding various dopants to the host material. In particular, addition of C_{60} at the concentration of 2 and 5 wt% to the ADT-TES-F resulted in a factor of ~3 and ~10 enhancement of the transient photocurrent amplitude, respectively. In constrast, addition of ADT-TIPS-CN at a concentration of 10 wt% reduced the amplitude and dramatically changed the photocurrent dynamics.

Continuous wave photocurrents with densities of ~1 A/cm^2 and photoconductive gains of up to 130 were achieved in ADT-TES-F films at 532 nm photoexcitation with light intensity of 0.58 mW/cm^2 at an average electric field of $4x10^4$ V/cm.

Availability of several high-performance solution-processable derivatives, such as compounds considered here, opens up new possibilities to tune optical and electronic properties of materials for specific applications that include photodetectors, light-emitting diodes, solar cells, and organic lasers, by synthetic modifications and by creating mixtures of various derivatives. Our ability to image several of the studied compounds on the single molecule level represents an exciting opportunity to relate microscopic properties of individual molecules to macroscopic properties that determine the device performance, which we are currently pursuing.

Acknowledgments

We thank Professors J. E. Anthony and R. J. Twieg for functionalized ADT, functionalized pentacene, and DCDHF derivatives. We also acknowledge technical assistance from Professor Y.-S. Lee and J. Tomaino. This work was supported in part by the Petroleum Research Fund, Office of Naval Research (grant #N00014-07-1-0457 via ONAMI Nanometrology and Nanoelectronics Initiative), and National Science Foundation via CAREER program (DMR-0748671).

References

1. Forrest, S. R. *Nature* **2004**, *428*, 911–918.
2. Park, S. K.; Jackson, T. N.; Anthony, J. E.; Mourey, D. A. *Appl. Phys. Lett.* **2007**, *91*, 063514.
3. Park, S. K.; Mourey, D. A.; Subramanian, S.; Anthony, J. E.; Jackson, T. N. *Appl. Phys. Lett.* **2008**, *93*, 043301.
4. Ostroverkhova, O.; Gubler, U.; Wright, D.; Moerner, W. E.; He, M.; Twieg, R. *Adv. Funct. Mater.* **2002**, *12*, 621–629.
5. Ostroverkhova, O.; Moerner, W. E.; He, M.; Twieg, R. J. *Appl. Phys. Lett.* **2003**, *82*, 3602–3604.
6. Willets, K. A.; Ostroverkhova, O.; He, M.; Twieg, R. J.; Moerner, W. E. *J. Am. Chem. Soc.* **2003**, *125*, 1174–1175.
7. Lord, S. J.; Lu, Z.; Wang, H.; Willets, K. A.; Schuck, P. J.; Lee, H. D.; Nishimura, S. Y.; Twieg, R. J.; Moerner, W. E. *J. Phys. Chem. A* **2007**, *111*, 8934–8941.
8. Borsenberger, P. M.; Weiss, D. S. *Organic Photoreceptors for Xerography*; Marcel Dekker, Inc.: New York, 1998; Vol. 59.
9. Ostroverkhova, O.; Singer, K. D. *J. Appl. Phys.* **2002**, *92*, 1727–1743.
10. Ostroverkhova, O.; Moerner, W. E. *Chem. Rev.* **2004**, *104*, 3267–3314.
11. Day, J.; Subramanian, S.; Anthony, J. E.; Lu, Z.; Twieg, R. J.; Ostroverkhova, O. *J. Appl. Phys.* **2008**, *103*, 123715.
12. Platt, A. D.; Day, J.; Subramanian, S.; Anthony, J. E.; Ostroverkhova, O. *J. Phys. Chem. C* **2009**, *113*, in press.
13. Anthony, J. E. *Chem. Rev.* **2006**, *106*, 5028–5048.
14. Ostroverkhova, O.; Cooke, D. G.; Shcherbyna, S.; Egerton, R. F.; Hegmann, F. A.; Tykwinski, R. R.; Anthony, J. E. *Phys. Rev. B* **2005**, *71*, 035204.
15. Pope, M.; Swenberg, C. E. *Electronic Processes in Organic Crystals and Polymers*, 2nd ed.; Oxford University Press: New York, 1999.
16. Subramanian, S.; Park, S. K.; Parkin, S. R.; Podzorov, V.; Jackson, T. N.; Anthony, J. E. *J. Am. Chem. Soc.* **2008**, *130*, 2706–2707.

17. Lim, S. H.; Bjorklund, T. G.; Spano, F. C.; Bardeen, C. J. *Phys. Rev. Lett.* **2004**, *92*, 107402.
18. Moses, D.; Soci, C.; Chi, X.; Ramirez, A. P. *Phys. Rev. Lett.* **2006**, *97*, 067401.
19. Ostroverkhova, O.; Cooke, D. G.; Hegmann, F. A.; Anthony, J. E.; Podzorov, V.; Gershenson, M. E.; Jurchescu, O. D.; Palstra, T. T. M. *Appl. Phys. Lett.* **2006**, *88*, 162101.
20. Ostroverkhova, O., Shcherbyna, S.; Cooke, D. G.; Egerton, R. F.; Hegmann, F. A.; Tykwinski, R. R.; Parkin, S. R.; Anthony, J. E. *J. Appl. Phys.* **2005**, *98*, 033701.
21. Hegmann, F. A.; Ostroverkhova, O.; Gao, J.; Barker, L.; Tykwinski, R. R.; Bullock, J. E.; Anthony, J. E. *Proc. SPIE* **2004**, *5352*, 196–207.
22. Day, J.; Platt, A. D.; Ostroverkhova, O.; Subramanian, S.; Anthony, J. E. *Appl. Phys. Lett.* **2009**, *94*, 013306.
23. Jurchescu, O. D.; Baas, J.; Palstra, T. T. M. *Appl. Phys. Lett.* **2004**, *84*, 3061–3063.
24. Hu, W.; Gompf, B.; Pflaum, J.; Schweitzer, D.; Dressel, M. *Appl. Phys. Lett.* **2004**, *84*, 4720–4722.
25. Lee, C. H.; Yu, G.; Heeger, A. J. *Phys. Rev. B* **1993**, *47*, 15543–15553.
26. *Introduction to Organic Electronic and Optoelectronic Materials and Devices*; Sun, S. S., Dalton, L. R., Eds.; CRC Press: Boca Raton, FL, 2008.
27. Munoz, E.; Monroy, E.; Garrido, J. A.; Izpura, I.; Sanchez, F. J.; Sanchez-Garcia, M. A.; Calleja, E.; Beaumont, B.; Gibart, P. *Appl. Phys. Lett.* **1997**, *71*, 870–872.
28. Gao, J.; Hegmann, F. A. *Appl. Phys. Lett.* **2008**, *93*, 223306.
29. Lehnherr, D.; Gao, J.; Hegmann, F. A.; Tykwinski, R. R. *Org. Lett.* **2008**, *10*, 4779–4782.
30. Troisi, A.; Orlandi, G. *Phys. Rev. Lett.* **2006**, *96*, 086601.

Chapter 16

Ultrathin Self-Assembled Organophosphonic Acid Monolayers/Metal Oxides Hybrid Dielectrics for Low-Voltage Field-Effect Transistors

Hong Ma,[1] Orb Acton,[1] Guy Ting,[2] Jae Won Ka,[1] Hin-Lap Yip,[1] and Alex K.-Y. Jen*,[1,2]

[1]Department of Materials Science and Engineering, University of Washington, Seattle, WA 98195
[2]Department of Chemistry, University of Washington, Seattle, WA 98195
***ajen@u.washington.edu**

By using organophosphonic acid self-assembled monolayers (SAMs) on metal oxides (MOs) such as AlO_x and HfO_2 as ultrathin gate dielectrics and templates, we have realized low-voltage organic field-effect transistors (OFETs) with low leakage currents and small subthreshold slopes. In the demonstrated OFETs, the following device characteristics has been achieved: 1) low leakage current density - down to few nA/cm^2; 3) large capacitance density - up to 760 nF/cm^2; 2) low operating voltage (<2 V); 3) small subthreshold slope - down to 85 mV/decade. This is achieved by: a) modification of MO dielectric surface with SAM to decrease charge carrier traps; b) tailoring of the surface energy of MO dielectric to control the molecular orientation and morphology of subsequently deposited semiconductor layer; c) well-packed and dense organophosphonic acid SAMs (<4 nm) on metal oxides (<4 nm) as ultrathin dielectrics.

Introduction

Organic thin-film transistors (OTFTs) based on π-conjugated materials are very attractive for a variety of large-area low-cost solution-processed/printed electronic applications (*1–4*), such as logic circuits, displays, sensors and electronic barcodes. It is critical to reduce the threshold voltage and the sub-threshold slope of the devices for low voltage operation of OTFTs. When combined with low gate leakage currents, OTFTs may also become a key element in high-end sensor applications, such as flexible touch pads and screens or thermal imaging tools for night vision, surveillance or for the detection of undesired heat loss paths in buildings. However, current OTFT devices still require rather high operating voltages, often exceeding 20 V. This is due to the low capacitance of thick gate dielectrics (usually less than 15 nF/cm^2) and the high density of defect states in organic semiconductor films and at the interface between the gate dielectric and organic semiconductor. To produce low-voltage operating OTFTs, there have been numerous attempts (*5*) to develop high capacitance gate dielectrics with reduced thickness, such as self-assembled molecular monolayers and multilayers and ultra-thin polymer layers, or high dielectric constant (*k*) metal oxides (TiO_2, ZrO_2), polymer electrolytes and high-*k* polymers.

Compared to the conventional silane-based molecules, there are several advantages to use organophosphonic acids to form self-assembled monolayers (SAMs) (*6*, *7*). These include: 1) better stability to moisture; 2) less tendency to form homocondensation between the phosphonic acids; and 3) the reaction between organophosphonic acid and the metal oxide substrate is not limited by the content of surface hydroxyl groups. These advantages enable organophosphonic acids to form dense, robust, and structurally well-defined functional phosphonate monolayers on metal oxide surface, which are ideal for their use as dielectrics in TFTs to ensure small leakage currents, operating voltages, and subthreshold slopes. Utilizing pentacene as the organic semiconductor and octadecylphosphonic acid (ODPA, σ-PA) SAM on AlO_x/Al as the dielectric, Klauk *et al.* has recently demonstrated OTFTs operating within 3 V, with a subthreshold slope of 100 mV/decade, and a leakage current density of 10^{-8} A/cm^2 at 2 V (*8*). The leakage currents of the device are one order of magnitude lower than the one fabricated with octadecyltrichlorosilane (OTS) SAMs on AlO_x/Al. Hill *et al* also fabricated pentacene-based OTFTs using 9-anthrylphosphonic acid (π-PA) SAMs as a buffer between the silicon dioxide (100 nm) gate dielectric and the active pentacene channel region. A substantial decrease of the subthreshold slope (from 1,500-1,700 mV/decade to 200 mV/decade) was observed compared to the control devices without using the SAM buffer (*9*). This is consistent with significantly reduced density of charge trapping states at the semiconductor-dielectric interface by introducing the anthracene-based SAM. It blocks surface hydroxy groups and provides an interfacial template with similar chemical structure yet still has a larger bandgap compared to the pentacene semiconductor. Recently, 1) we have developed anthryl-alkyl-phosphonic acid (π-σ-PA) SAMs which combine the advantages of σ-PA dielectric with π-PA interfacial modification to further enhance the performance of OTFTs. By using pentacene as the organic semiconductor and π-σ-PA SAMs on AlO_x/Al as the dielectrics (Figure 1), we have fabricated OTFTs operating within -3 V, with a subthreshold slope down to 85 mV/decade, and a leakage current density of 10^{-8} A/cm^2 at -2 V (*10*). 2) Solution processed phosphonic acid SAMs/hafnium oxide (HfO_2) hybrid dielectrics have been integrated into OTFTs to achieve operating voltages under -1.5 V (Figure 3). Using anthryl terminated phosphonic acid SAMs on solution processed HfO_2, smooth and pinhole free hybrid dielectrics are realized with high capacitances (0.69 μF/cm^2) and low leakage current densities (down to 6x10^{-9} A/cm^2). These hybrid dielectrics allow pentacene based OTFTs to operate at drain-source and gate-source voltages under -1.5 V with very low subthreshold slopes (100 mV/dec), high on-off current ratios (10^5-10^6), and hole mobilities as high as 0.22 cm^2/(V s) (*11*). (3) C_{60}-based OTFTs have been fabricated using an ODPA SAM/sol gel processed hafnium oxide hybrid dielectric (Figure 5). With the combination of high capacitance (580 nF/cm^2) and low leakage current density (8x10^{-9} A/cm^2), this hybrid dielectric yields C_{60} OTFTs operating under 1.5 V with an average n-channel saturation field-effect mobility of 0.28 cm^2/(V s), high on-off current ratio of 10^5 and low sub-threshold slope of 100 mV/dec (*12*).

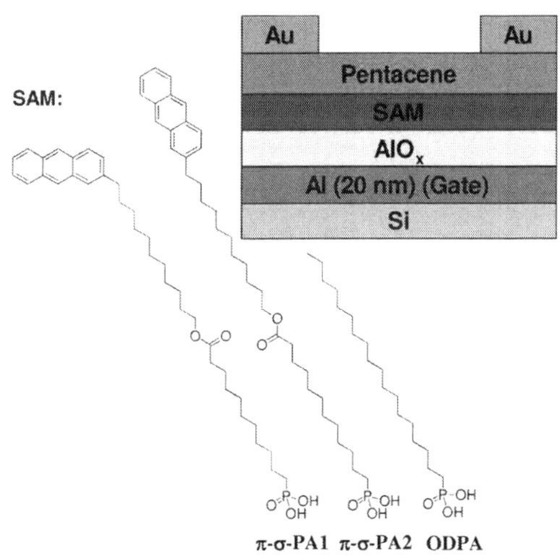

Figure 1. Schematic view of the pentacene TFTs with different phosphonic acid SAMs on AlOx/Al as the dielectrics. (Reproduced with permission from reference (10). Copyright 2008.)

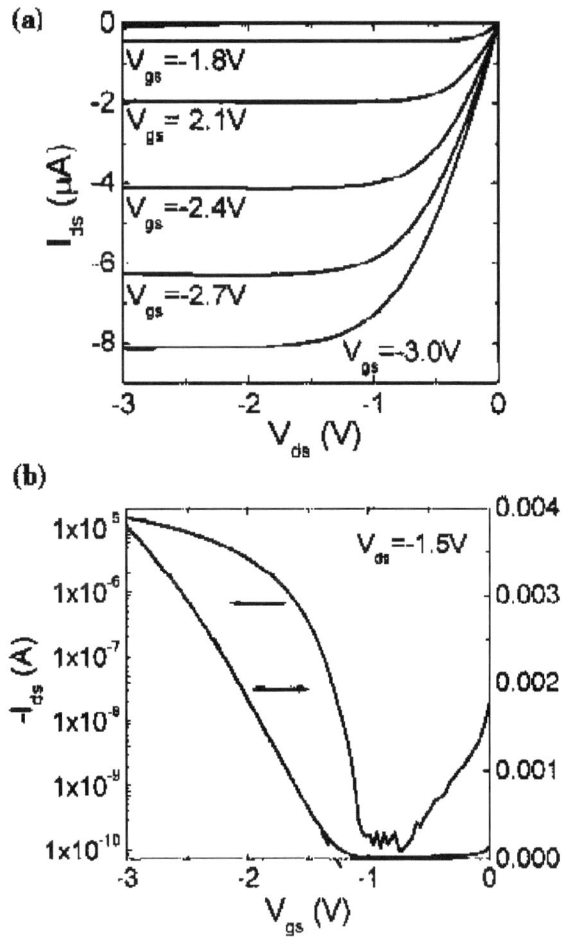

Figure 2. (a) Typical output current-voltage characteristics of a 90 μm channel length and 9000 μm width pentacene TFT with π-σ-PA1 SAM on AlO_x/Al as the dielectric. (b) Transfer characteristics [$-I_{ds}$ vs V_{gs} at V_{ds}=-1.5 V], and [$(-I_{ds})^{1/2}$ vs V_{gs} at V_{ds}=-1.5 V]. (Reproduced with permission from reference (10). Copyright 2008.)

Results and Discussion

π-σ-PA SAMs on AlO_x as Dielectrics for Low-Voltage Pentacene TFTs

A summary of the electrical parameters for pentacene TFTs with different SAMs on AlO_x/Al as the dielectrics is listed in Table I. The OTFTs using pentacene vapor-deposition on room-temperature substrates with these SAM/AlO_x dielectrics exhibit mobilities of 0.14-0.30 cm^2/V s, on/off current ratios of 10^5, threshold voltages of –(1.3-1.5) V and leakage current densities of 10^{-8} A/cm^2.

With ultra-thin SAM/AlO_x dielectrics, high capacitances up to 760 nF/cm^2 at 10 kHz have been obtained, allowing operation of OTFTs within -3 V (Figure 2). It is noteworthy that vast improvements in the gate leakage current (~ 2 orders), on/off current ratio (1 order) and sub-threshold slope down to 85 mV/decade are achieved compared to the control devices without SAMs. This is the best sub-threshold slope reported so far for an organic transistor, and close to the theoretical room-temperature minimum of ~58 mV/decade [$kT/q\ln(10)$], which is independent of dielectric thickness in the trap free limit. It is very encouraging that improvements of device parameters in leakage current densities, capacitance densities and subthreshold slopes could be simultaneously achieved for π-σ-PA1 and π-σ-PA2 SAM modified AlO_x dielectrics compared to the ODPA (σ-PA) SAM/AlO_x dielectrics. The π-σ-PA1 and π-σ-PA2 SAMs are thicker than ODPA SAM so their leakage current densities also decrease. The π-conjugated anthracene terminal group is more polarizable than methyl terminal group so thicker π-σ-PA1 and π-σ-PA2 SAMs still possess larger capacitance densities than ODPA SAM ($C_i=k\varepsilon_0/t$, where t and k are the thickness and relative permittivity of the dielectric, respectively). The π-σ-PA1 and π-σ-PA2 SAMs may also efficiently passivate the hydroxy groups on AlO_x/Al surface and provide a favorable anthracene interface with similar chemical structure yet still a larger bandgap than pentacene. Compared to the ODPA SAM, the lower sub-threshold slopes for devices made from π-σ-PA1 and π-σ-PA2 SAMs could be due to smaller density of charge trapping states (surface hydroxyl groups and interaction-induced trapping states (*13*)) at the semiconductor-dielectric interface.

Table I. Summary of the Electrical Parameters for Pentacene TFTs with Different SAMs on AlO_x/Al as the Dielectrics[a]

Dielectric	*rms (nm)*	*Contact angle (°)*	*J at 2V (A/cm^2)*	*C_i (nF/cm^2)*	*μ (cm^2/V s)*	*V_t (V)*	*S (mV/dec)*	*I_{on}/I_{off}*
π-σ-PA1 /AlO_x	0.65	83	5×10^{-8}	760	0.18	-1.3	85	10^5
π-σ-PA2 /AlO_x	0.75	84	5×10^{-8}	700	0.14	-1.4	85	10^5
ODPA /AlO_x	0.75	107	8×10^{-8}	600	0.30	-1.5	110	10^5
AlO_x	0.84	<10	2×10^{-6}	950	0.30	-1.7	200	10^4

[a] rms, root-mean-square roughness of the gate dielectrics; *J*, leakage current density; C_i, capacitance density at 10 kHz; *μ*, field-effect mobility; V_t, threshold voltage; *S*, sub-threshold slope; I_{on}/I_{off}, on/off current ratio.

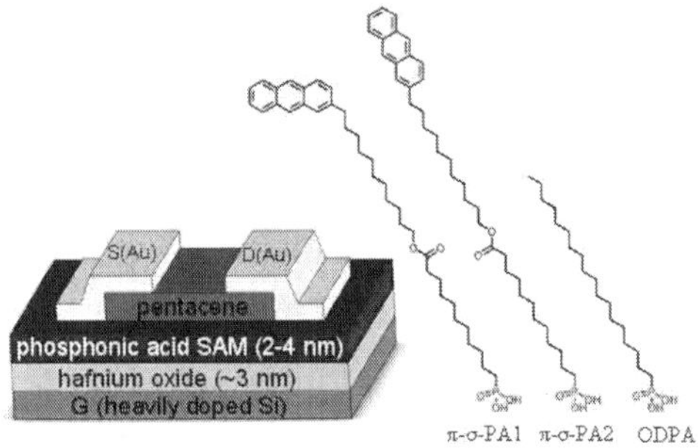

Figure 3. Schematic view of top contact OTFT using different PA SAM/hafnium oxide hybrid gate dielectrics. (Reproduced with permission from reference (11). Copyright 2008.)

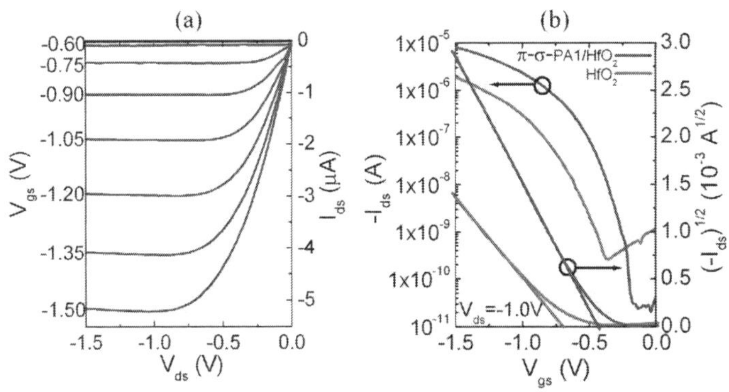

Figure 4. Electrical characteristics of a typical pentacene based OTFT using π-σ-PA1/HfO_2 hybrid dielectric. (a) Drain current (I_{ds}) vs drain-source voltage (V_{ds}). (b) Transfer curve ($-I_{ds}$ vs V_{gs}, $(-I_{ds})^{1/2}$ vs V_{gs}) at a constant V_{ds} of -1.0V with equivalent plot for bare HfO_2 included for comparison. (channel length = 90 μm; channel width = 9000 μm). (Reproduced with permission from reference (11). Copyright 2008.)

π-σ-PA SAMs/Sol-Gel HfO_2 Hybrid Dielectrics for Low-Voltage Pentacene TFTs

Low-voltage top contact OTFTs were fabricated by sequentially thermally evaporating pentacene and Au through shadow masks onto hybrid dielectrics (Figure 3). Device testing was performed under ambient conditions and completed within six hours of initial exposure. As can be seen from the summarized results in Table II, high quality devices with operating voltages under -1.5 V are achieved by employing SAM/HfO_2 dielectrics (Figure 4).

The systematic improvements in the device characteristics including I_{on}/I_{off}, V_t, S, and μ_{sat} using the π-σ-PA1 SAM compared to bare HfO_2 are evident (Table

II). It is notable that the improvement of the I_{on}/I_{off} using either π-σ-PA1 or π-σ-PA2 compared to bare HfO_2 or ODPA correlates to the observed decreased leakage current density using the anthryl terminated SAMs.

The improvement in the charge carrier mobility using any one of the SAMs compared to the bare HfO_2 can be explained in part from (1) the lower surface energy of the SAM allowing better interconnection between grains during pentacene deposition (*14*) and (2) the low-*k* monolayer surface physically buffering the charge carriers from the highly polar underlying high-*k* surface suggested to reduce charge carrier mobility in single crystal transistors (*15*). That the charge carrier mobility using π-σ-PA1 versus that of π-σ-PA2 and ODPA is slightly improved suggests that the chemical compatibility at the pentacene-dielectric interface is important not only by the surface group (changing from methyl to anthryl), but also possibly the orientation of the surface group.

Using π-σ-PA1, a subthreshold slope as low as 100 mV/dec has been achieved, which is close to the theoretical room-temperature limit of ~58 mV/decade [$kT/q\ln(10)$]. The low subthreshold slope and V_t using either π-σ-PA1 or π-σ-PA2 is consistent with reducing charge traps by eliminating surface hydroxyl groups with a SAM on an inorganic oxide (*16*), while simultaneously improving the chemical compatibility at the pentacene-dielectric interface using the anthryl surface group (*13*) compared to the methyl surface group of ODPA.

Table II. Dielectric and Pentacene TFT Data for Different SAMs on HfO_2

Dielectric	*J at 1.5V (A/cm²)*	C_i *(nF/cm²)*	μ_{sat} *(cm²/V s)*	V_t *(V)*	*S (mV/dec)*	I_{on}/I_{off}
π-σ-PA1 /HfO_2	6×10^{-9}	690	0.22	-0.41	100	10^5-10^6
π-σ-PA2 /HfO_2	6×10^{-9}	640	0.15	-0.41	110	10^5-10^6
ODPA /HfO_2	4×10^{-8}	580	0.15	-0.53	130	10^5
HfO_2	8×10^{-8}	1040	0.08	-0.72	160	10^4

ODPA SAM/Sol-Gel HfO_2 Hybrid Dielectrics for Low-Voltage C_{60} TFTs

Small angle X-ray reflectivity (XRR) and ellipsometry thickness measurements indicate a 3.1 nm thick HfO_2 on 1.7 nm thick SiO_2. AFM images indicate that the hybrid dielectric surfaces are smooth and pinhole free with a root-mean-square (rms) roughness of 0.19 nm before and after SAM preparation (Figure 5b). Aqueous advancing (and receding) contact angles increase from <10° before ODPA SAM preparation to 110° (105°) after. ATR-FTIR spectrum gave the chemical functionality of ODPA on HfO_2 with characteristic methylene ($-CH_2$) stretching vibrations at 2851 and 2919 cm^{-1} indicative of well-ordered densely packed monolayers (Figure 5c).

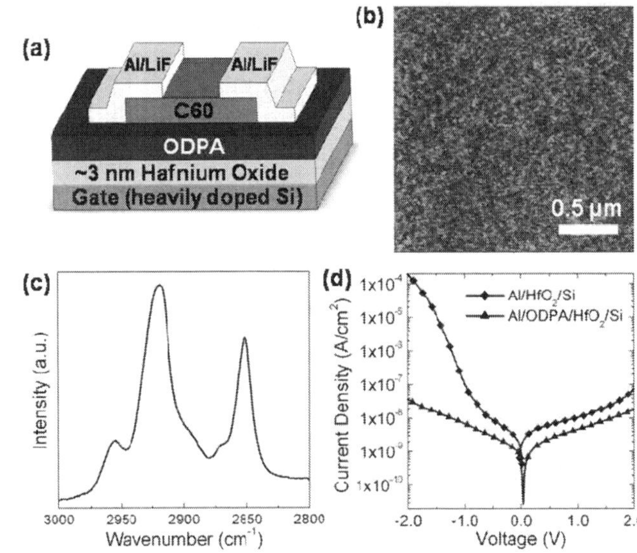

Figure 5. (a) Schematic view of top contact OTFT using ODPA SAM/HfO_2 hybrid gate dielectric. (b) AFM height image of ODPA on HfO_2 (RMS = 0.19 nm). The height in the AFM image ranges from 0 (dark) to 2 nm (light). (c) ATR-FTIR spectrum of ODPA on HfO_2. (d) Leakage current density versus applied voltage of ODPA/HfO_2 hybrid dielectric versus bare HfO_2. (Reproduced with permission from reference (12). Copyright 2008.)

The leakage current density of HfO_2 is $3x10^{-8}$ A/cm^2 at an applied voltage of 1.5 V, while ODPA decreases the leakage current further to $8x10^{-9}$ A/cm^2 at the same applied voltage (Figure 5d). The change from asymmetric to symmetric *J-V* characteristics with the formation of ODPA on HfO_2 shows that ODPA effectively blocks the leakage current associated with electrons passing from heavily doped Si to the Al electrode through the HfO_2 (*17*). As expected, with the formation of an ODPA SAM on HfO_2, the capacitance density decreases due to a larger total dielectric thickness from 1.04 μF/cm^2 to 0.56 μF/cm^2 for bare HfO_2 and ODPA/HfO_2, respectively. With the combination of ODPA and HfO_2, low leakage current and high capacitance is achieved rendering this hybrid dielectric suitable for low-voltage OTFTs.

As can be seen from the results in Table III, high quality C_{60} OTFTs with operating voltages under 1.5 V are achieved by employing the ODPA SAM/HfO_2 hybrid dielectric. The systematic improvements in the device characteristics including two orders of magnitude in the on-off current ratio (I_{on}/I_{off}), lower threshold voltage (V_t), lower subthreshold slope (S), and *n*-channel saturation field-effect mobility (μ_{sat}) nearly 3x greater using the ODPA/HfO_2 compared to bare HfO_2 are evident. A sub-threshold slope of 100 mV/dec was achieved which is close to the theoretical room-temperature limit of ~58 mV/dec [kT/qln(10)]. These improvements can be explained in part from a combination of the low surface energy and compatible chemical functionality at the OSC/dielectric interface. The surface energy (γ_s) of the different dielectrics are 22.5, and 72.6 mJ/m^2 for ODPA/HfO_2 and bare HfO_2, respectively. The reduction of surface energy using ODPA promotes greater diffusion of C_{60} on the dielectric surface during thin film deposition resulting in large grains with an average size of ~2.3 μm versus only ~0.1 μm for bare HfO_2 (Figure 6). The small grains on bare HfO_2 may severely limit the charge carrier mobility due to the high density of grain boundaries acting as trap sites while the large grains on ODPA/HfO_2 provide a pathway for high mobility with substantially fewer grain boundaries. In addition, the ODPA SAM not only effectively passivates the surface hydroxyl groups known to trap electrons in n-type OSCs (*16*), but may also act to physically buffer charge carriers in the OSC from the highly polar underlying high-k HfO_2 surface suggested to reduce charge carrier mobility in single crystal transistors (*15*).

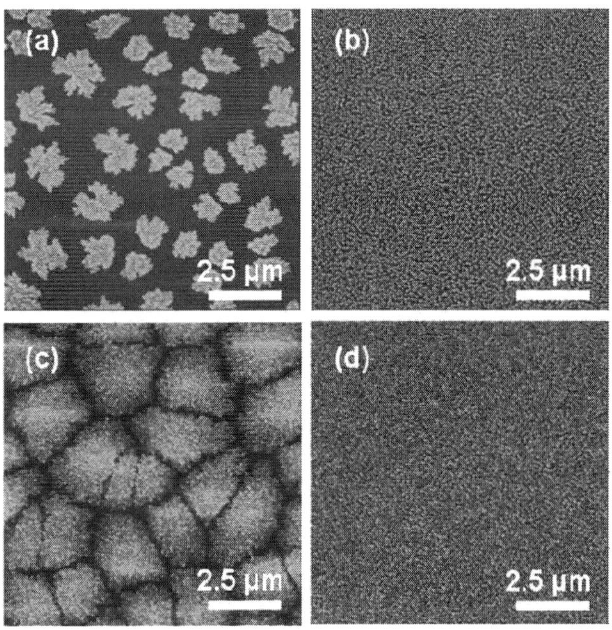

Figure 6. AFM height images of initial stages of C_{60} growth on (a) ODPA/HfO_2 and (b) bare HfO_2. 50 nm of C_{60} on (c) ODPA/HfO_2 and (d) bare HfO_2 with mean grain size of ~2.3 μm and ~0.1 μm, respectively. The height in the AFM images ranges from 0 (dark) to 80 nm (light). (Reproduced with permission from reference (12). Copyright 2008.)

Table III. C_{60} OTFT Data for ODPA/HfO_2 Hybrid Dielectric[a]

Dielectric	*J at 1.5V (A/cm²)*	C_i *(μF/cm²)*	γ_s *(mJ/m²)*	μ_{sat} *(cm²/V s)*	V_t *(V)*	*S (mV/dec)*	I_{on}/I_{off}
ODPA /HfO_2	8×10^{-9}	0.58	22.5	0.28	0.35	100	10^5
HfO_2	3×10^{-8}	1.04	72.6	0.097	0.40	170	10^3

[a] γ_s, dielectric surface energy.

Conclusions

We have demonstrated phosphonic acid based SAM/metal oxide hybrid dielectrics combining the strengths of molecular π-σ SAM with high-*k* metal oxide for low-voltage OTFTs. By using anthryl terminated PA SAMs compared to a methyl terminated one, higher capacitances and lower leakage current densities are achieved while simultaneously improving pentacene based OTFTs performances including lower subthreshold slopes, higher I_{on}/I_{off}, and improved hole mobilities. We have also realized low-voltage high-performance C_{60}-based OTFTs using an ODPA SAM/sol-gel processed HfO_2 hybrid dielectric. We believe that this approach of hybrid dielectrics is generally applicable for low-voltage OTFTs.

Acknowledgments

This work is supported by the National Science Foundation (the NSF-STC Program under the Agreement Number DMR-0120967) and the Institute of Advanced Materials & Technology, University of Washington. A. K. Y. Jen thanks the Boeing-Johnson Foundation for financial support.

References

1. Forrest, S. R. *Nature* **2004**, *428*, 911.
2. Jackson, T. N. *Nat. Mater.* **2005**, *4*, 581.
3. Klauk, H. *Nat. Mater.* **2007**, *6*, 397.
4. Sirringhaus, H. *Nat. Mater.* **2003**, *2*, 641.
5. Facchetti, A.; Yoon, M. H.; Marks, T. J. *Adv. Mater.* **2005**, *17*, 1705.
6. Hanson, E. L.; Schwartz, J.; Nickel, B.; Koch, N.; Danisman, M. F. *J. Am. Chem. Soc.* **2003**, *125*, 16074.
7. Kang, M. S.; Ma, H.; Yip, H. L.; Jen, A. K.-Y. *J. Mater. Chem.* **2007**, *17*, 3489.
8. Klauk, H.; Zschieschang, U.; Pflaum, J.; Halik, M. *Nature* **2007**, *445*, 745.

9. McDowell, M.; Hill, I. G.; McDermott, J. E.; Bernasek, S. L.; Schwartz, J. *Appl. Phys. Lett.* **2006**, *88*, 073505.
10. Ma, H.; Acton, O.; Ting, G.; Ka, J. W.; Yip, H. L.; Tucker, N.; Schofield, R.; Jen, A. K.-Y. *Appl. Phys. Lett.* **2008**, *92*, 113303.
11. Acton, O.; Ting, G.; Ma, H.; Ka, J. W.; Yip, H. L.; Tucker, N. M.; Jen, A. K.-Y. *Adv. Mater.* **2008**, *20*, 3697.
12. Acton, O.; Ting, G.; Ma, H.; Jen, A. K.-Y. *Appl. Phys. Lett.* **2008**, *93*, 083302.
13. Hill, I. G.; Hwang, J.; Kahn, A.; Huang, C.; McDermott, J. E.; Schwartz, J. *Appl. Phys. Lett.* **2007**, *90*, 012109.
14. Yang, S. Y.; Shin, K.; Park, C. E. *Adv. Funct. Mater.* **2005**, *15*, 1806.
15. Hulea, I. N.; Fratini, S.; Xie, H.; Mulder, C. L.; Iossad, N. N.; Rastelli, G.; Ciuchi, S.; Morpurgo, A. F. *Nat. Mater.* **2006**, *5*, 982.
16. Chua, L. L.; Zaumseil, J.; Chang, J. F.; Ou, E. C. W.; Ho, P. K. H.; Sirringhaus, H.; Friend, R. H. *Nature* **2005**, *434*, 194.
17. Shaimeev, S.; Gritsenko, V.; Kukli, K.; Wong, H.; Lee, E-H.; Kim, C. *Microeletron. Reliab.* **2007**, *47*, 36.

Chapter 17

Ultrafast Photoinduced Carrier Dynamics of Organic Semiconductors Measured by Time-Resolved Terahertz Spectroscopy

Okan Esenturk,[*,1,2] Joseph S. Melinger,[3] Paul A. Lane,[4] and Edwin J. Heilweil[2]

[1]Dept. of Chemistry and Biochemistry, Bldg 091, University of Maryland, College Park, MD 20742
[2]Optical Technology Div., NIST, Gaithersburg, MD 20899-8443
[3]Electronics Science & Technology Div., NRL, Washington, DC 20375
[4]Optical Sciences Div., NRL, Washington, DC 20375
[*]esenturk@umd.edu

Intrinsic properties of organic semiconductors are investigated by Time-Resolved Terahertz Spectroscopy (TRTS) to assess their relative mobilities and efficiencies. Our results are well correlated with device measurements and show the effectiveness and advantages of using this non-contact optical technique to rapidly identify prospective materials. After a brief introduction of the TRTS technique, we summarize our results from relative mobility measurements of the organic semiconductor polymers poly(3-hexylthiophene) (P3HT) and poly(2,5-bis(3-tetradecylthiophen-2yl)thieno[3,2-b]thiophene (PBTTT) as model active layers in transistor devices. We also synopsize our recent results from relative efficiency studies of blend and alternating multilayer films of zinc phthalocyanine (ZnPc) and buckminsterfullerene (C_{60}) as model active nanolayers for efficient solar cells.

Introduction

Relatively inexpensive organic semiconductors are promising candidates to augment currently applied silicon-based systems because of their increasing efficiencies and mobilities in addition to being potentially inexpensive materials for device applications. A major component of all organic based devices is the active layer where carrier generation and transfer occurs. Depending on the device structure and purpose, either the carrier mobility (e.g. for a transistor) or generation efficiency (e.g. for solar cells) of the organic semiconductors becomes very important. Understanding the fundamentals of carrier generation, recombination, and transport properties at early times (< ns) will help in the characterization and optimization of the active layer and hence, on the final efficiency of devices.

In this chapter we briefly describe our application of the all-optical Time-Resolved Terahertz Spectroscopy (TRTS) technique and demonstrate its use for measuring and comparing transient dynamics of photo-generated carriers in model organic semiconductors. With its unique frequency range in the far-infrared (IR) region, which has been demonstrated to directly characterize carrier dynamics (*1–13*) and being a non-contact optical measurement approach, TRTS studies reveal intrinsic photo-carrier properties of photoconducting films free from external perturbations such as electrode contacts that occur in device measurements. Differences between the two measurement techniques (TRTS and direct device electrical contact measurements) arise from their time and structural sensitivity. The ultrafast time-resolved THz technique mainly focuses on very short-time dynamics (up to a few ns) that probes "local" carrier migration phenomena. In contrast, device measurements measure "long-time" (μs or longer) carrier transport processes over length scales of tens to hundreds of micrometers and are highly dependent on the device structure (*14*). Depending on the nature of the measured property, both measurements may correlate well or they could be complementary techniques towards developing a better understanding of the active layers and entire system behavior.

First, we present examples from recent studies of high mobility organic semiconductor materials poly-3-hexylthiophene (P3HT) and poly(2,5-bis(3-tetradecylthiophen-2yl)thieno[3,2-b]thiophene (PBTTT). In this work we investigated internal and external effects on the carrier mobility of these polymers to better understand how the mobility is affected by structural changes and film processing conditions (*10*, *12*). In these cases, the cast polymer samples represent model active layers of a device used in applications such as transistors, where attaining the highest mobility of the film is a prime objective. We also discuss the dependence of carrier mobility on i) polymer molecular weight or chain length (*10*) and ii)_monomeric unit differences (*12*).

A brief summary of our most recent work on thin films of blended and nanometer (nm) scale alternating multi-layer films of zinc-phthalocyanine (ZnPc) and Buckminsterfullerene (C_{60}) is also presented (*15*). In these cases, the films are model active layers for solar cell systems and the THz transient carrier dynamics are analyzed to extract their relative photon-to-carrier efficiencies. Through this study we aim to address processes such as exciton diffusion, charge separation, and relative photon-to-carrier efficiencies, which are important factors for efficient operation of organic solar cells. We discuss the TRTS transients in terms of the role played by the ZnPc:C_{60} interface and bulk layer content in photo-carrier generation and subsequent photo-carrier relaxation processes.

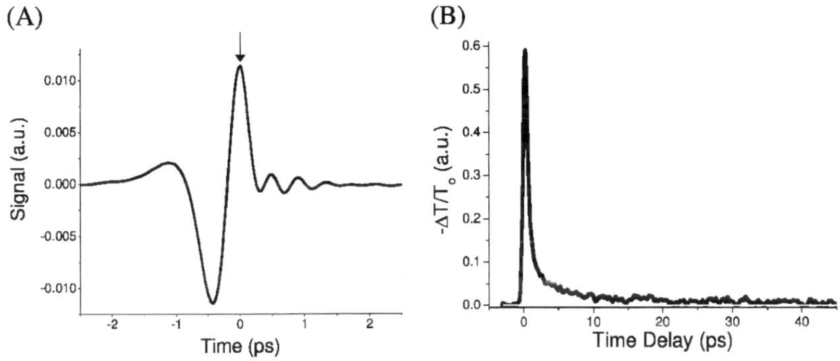

Figure 1. (A) A typical transmitted THz electric field waveform detected after a sample (vapor deposited C_{60}).(B) UV-excited transient differential THz transmission response of the same sample.

Experimental

P3HT polymers used in molecular weight studies were purchased from Sigma-Aldrich Co. and EMD Chemicals Inc. (*16*) P3HT and PBTTT polymers used in structural investigations were received from EMD Chemicals Inc. All polymers were used as received without further purification. Details on sample preparation methods may be found in references (*10*) and (*12*). ZnPc and C_{60} were purchased from Sigma-Aldrich Inc. and purified via vacuum train sublimation. The blend and layered ZnPc/C_{60} thin films were vapor deposited under high vacuum. The blend films were deposited by varying weight ratios of ZnPc and C_{60} between 0 % and 100 %. A multi-layered film structure is formed by vapor deposition of alternating layers of ZnPc and C_{60}. The alternating layer thickness ranged between 5 nm and 40 nm for the investigated multilayer films.

Details of the Ti:Sapphire kHz laser amplifier-based TRTS apparatus were reported elsewhere (*10*). Unless specified otherwise, the thin films were optically excited with 400 nm (≈3 eV, fluence of $5.7 \pm 0.2 \times 10^{14}$ photons/cm^2), 60 femtoseconds pulses and interrogated with optically gated and synchronized time-delayed ultrafast THz probe pulses (about 0.5 ps FWHM with center frequency at ~1 THz). The strongest observed pump-induced modulation in the THz field is less than 3 % of the peak transmitted THz field amplitude when all the excitation light is absorbed. All data are corrected for optical density (OD) variations of the films at the excitation wavelength. Time-resolved data was obtained by averaging twenty to sixty time-delay sweeps, where each sweep takes approximately 1 minute to collect. The TRTS spectrometer was contained inside a plastic housing purged with water and CO_2-free air to minimize water vapor interference effects. We did not observe photo-oxidation or photo-bleaching of the films over the course of the experiments. Many hour long exposure of the films to 400 nm excitation light optically damages the films.

Time-dependent dynamics of the photo-generated charge carriers were measured by measuring the UV excitation/pump-pulse induced modulation of THz probe pulses transmitted through the sample after photoexcitation as a function of time delay. The longest accessible time delay of about 500 ps was limited by the length of the optical delay line employed in our apparatus. Time zero was measured using a double-side polished silicon wafer as a reference sample (in dry-air) and is defined as the half-rise of the observed single-sided modulation.

TRTS Technique in Brief

Typical time-domain THz waveform and photoexcitation induced transient differential THz transmission signals ($\Delta T/To$) are shown in Figure 1 (a) and (b), respectively. The differential THz transmission signal refers to a change in THz peak transmission amplitude (shown by an arrow) and is recorded as the relative time delay between the THz probe pulse and pump pulse varies. Therefore, the differential THz transmission signal is zero before excitation of the sample with the pump pulse (time delay $t < 0$). Once the sample is excited by pump pulses, THz probe pulses detect the charge generation ($t \approx 0$) and subsequent evolution of the generated carriers ($t > 0$). Hence, the differential transmission signal contains information about the ultrafast photo-induced carrier dynamics during carrier generation, evolution and recombination. The TRTS instrument has ~0.5 ps time resolution based on response measurements of the double-side polished silicon wafer. Therefore, the initial response of the instrument (peak amplitude of the differential transmission around $t = 0$ ps) may not be directly proportional to the instantaneous photoconductivity but reflects an integrated value detected by the system.

At any given time delay the measured differential THz transmission is proportional to the effective mobility and carrier density which is proportional to the carrier generation efficiency. For further details please refer to reference (*6*) and references therein. The exact carrier density and/or mobility cannot be estimated from the present measurements without independent measurements of either the effective carrier mobility or generation efficiency of the sample. Modified Drude conductivity models have been employed (*6*, *8*, *11*, *17*) as a solution to estimate carrier density and effective carrier mobility. However, the approximately limited 2.5 THz bandwidth of our THz probe pulses limits the accuracy and reliability of available models that can be applied to our results (*13*). Instead, we compared the experimentally measured differential THz transmission signals from a set of samples that have similar molecular and morphological structural properties in order to assess the relative mobilities and efficiencies of the samples.

For these photoconductivity studies of semiconducting polymers, we assumed that they have similar carrier generation efficiencies since:

i) they are based on the same or similar monomer unit,

ii) they have similar stereo-regularity,

iii) only the chain length varies as the molecular weight increases
iv) the chain lengths are very long
v) they have similar morphology when cast on a substrate.
vi) reported carrier efficiencies of two different polymer structures are similar (*7*)

$$\frac{\mu_1 \cancel{\eta} = \cancel{C}\,(\Delta T/To)_1}{\mu_2 \cancel{\eta} = \cancel{C}\,(\Delta T/To)_2} \qquad 1$$

Under these conditions the constants (C) and the carrier generation efficiency in the above equation (η) cancels out when the measured differential THz transmission signals of the films are ratioed against each other at the same time delay (see equation 1). The ratio of the differential transmission signals is then directly proportional to the ratio of the effective mobility (μ) of the carriers in the measured films. A similar argument is applied for the comparison between carrier mobilities of P3HT and PBTTT thin films.

In the case of multilayered ZnPc and C_{60} super structures, we assume similar mobilities for the generated carriers since the layers were composed of the same neat materials and only the individual layer thicknesses were changed. This implies that the generated carrier concentration is the only factor changing with a change in thickness of the layers. Therefore, the ratio of differential THz transmission signals of alternating multilayer films at a fixed delay time corresponds to the ratio of the carrier density resulting from photo-excitation of the films. The decay rates and signal levels are used to evaluate the efficiencies of the multilayer films relative to each other. Comparing blended films is a bit more complicated since the chemical composition of the film is constantly changing as the deposited ratio of ZnPc and C_{60} changes. This makes it difficult to assume that either one of the variables (mobility or generation efficiency) is constant. Despite these complications, comparison of the measured differential transmission data still shows significant correlation to the chemical composition of the measured film.

The carrier dynamics measured by TRTS for all types of films are complex, exhibiting system dependent multiple lifetime components with decay times on the sub-picosecond (ps) to nanosecond (ns) timescales. The relative strengths of the individual components and their decay times are found to depend in a sensitive way to the local film structure. Investigating the ultrafast carrier dynamics of these films offer unique opportunities for improving the photoactive characteristics of the materials by better understanding the carrier properties. The following sections provide a synopsis of our work on the photoconductive properties of organic semiconductors.

Relative Carrier Mobility of Semiconductor Polymers Using TRTS

TRTS is a powerful technique to assess "local" conductive properties of organic semiconductors. For example, on the picosecond timescale, free carriers can only migrate on the nanometer length scale in organic semiconductors (*8*, *10*). One of the most significant results we observed was from the comparison of regio-random and regio-regular P3HT polymers of same molecular weights shown in Figure 2 (*10*). The TRTS measurements of drop-cast films of these polymers demonstrated the importance of molecular structural regularity and how it potentially affects the final mobility of the polymer. The differential transmission signal amplitudes from regio-regular P3HT is almost an order of magnitude larger than that of the regio-random P3HT of the same molecular weight. In addition, the ultrafast dynamics data showed that the photo-generated carriers recombine extremely fast (within a few ps) for regio-random samples but a significant amount of carriers generated in regio-regular polymers live longer (>250 ps, not shown). The very short carrier lifetime observed for regio-random polymer films is an indication of short diffusion length, suggesting that the carriers do not travel along the regio-random P3HT films. On the other hand, the long-lived carriers of regular polymer shows that once the carriers leave the generation site they diffuse trough the film via percolation paths made by the overlapping regular regions of the P3HT polymer. This study suggested that longer conjugation lengths results in higher mobility as long as percolation paths for the carriers exist. Study of varying molecular weight P3HT films is the first example of such an effect on the effective mobility of the films. Our results showed that the mobility increases with molecular weight and also correlated well with device measurements of similar molecular weight P3HT polymers films, suggesting that the local film properties measured by TRTS extends throughout the entire morphology of the film.

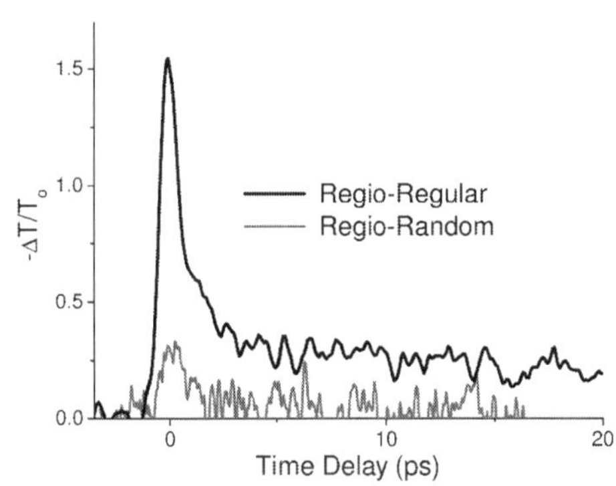

Figure 2. Transient decay dynamics of regio-random and regio-regular 55kDa P3HT polymers.

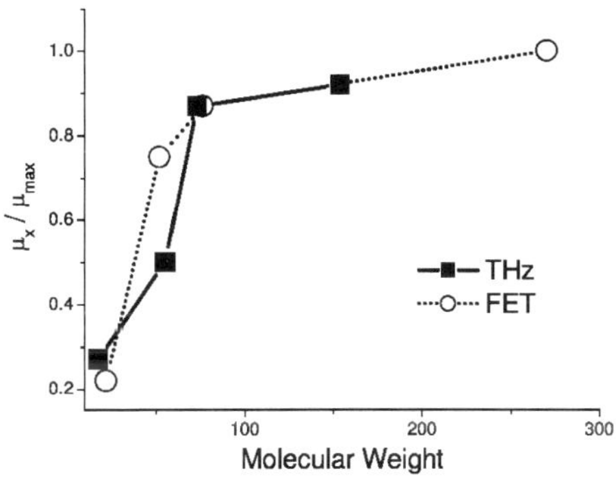

Figure 3. Relative mobilities measured by TRTS and device measurements. Lines to guide eye. Reproduced with permission from reference (10). Copyright AIP 2008.

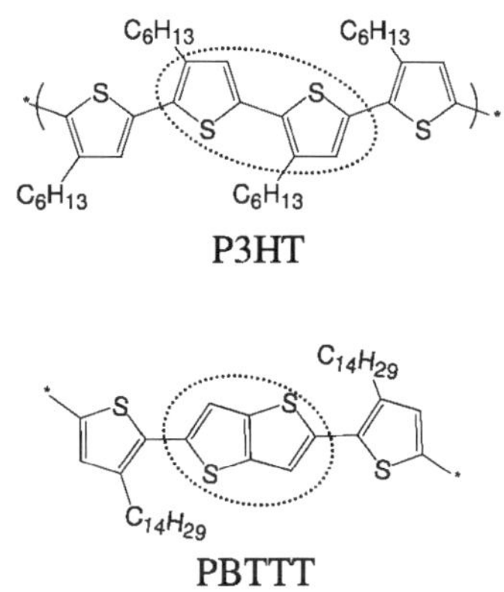

Figure 4. Molecular structures of P3HT and PBTTT monomers.

Figure 3 compares the relative mobility ratios of the drop-cast films for four different molecular weight P3HT polymers by TRTS. The relative mobility increases sharply with increasing molecular weight of the polymers to an asymptotic value for the highest molecular weight P3HT film (*10*). The data showed a four-fold increase in the relative mobility of the P3HT films for the manufacturer's supplied weight range between ~20 kDa and ~150 kDa. The results suggest that the local structure improves as the molecular weight of the polymer increases. The correlation of our results with device measurements shown in Figure 3 illustrates that this local improvement extends throughout the film morphology and is an important feature for the final device. Details about the comparison to device measurements, discussion concerning minor deviations and the effect of temperature and casting conditions are found in reference (*10*)

To further investigate the role of the conjugation length and the structural effects that may improve the mobility of carriers in polymer thin films, we examined and compared the relative mobilities of the materials, P3HT and PBTTT (*12*). These two polymers have very similar monomeric structures. The only differences in their local and global film structures arise from the removal of alternating single bonds and their side chain between adjacent thiophene units, resulting in the fusion of two central thiophene molecules in the monomer unit, and having longer side chains (Fig 4). This makes the PBTTT polymer more rigid and permits a higher degree of side chain interdigitation compared to P3HT (*18*). Additionally, this modification also creates a longer effective conjugation length for individual polymer strands and a higher probability of overlapping conjugation segments between PBTTT polymers. The increased intermolecular interactions leads to better percolation paths for the carriers through a deposited film. As a result, the PBTTT polymer has a more conductive local and global organization compared to P3HT. P3HT and PBTTT films were spin-cast to thicknesses of only about 30 nm for PBTTT films and about 50 nm for P3HT films. The TRTS measurements of few tens of nanometer thick polymer films are unique and enables us to compare our results to device measurements with greater confidence. TRTS measurements revealed that the mobility of carriers in an annealed PBTTT film is approximately an order of magnitude higher than the mobility of the carriers in P3HT films (*12*). This result agrees with the expectation that the local organization of the films significantly improves carrier diffusion. These findings also correlate well with electrical contact device measurements indicating that the local structure is again a significant contributor to the final effective mobility of the carriers in these thin films.

Ultrafast photo-carrier dynamics studies of thin polymer films illustrated the importance of conjugation length and local order on the final carrier mobility of the films. The studies suggest that longer conjugation length of polymers increases the travel distance of the carriers with fewer intermolecular hopping events, and an increased local ordering increases the possibility of a successful hopping process by increasing chances of overlapping conjugation segments of the polymers. Overall, longer conjugation with higher local order (e.g., in PBTTT) results in better percolation paths for the carriers to travel. Such improvement of polymer structure in thin films results in an overall two orders of magnitude increase in mobility. This study also showed that the difference in sensitivity to measured conduction length scales makes both TRTS and device techniques complementary to each other and provides the opportunity to identify possible limiting causes of impeded carrier conduction by directly comparing measurements of thin films.

Relative Carrier Generation Efficiencies of ZnPc and C_{60} Blended and Multilayer Films Studied by TRTS

We recently extended our TRTS measurements to study the transient decay dynamics of photogenerated carriers in blended and multi-layered ZnPc/C_{60} films that are model active layers for solar cell devices (*15*). Investigation of ZnPc and C_{60} films was conducted by the following approaches: (1) we measured the transient response of various co-vapor deposited blended films to determine the optimum blend ratio to achieve the maximum photoconductivity of the films and (2) we assessed the results and checked for trends with the blending ratio to achieve a better understanding of the roles played by the each component. Results from these blended film studies guided us towards examining alternating multilayered films to find more efficient photon harvesting scenarios, which is the second part of the study. The alternating multilayer results signified the importance of the interface in the charge generation and lifetime of generated carriers. The following sections discusses TRTS results of blend and multilayer thin films.

i. Blended Film Studies

Figure 5 shows the measured and OD-corrected differential transmission data for vapor-deposited blend films of ZnPc and C_{60} with blend ratios ranging from 0 to 100 weight percent. As in the polymeric film studies, we observed complex decay dynamics for the photo-generated carriers that depend on the composition of the films. The figure exhibits two component responses to compare and contrast the data from these films; (1) measurements at the response maximum and (2) data from the longer time tail (especially beyond 30 ps where the decay time slows down). The maximum signal peak response is dominated by initially generated carriers. The highest peak amplitude was observed for a neat C_{60} film. However, over 90% of the carriers generated in the neat C_{60} film disappear within only a few ps (< 5 ps). This decay rate is extremely fast resulting in a very small signal level at later times. On the other hand, 400 nm photo-excitation of neat ZnPc produces almost no detectable signal. The very weak TRTS response from both neat samples beyond 20 ps is consistent with their reported low conductivities (*19*, *20*). The peak amplitude increases linearly with the amount of C_{60} content of the blended films, suggesting that either C_{60} is the main source of carrier generation or the increased interfacial interactions results in higher carrier generation when the films are excited with 400 nm excitation (black triangles, Figure 5 inset). This result is surprising given that ZnPc and C_{60} have comparable absorption coefficients at 400 nm. The longer-lived carriers (which generate signal at > 30 ps in this case) deserve the most attention in this study since they are directly responsible for the photo-conversion efficiency a device. The long-time tail amplitude increases as the C_{60} content of the film increases up to 50 weight percent. At higher loadings, the amplitude decreases (red squares in inset of Figure 5). The data suggests that the highest photoconductivity and possibly best material efficiency is achieved around 50:50 weight percent blend ratio of ZnPc to C_{60}.

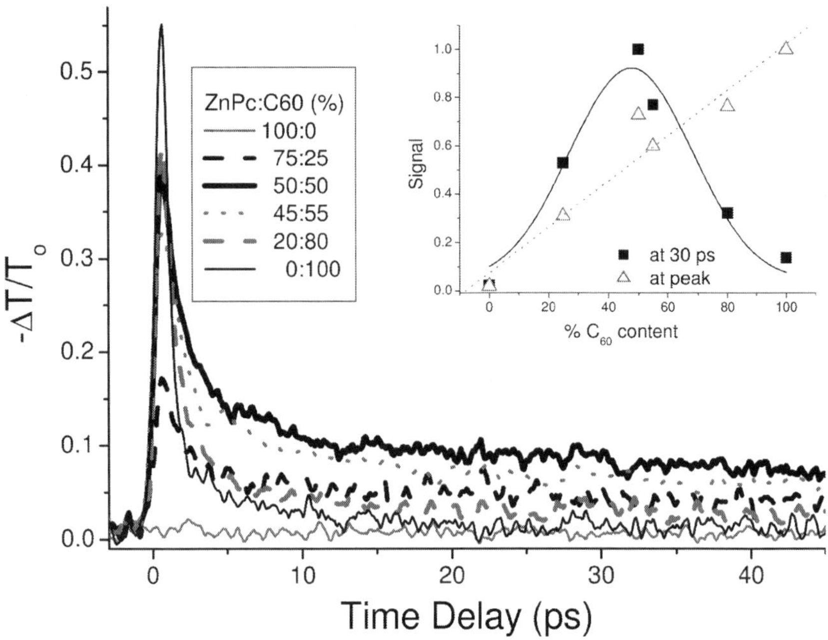

Figure 5. Transient signal decays from ZnPc and C_{60} blended films measured by TRTS. Data are corrected for optical density variations at the 400 nm excitation wavelength. The inset shows amplitude values at peak (triangles) and at 30 ps (squares) of transient data as a function of %C_{60}.

Despite exhibiting high carrier population generation, the long-lived free carrier fraction is only about 1/5 of the peak amplitude for the best blend film. This limit may be due to the chemical and physical properties of the films themselves. At 50% blending, the individual components of the blended films are presumably optimally mixed, providing the best interaction between ZnPc and C_{60} for charge separation. On the other hand, this condition may preclude the best pathway for charge transport through the film. Charge diffusion requires percolation paths or a connected network of the same type of molecule to create a band-like structure for efficient charge transfer. However, a well-mixed system may not provide that condition. A solution to this argument would be to generate as many conducting interfaces as possible while keeping the charges (electrons and holes) separated. We therefore surmised that alternating multi-layered films of ZnPc and C_{60} comprised of a few nanometer-thick individual layers might be a better solution to achieve higher efficiency.

ii. Alternating Multilayer Films with Nanometer Thick Individual Layers

We then measured the transient dynamics of a series of alternating ZnPc and C_{60} multilayer films ranging from 5 nm to 40 nm in alternating layer thickness to assess the relative photon-to-carrier efficiencies and better understand carrier generation and evolution dynamics in these controlled films. Below we briefly outline some of the more interesting results, while the details of the study can be found in reference (*15*). Figure 6 compares differential Terahertz transmission decays ($\Delta T/T_0$) of alternating multilayer films, a 1:1 blend film, and neat ZnPc and C_{60} films as a function of delay time between the 400 nm pump and THz probe pulses. The carrier generation and subsequent transport properties are complex and highly dependent on the relative fractions of ZnPc and C_{60} for blended films (as discussed above) and on the thickness of ZnPc and C_{60} layers for the multilayer films. The transient signal of the multilayered films shows a very rapid initial decay (especially for the multilayer films with thickest alternating nanolayers) similar to that of the neat C_{60} film for delay time < 5 ps, which likely arises from sub-ps geminate recombination of generated carriers in bulk C_{60}. This rapid decay appears to be slightly slower for the 1:1 blended film. After the initial fast decay, the relaxation of the carriers in the blended film appears to be comparable to that observed for the multilayer films (i.e 20 nm alternating multilayer film vs blend film), possibly due to a higher recombination rate of the carriers and/or increased trapping in the blend film. Amplitudes at the peak response (~0 ps) and tail (>30 ps, where the neat C_{60} response diminishes) appears to depend on the nanostructure of the films. Both the peak amplitude and the tail increases as the alternating nanolayer thickness decreases from 40 nm to 5 nm. Simple signal estimates showed that the peak amplitude response for the alternating multilayer films mimic the sum of half of the neat C_{60} film peak amplitude response (since approximately half of the multilayered films are C_{60}) and the response at 45ps (where the neat C_{60} contribution is negligible). This finding and 800 nm excitation studies (*15*) (not discussed) suggest the possibility that two independent contributions from bulk C_{60} and interfacial charge generation are contributing to the observed response of the multilayered films. As the alternating nanolayer thickness is reduced from 40 nm to 5 nm, the interface number (or total interfacial interaction area) of a multilayer film increases as the total film thickness is kept constant. This results in an increase in the interfacial charge generation contribution to both the peak and the tail amplitudes of the observed differential transmission signal and may also explain the slight peak shift (~0.2 ps) to later time delay. In addition, the large increase in differential transmission tail amplitude is linearly proportional to the increase in interface number of the multilayer films (where bulk C_{60} contribution is almost zero). Such a correlation suggests that exciton diffusion is either nonexistent or very slow compared to the measurement time. This study also indicated that the active region at the interface is smaller than 2.5 nm since a deviation from signal linearity with layer thickness was not observed, even for the multilayer films composed of 5 nm thick alternating layers. These results signifies the importance of controlling the molecular component interactions and interface for successful photon-to-carrier conversion and also of the required co-existence of bulk components in each phase for retention of the generated carriers and their transport.

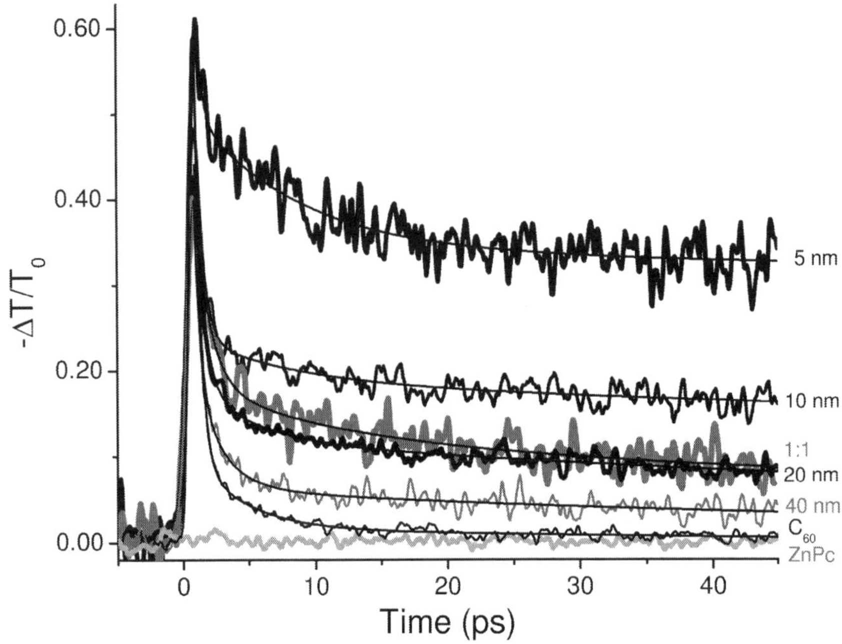

Figure 6. Transient decay dynamics of blend and layered ZnPc/C_{60} films. Data are corrected for OD variations at 400 nm. Reproduced with permission from reference (15).

Conclusions

In this chapter we presented a brief overview of the TRTS technique and its application to newly synthesized organic semiconductors that are model active layers for organic-based photoconductive devices. Conducting polymer film studies revealed that the effective mobility of carriers in thin films is increased when molecular-scale local structure is better organized and the conjugation length is lengthened.. A comparison between TRTS and device measurements signified the strength of the fully optical TRTS technique to further understand the carrier properties of photo-active thin polymer films without introducing potential device artifacts.

Blend and multi-layered ZnPc and C_{60} nano-films investigated using the TRTS technique showed a strong correlation between the composition and nanolayer thickness to photon-to-carrier efficiency. Neat films of both species produces immeasurable amounts of long-lived carriers. The results suggest that two independent factors contribute to the charge generation and conduction processes in nanolayered films, namely bulk and interfacial charge separation. These studies suggest that multi-layer systems composed of a few nm thick alternating layers have significantly better efficiency compared to blended component films and may ultimately be useful for high-efficiency photo-conversion applications.

Acknowledgments

This work was partially funded by NIST Physics Laboratory internal Scientific and Technical Research Support (STRS) for O. Esenturk (NIST Associate and Guest Researcher) and E. J. Heilweil.

References

1. Grischkowsky, D.; Keiding, S.; Vanexter, M.; Fattinger, C. *J. Opt. Soc. Am. B* **1990**, *7*, 2006.
2. Beard, M. C.; Turner, G. M.; Schmuttenmaer, C. A. *J. Appl. Phys.* **2001**, *90*, 5915.
3. Jeon, T. I.; Grischkowsky, D.; Mukherjee, A. K.; Menon, R. *App. Phys. Lett.* **2001**, *79*, 4142.
4. Schmuttenmaer, C. A. *Chem. Rev.* **2004**, *104*, 1759.
5. Hendry, E.; Koeberg, M.; Schins, J. M.; Siebbeles, L. D. A.; Bonn, M. *Phys. Rev. B* **2004**, *70*.
6. Ostroverkhova, O.; Shcherbyna, S.; Cooke, D. G.; Egerton, R. F.; Hegmann, F. A.; Tykwinski, R. R.; Parkin, S. R.; Anthony, J. E. *J. Appl. Phys.* **2005**, *98*.
7. Hendry, E.; Koeberg, M.; Schins, J. M.; Siebbeles, L. D. A.; Bonn, M. *Chem. Phys. Lett.* **2006**, *432*, 441.
8. Ai, X.; Beard, M. C.; Knutsen, K. P.; Shaheen, S. E.; Rumbles, G.; Ellingson, R. J. *J. Phys. Chem. B* **2006**, *110*, 25462.
9. Lloyd-Hughes, J.; Richards, T.; Sirringhaus, H.; Castro-Camus, E.; Herz, L. M.; Johnston, M. B. *App. Phys. Lett.* **2006**, *89*, 112101.
10. Esenturk, O.; Melinger, J. S.; Heilweil, E. J. *J. App. Phys.* **2008**, *103*, 023102.
11. Cunningham, P. D.; Hayden, L. M. *J. Phys. Chem. C* **2008**, *112*, 7928.
12. Esenturk, O.; Kline, R. J.; Deongchamp, D. M.; Heilweil, E. J. *J. Phys. Chem. C* **2008**, *112*, 10587.
13. Parkinson, P.; Lloyd-Hughes, J.; Johnston, M. B.; Herz, L. M. *Phys. Rev. B* **2008**, *78*, 11.
14. Kline, R. J.; McGehee, M. D. *Polym. Rev.* **2006**, *46*, 27.
15. Esenturk, O.; Melinger, J. S.; Lane, P. A.; Heilweil, E. J. *J. Phys. Chem. C* **2009**, submitted.
16. Certain commercial equipment, instruments, or materials are identified in this paper to adequately specify the experimental procedure. In no case does identification imply recommendation or endorsement by NIST.
17. Hendry, E.; Schins, J. M.; Candeias, L. P.; Siebbeles, L. D. A.; Bonn, M. *Phys. Rev. Lett.* **2004**, *92*.

18. Kline, R. J.; DeLongchamp, D. M.; Fischer, D. A.; Lin, E. K.; Richter, L. J.; Chabinyc, M. L.; Toney, M. F.; Heeney, M.; McCulloch, I. *Macromolecules* **2007**, *40*, 7960.
19. Siebentritt, S.; Gunster, S.; Meissner, D. *Synth. Met.* **1991**, *41*, 1173.
20. Walzer, K.; Maennig, B.; Pfeiffer, M.; Leo, K. *Chem. Rev.* **2007**, *107*, 1233.

Chapter 18

Photoinduced Carrier Dynamics in Organic Bulk Heterojunctions Studied by Sub-Picosecond Optical Pump-THz Probe Time Domain Spectroscopy

Yi-Hsing Peng,[1] Weilou Cao,[1] Mihaela Ballarotto,[1] Danilo Romero,[1] Warren N. Herman,[2] and Chi H. Lee*,[1]

[1]Department of Electrical and Computer Engineering, University of Maryland, College Park, MD 20742
[2]Laboratory for Physical Sciences, College Park, MD 20740
***chlee@umd.edu**

The purpose of this work is to validate an analysis procedure proposed by Nienhuys and Sundstrom (*1*) for extracting transient complex conductivity from optical pump-THz probe time domain spectroscopy measurements and to use this method to investigate the photoinduced carrier dynamics in P3HT/PCBM blends. The conductivity artifacts predicted in reference (*1*) were verified in experiments on GaAs, well-known to obey the Drude model, and corrected by the appropriate time and frequency domain transformations (*1*). Unlike the case for GaAs, the complex conductivities of the P3HT/PCBM blends were described by the Drude-Smith model (*2*). A significant drop in conductivity in the first pico-second after carrier generation was found to be mainly due to the decrease in mobility associated with polaron formation.

Introduction

Photovoltaic devices based on bulk heterojunctions consisting of conjugated polymer and fullerene composites have been intensively investigated recently as alternative renewable energy sources. The development of this technology promises to produce low-cost solar energy harvesting devices capable of being implemented on large area surfaces with high form factor. The current state-of-the-art single-cell power-conversion efficiency of 5% has been reported in devices using regio-regular poly(3-hexylthiophene) (P3HT) as donor and [6,6]-phenyl-C_{61}-butyric acid methyl ester (PCBM) as acceptor molecules (*3*, *4*). Higher efficiency above 6% has been reported in a tandem cell device architecture (*5*). The efficiency of these photovoltaic devices is governed by a series of optoelectronic processes starting with light absorption, followed by ultrafast charge-transfer from donor to acceptor and subsequent exciplex dissociation into electron and hole, and concomitant charge transport and collection to the respective electrodes (*6*). Understanding the charge dynamics in each of these processes is critical towards the development of high efficiency polymer-based solar cells.

An ideal probe of the above-described processes is the time-domain terahertz (THz) spectroscopy technique. This technique involves the measurement of freely propagating THz pulses transmitted through a sample under investigation. Because the amplitude and phase informations are contained in the time-resolved phase-sensitive probe of the electric field, one can obtain the frequency-dependent complex transmission coefficient by means of Fourier analysis. From modeling of the transmission coefficient, optical quantities such as the complex index of refraction, absorption coefficient, and complex conductivity spectra in the THz range can be extracted (*7*). In conjunction with the use of femtosecond optical pulses to photoexcite a sample, optical pump - THz probe time domain spectroscopy (OPTP-TDS) has been reported to study transient dynamics of charge carriers in semiconductors (*8*) and semiconducting polymers (*9–12*).

Recently, Nienhuys and Sundstrom's (*1*) theoretical analysis of the optical-pump terahertz-probe measurement showed that the conductivity obtained directly from experiment contains both the impulse response of the charge carrier and the kinetics of the carrier population. In this case, the determination of the conductivity is ambiguous when the carrier population varies abruptly during the time of THz probing, particularly at the early stage of the photoexcitation. Also, the fast dynamics determined from measurement of the transient THz peak absorption may not necessarily reflect the true charge carrier dynamics in a sample for the same reason. To obtain the correct form of the frequency-dependent complex conductivity, a series of Fourier and time shifted transformations were suggested for these scenarios (*1*).

The purpose of this paper is to validate a procedure to analyze the results of OPTP-TDS measurements in order to extract the transient complex conductivity, and to use this method to investigate the photoinduced carrier dynamics in P3HT:PCBM blends. As a test of the validity of this procedure, GaAs is used as a benchmark sample to perform the OPTP-TDS measurement since carrier dynamics in this compound is well-known to obey the Drude model. The conductivity artifacts predicted by Nienhuys and Sundstrom have been verified in the experiments, and corrected by the appropriate time and frequency domain transformations (*1*). Then, we present studies of the charge-carrier dynamics in P3HT:PCBM blends using the established OPTP-TDS measurement method. Unlike the case for GaAs, the complex conductivity of the P3HT:PCBM blends are better described by the Drude-Smith model that was developed to explain the unconventional low-frequency conductivity of disordered granular systems (*2*). We present the transient photoconductivity of P3HT:PCBM blends with sub-picosecond resolution. We find that the significant drop in conductivity in the first pico-second after carrier generation is mainly due to the decrease in mobility associated with polaron formation.

Experiments and Analysis Methods

Organic bulk heterojunctions consisting of blends of regioregular P3HT (Rieke Metals, 90~93% regioregular) as donor and PCBM (Nano-C) as acceptor molecules were investigated for this work. The photovoltaic characteristics of devices fabricated using these materials as the active layer are known to depend on the blend composition. To study the carrier dynamics that are responsible for the photovoltaic properties, a series of P3HT:PCBM solutions of varying ratio by weight (weight ratio = 1:0, 4:1, 1:1, 1:4, and 0:1) were prepared. The solutions were stirred overnight at 50°C to uniformly disperse the organic components in dichlorobenzene. Films, approximately 10 μm thick, were prepared by drop-casting the solution onto a quartz substrate and allowed to dry slowly at room temperature to promote the ordering of the P3HT chains. We estimate the optical densities (OD) of the samples at 400 nm wavelength to be larger than 3.0 which ensures that the sample absorbs nearly all of the excitation light. The films were prepared under nitrogen atmosphere inside a glovebox system to minimize photo-oxidative degradation.

The setup for the OPTP-TDS, similar to that in reference (*8*), is shown in Figure 1. A 1 mJ, 800 nm, 100 fs, 1 kHz repetition rate amplified Ti:sapphire laser was used as the fundamental laser source for the THz system. The THz probe pulses, about 1 ps duration single cycle electromagnetic pulses, are generated via optical rectification by shining 350 μJ of 800 nm pulses onto a 1-mm-thick [110] ZnTe crystal. The THz pulses are collected, collimated, and focused onto the sample with 90 degree off-axis gold coated parabolic mirrors to generate the THz peak electric field of about 0.8 kV/cm. At the same time, the samples are pumped with 300 μJ, 400 nm pulses (fluence = 3×10^{19} photons/m^2) obtained from frequency doubled 800 nm pulses. The 400 nm pump pulse has a beam diameter of 5 mm, which is much larger than the 1 mm diameter of the THz probe pulse at the sample ensuring a uniform excitation of the probed region. The THz probe pulses are transmitted through the photoexcited sample to be modulated by the carrier density. The pump-induced modulation of the THz electric field is less than 1% in the measurement, as shown in Figure 2. The transmitted THz pulses are again collected, collimated and focused onto another 1-mm-thick [110] ZnTe crystal and collinear with the 10 nJ of 800 nm gate pulses to measure the electric field amplitude using free space electro-optic sampling.

The experimental data are collected using a two-dimensional (2D) scan method as described in references (*8*) and (*13*). A series of THz difference waveforms are obtained by chopping the optical pump pulses and delaying the

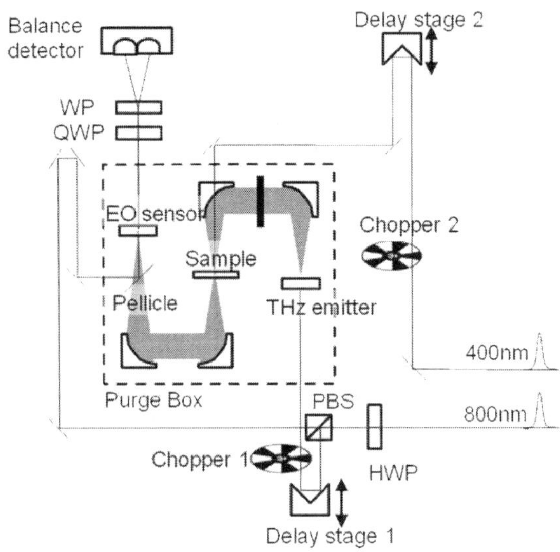

Figure 1. The setup of the OPTP-TDS. The THz path is contained inside a nitrogen filled purge box to eliminate water vapor absorption.

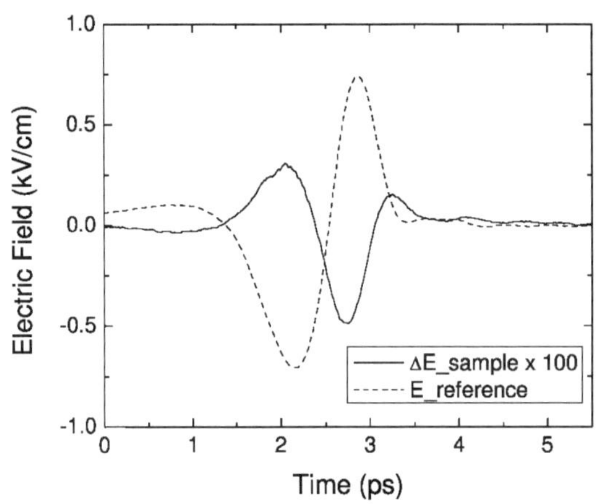

Figure 2. The electric field E(t) of the THz pulse (dotted line) transmitted through the P3HT:PCBM 1:1 blend sample with the modulation ΔE(t) measured 2 ps after photoexcitation (solid line)

optical pump arrival time in increments of 0.5 ps. The distortion from the detector response of the 1-mm-thick ZnTe crystal is deconvolved numerically from the measured wave form (*14*), and the data are processed so that each point in the measured waveform has passed through the sample after the same amount of time from the photoexcitation event (*13*) , as shown in Figure 3.

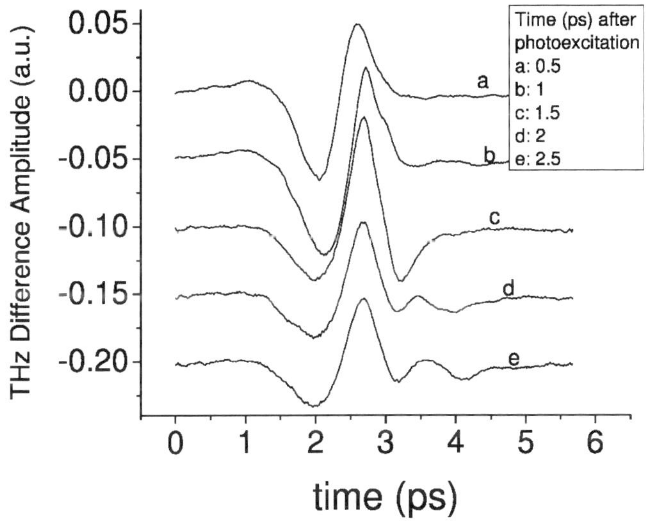

Figure 3. Several THz difference scans in photoexcited P3HT measurement for different probing times. The data are shifted in y-axis for clear view.

To extract the conductivity, $\sigma_M(\omega,\tau)$, from the OPTP-TDS measurement, first we consider the conductivity in the frequency domain,

$$\sigma_M(\omega) = \frac{F[J(t)](\omega)}{F[E_0(t)](\omega)} , \qquad (1)$$

where the notation $F[x(t)](\omega)$ refers to the Fourier transformation from time to frequency domain of the function $x(t)$, and $J(t)$ is the current density due to the applied electric field $E_0(t)$. If the thickness L of the photoexcited material is small compared to the THz wavelength (the thickness here is that of the photoexcited layer), one can show that $J(t)$ is obtained from the difference of the THz electric fields (1) as,

$$J(t) = -\frac{\varepsilon_0 c(n_{THz} + n_{quartz})}{L}[E(t) - E_0(t)] , \qquad (2)$$

where ε_0 is the permittivity of vacuum, c is the speed of light, n_{THz} and n_{quartz} are the THz refractive indices for the unexcited sample and quartz substrate, respectively. $E(t)$ and $E_0(t)$ are the electric field of the THz wave after passing through the excited sample and the unexcited sample, respectively. $E(t)$-$E_0(t)$ and $E_0(t)$ are the THz fields measured directly from experiment. By substituting eq (2) into eq (1), one can calculate the *measured* conductivity, $\sigma_M(\omega,\tau)$.

According to reference (*1*), under non steady state conditions in which the carrier density changes during the THz probe pulse duration, the conductivity, $\sigma(\omega,\tau)$, can be expressed as:

$$\sigma(\omega,\tau) = F[N(\tau - t)\, j_0(t)] \quad , \tag{3}$$

where τ is the delay time after excitation, N is the time dependent carrier density function, and $j_0(t)$ is the single charge impulse response. Note that if $N(\tau\text{-}t)$in eq 3 is constant, it is easy to show that $\sigma_M(\omega,\tau)$ derived from eqs 1 and 2 represents the correct measured conductivity given by eq 3. However, because $N(t)$ is in general time dependent in the OPTP-TDS measurement, the experimentally derived conductivity will be influenced by the rapidly changing $N(t)$ and as such will not have a straightforward physical meaning. In order to recover the *true* conductivity, $\sigma(\omega,\tau)$, one can apply the following sequence of transformations (*1*)

$$\begin{aligned}
&\sigma_M(t,\tau) = F^{-1}[\sigma_M(\omega,\tau)](t) = N(\tau - t)\, j_0(t) \ , \\
&\sigma(t,\tau) = \sigma_M(t,\tau + t) = N(\tau)\, j_0(t) \ , \\
&\sigma(\omega,\tau) = F[\sigma(t,\tau)](\omega) = N(\tau) F[j_0(t)](\omega) \ .
\end{aligned} \tag{4}$$

Clearly, the conductivity derived above is not coupled with the dynamics of the carrier population and represents the *true* conductivity. Indeed, because of the limited bandwidth of the THz pulses and the detection system, the Fourier transformation in eq (4) suffers some data loss. However, the frequency-dependent conductivity that we desire can be recovered reasonably well even using limited bandwidth data (*1*). We point out here that, in order to perform the transformation in eq (4), two dimensional scan is necessary to obtain sufficient information of conductivity in both pump and probe time durations.

Results and Discussion

1. GaAs Benchmark Measurement

As a test of the validity of the analysis discussed above, we used GaAs as a benchmark sample in our OPTP-TDS measurement. This is because GaAs is a material in which the carrier dynamics is well-described by the simple Drude model. The sample used in our measurement is double-side polished 2" semi-insulating GaAs wafer from AXT. The thickness is 0.35 mm and the surface orientation is [100] to ensure that it does not generate THz radiation upon normal incident photoexcitation (*15*). In the OPTP-TDS measurement, the optical pump wavelength is 800 nm (1.55 eV), which is just above the room temperature energy gap (1.43 eV) of GaAs (*16*), eliminating the possibility of higher energy band optical transitions (*8*). The pump beam energy is attenuated to 0.12 μ J/pulse spread over a 5 mm diameter cross-sectional area (fluence: 2.46×10^{16} photons/m^2)

to maintain a low photoexcited carrier density to avoid effects associated with non-parabolicity.

The real part of $\sigma_M(\omega,\tau)$ measured for photoexcited GaAs is shown in Figure 4(a). For times later than 2 ps after initial photoexcitation, the conductivity is like that in the steady state THz experiment, and can be fitted well using a simple Drude model, as shown in Figure 5. The fitting results show that the photoexcited carriers in GaAs have the mobility 5750 $cm^2V^{-1}s^{-1}$ and the carrier density 1.9×10^{16} cm^{-3}. These values agree with the mobility measured in reference (*8*), and with the absorbed photon density of 2.5×10^{16} cm^{-3} calculated using $N_{absorb} = FT/d$, where F is the fluence, T is the transmissivity, and d is the skin depth. However, from the time 0 ps to 2 ps, the conductivity deviates from the Drude model. Figure 6a displays $\sigma_M(\omega,\tau)$ at time delay of 0.8 ps. The conductivity initially rises to a maximum and then decreases at higher frequencies, unlike a Drude behavior. Similar behavior has been reported in reference (*8*). According to Nienhuys and Sundstrom's theoretical analysis, this behavior is an artifact associated with strongly time-dependent carrier density during the short temporal probe of the GaAs sample (*1*). This is simulated using eq (3) assuming a Drude like carrier response and a step function with a 100 fs rise time for the carrier density. The results are shown in Figure 4(b). Note the initial rise in the simulated conductivity at early times. Because we used only the Drude model to describe the charge carrier dynamics in the simulation, this non-Drude response at the beginning of the photoexcitation is an artifact introduced by the temporal change in the carrier density.

To retrieve the *true* conductivity from the measurement, we apply the transformations in eq (4) to the *measured* conductivity. The result is shown in Figure 4(c). Comparing Figure 4(a) and (c), we find that the conductivity artifacts at early times disappear after the correction transformations. Figure 6 shows the *measured* (before correction) and *true* (corrected) complex photo-induced conductivity of photoexcited GaAs at the time 0.8 ps after the initial photoexcitation. The lines shown here are the results of Drude model fitting. For the uncorrected data in Figure 6(a), the conductivities deviate from the simple Drude model. However, after the data correction transformations, the conductivities at 0.8 ps basically follow the Drude model, and are described by the same fitting parameters as what we get for the conductivities measured at later time ($t > 2$ ps).

From this GaAs measurement, we establish the measurement and analysis methods for OPTP-TDS to obtain the correct time-resolved photo-induced conductivity. The photo-induced conductivities of photoexcited GaAs from a time just after excitation to the time 10 ps after excitation basically follow the simple Drude model. The conductivity artifacts at the beginning of the excitation predicted by Nienhuys and Sundstrom are observed (*1*), and the *true* transient conductivities are obtained by applying the correction transformations to the conventional OPTP-TDS analysis method (*13*).

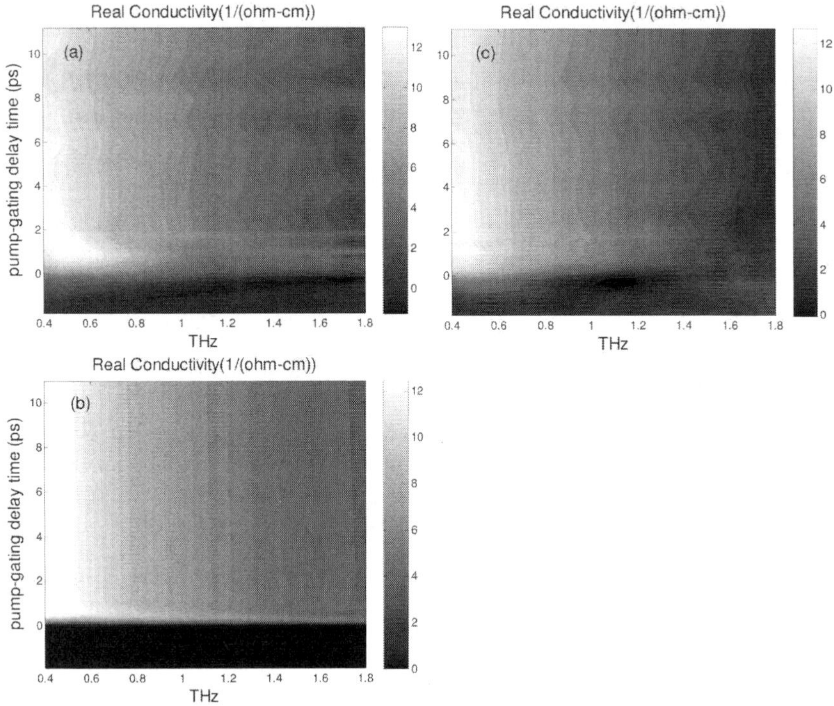

Figure 4. Two-dimensional (2D) contour plot of real part of (a) the measured (before correction) conductivity; (b) the simulated conductivity using eq (3); (c) the true (corrected) conductivity of photoexcited GaAs at room temperature.

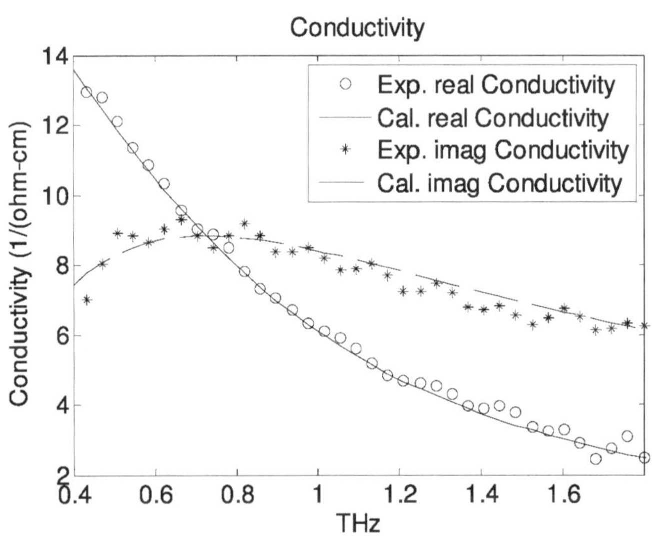

Figure 5. Real and imaginary conductivity of the photoexcited GaAs at a time 10 ps after the initial pump. The lines are the results of fitting using a Drude model.

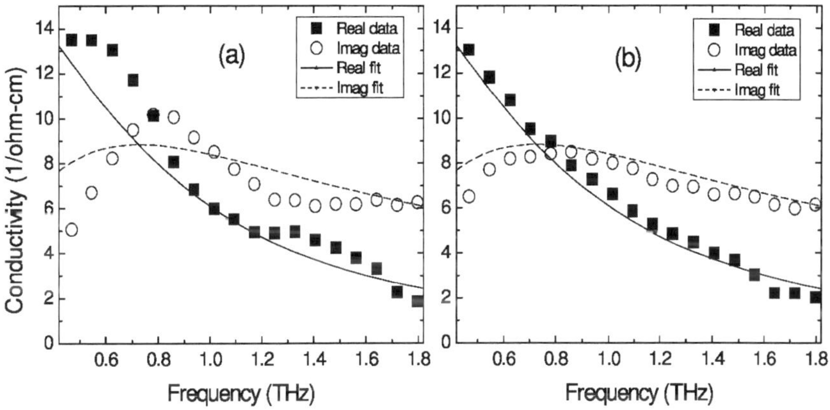

Figure 6. (a) Measured (uncorrected) and (b) true (corrected) complex photo-induced conductivity of photoexcited GaAs at the time 0.8 ps after the initial photoexcitation. The lines show the Drude model fit.

2. P3HT/PCBM Measurement

Figure 7 shows the real part and imaginary part of the frequency dependent conductivity of a photoexcited P3HT/PCBM 1:1 blend before and after the correction transformations for 0.5 ps after photoexcitation. The solid lines are the results of a fit of the data using the Drude-Smith model. Comparing the uncorrected conductivities with the corrected conductivities, some obvious differences are observed. In the real part of the conductivities, the uncorrected data are underestimated at the lower frequencies. On the other hand, the uncorrected data are overestimated at lower frequency and underestimated at higher frequency in the imaginary part of the conductivities.

There are two main differences between the photo-induced complex conductivities of GaAs and P3HT:PCBM blends. First, the real part of the conductivity at THz frequencies increases with frequency in the P3HT:PCBM blends, while it decreases with frequency in GaAs. Second, the imaginary part of the conductivity is negative in P3HT:PCBM blends and positive in GaAs. As we have shown above, the low frequency carrier dynamics in GaAs are well-described by a simple Drude conductivity. On the other hand, the behavior we observe in the P3HT:PCBM blends cannot be explained by a simple Drude model. Instead, they are characteristics of localized carriers. To describe the carriers in the semiconducting polymer, which has strong carrier localization due to nanoscale inhomogeneity and disorder, we use the Drude-Smith model (*2*):

$$\sigma_{DSM}(\omega) = \frac{ne^2\tau / m^*}{1 - i\tau\omega}\left[1 + \frac{c}{1 - i\tau\omega}\right] \tag{5}$$

where n is the carrier density, τ is the scattering time, m^* is the effective mass. The coefficient c represents that fraction of the charge carrier's original velocity that is retained after the first collision. Unlike the simple Drude model, it has a memory or persistence of velocity effect after the scattering. The truncated single-scattering approximation suggested by Smith has well described the semimetals near the metal-insulator transition, such as liquid mercury (*2*). In our data analysis, the terahertz refractive indices of the P3HT:PCBM blends are approximated by the weighted average of the P3HT, having index = 2.22, and PCBM, having index = 1.35 (*10*). The average effective mass of carriers in the P3HT:PCBM blends is held constant at 1.7 m_e (*17*). When we apply the Drude-Smith model to fit the data, the corrected conductivities allow better fits than the uncorrected data, as shown in Figure 7. It is similar to the GaAs measurements where the corrected data fit better to a pure Drude model than the uncorrected data in the early times after photoexcitation.

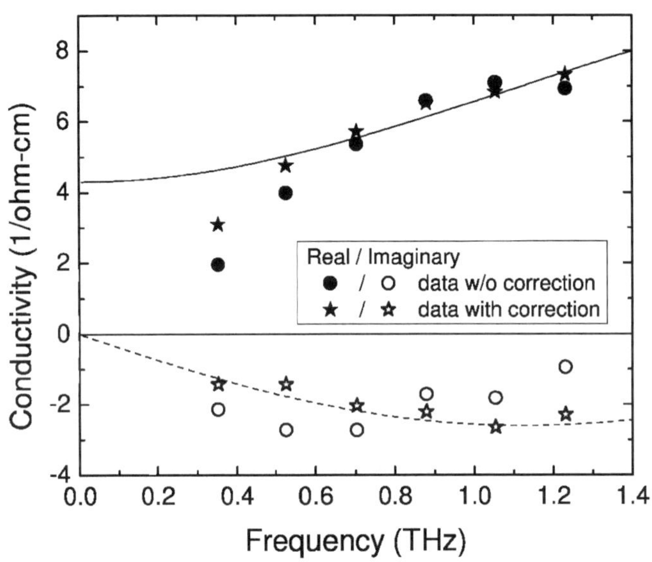

Figure 7. The real part and imaginary part of the frequency dependent conductivity of photoexcited P3HT/PCBM 1:1 blend before and after the correction transformations for times 0.5 ps after photoexcitation. The lines are the results of Drude-Smith model fit.

Figure 8 shows the corrected real (σ_{real}) and imaginary (σ_{imag}) parts of the time-resolved conductivity of P3HT/PCBM blends at 1 THz for delay times from -2 to 10 ps. The data are obtained by slicing the corrected 2D time resolved conductivity spectrum at frequency 1 THz for different blends. This data presenting method for carrier dynamics is different from the traditional method, which is to measure the time resolved transmittance changes of the peak of the THz probe (*9–11*). This method avoids the influence of the time profile of the THz probe pulse to the THz peak transmittance measurement, because the peak transmittance is the result of a convolution of a single charge impulse response, carrier density function, and electric field of THz pulse (*1*). For different blend ratios, the P3HT/PCBM 1:1 blend has the highest conductivity among other blends. Figure 9 shows the results of comparing the real conductivity of different P3HT/PCBM blends for a frequency 1THz at the time 2.5 ps after photoexcitation. It agrees with the results of the P3HT/PCBM photocell measurements that a 1:1 blend has the best performance (*18*). The data also show two different temporal behaviors for the real part of the conductivity for all blends. The first one is the very fast rise and fast decay in the first pico-second. The second one is a slower gradual decay after 2 ps. For the time resolved imaginary parts of the conductivities, the mixtures (0.8, 0.5, and 0.2 blends) do not change much after photoexcitation, while the pure materials (P3HT and PCBM) change like their real parts do.

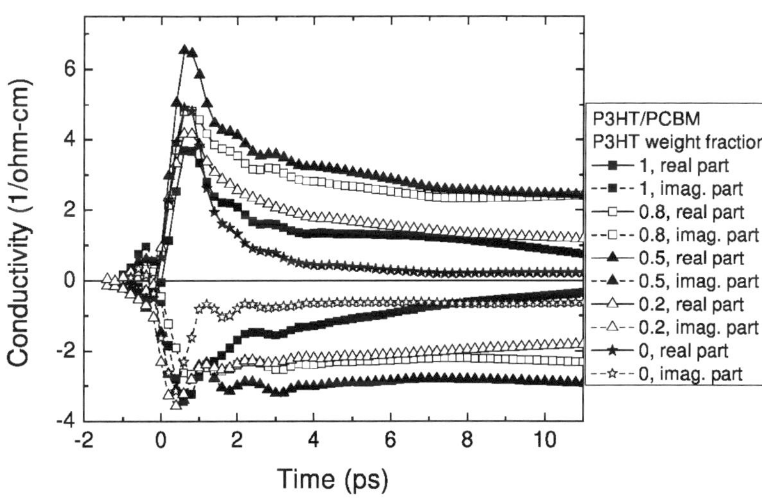

Figure 8. Time-resolved complex conductivity of P3HT/PCBM blends at frequency = 1 THz.

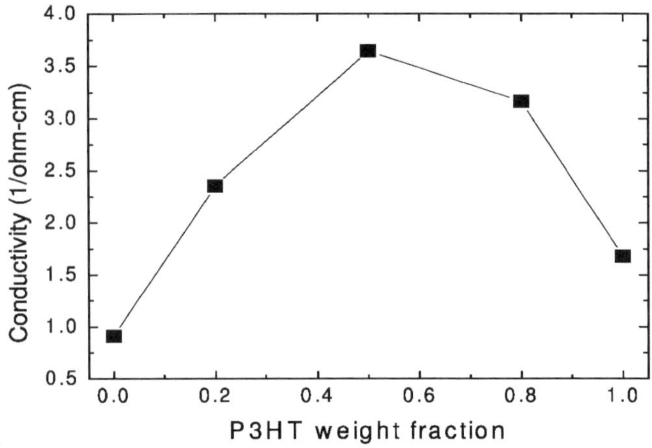

Figure 9. Real conductivities for 1 THz at 2.5 ps after excitation versus different P3HT/PCBM blend ratio.

In order to analyze the dynamics, model fitting to the time resolved conductivity spectrum is used to obtain other physical properties. Figure 10 shows the complex conductivity at the probe delay times 0.5, 1.5, and 8.5 ps for the different weight ratio of the P3HT/PCBM blends. The lines are the results of fitting using the Drude-Smith model, and the extracted parameters are shown in Table 1. The scattering time, τ, plasma frequency, ω_p, and the parameter of persistence of velocity, c, can be extracted directly from the fitting. The dc mobility, $\mu\_DC = e\,\tau / m^{*}\left[1+c\right]$, and carrier density, $N = \varepsilon_0 \omega_p^2 m^{*} / e^2$, can be obtained by assuming 1.7 m_e as the effective mass (*17*). Indeed, the uncertainties in the effective mass directly produce uncertainties in the values of the mobility and carrier density. Nevertheless, we can still compare the relative changes of the time resolved properties for the same material.

The photon-to-carrier yields we observed for all blends are about 1~2%. These carrier yields agree with other OPTP-TDS reports (*11*, *12*), but are much lower than the quantum efficiency obtained from the polymer photocell measurements (*19*). A possible explanation is the built-in potential. In photocell measurements, the polymer is sandwiched by two different electrodes. Because of the different work functions of the two electrodes, the polymer experiences a built-in electric field. The field can be as high as 10^5 V/cm, which is 2 orders of magnitude higher than the electric field of our THz probe. The high built-in electric field may help to dissociate the excitons generated by photoexcitation into free carriers. In contrast, the contactless polymer sample in OPTP-TDS measurement experiences no other electric field except the weak THz probe.

The dc mobilities of P3HT/PCBM blends are on the order of several cm^2/Vs, as shown in Table 1. These values are several orders of magnitude larger than those measured at GHz frequencies (*20*). The reason for the difference is that the dc mobilities we obtained here are the results of prediction from THz measurements. Because of the high frequency nature of the THz wave, the carriers driven by the THz field move only on a small length scale. The factors limiting the carrier mobility for long range transportation (e.g., disorder in the film) will not contribute to the THz mobility. It has been shown that the measured mobilities of the conducting polymers depend strongly on the probing frequency for the low frequency regime (*21*).

The dc mobility and carrier density for 1:1 blend P3HT/PCBM for a probe delay time from 0 ps to 10 ps are shown in Figure 11. The dc mobility drops significantly during the first pico-second and stays at approximately the same value afterward, while the carrier density is almost constant all the time after photoexcitation in this time period. Similar results were found for P3HT/PCBM 4:1 and 1:4 blends. The results indicate that the significant drop in conductivity during the first pico-second after carrier generation is mainly due to the time dependence of the mobility.

To understand these differences, we recall that the free-carrier Drude conductivity for $\omega\tau \ll 1$ is $\sigma_{real} \approx ne^2\ \tau/m^*$ where n is the carrier density, m^* and τ are the carrier effective mass and scattering time, respectively. Prior to 0.5 ps, the temporal conductivity response is due to increasing n, most likely associated with the rise time of the pump pulse intensity and exciton dissociation. After 0.5 ps, when free carriers are generated, charge-carrier transport is renormalized by interaction effects. In this case, the temporal conductivity response is governed by the mobility. In the Drude-Smith model, the dc mobility is $\mu\_DC = e\ \tau/m^*[1+c]$ and we assume that m^* is a constant. In our measurement results, the parameter c for the P3HT/PCBM blends decreases and is closer to -1 after 0.5 ps. Recall that the parameter c is a persistence of velocity effect in the Drude-Smith model. A more negative value for the parameter c means more velocity damping. When the free carriers are generated, their electric field starts to influence the polarization of the nearby atoms, and forms polarons. The formation time of the polaron is about sub-picosecond. The resulting polaron formation acts as a potential well that hinders the movements of the charge. In another words, the carriers become heavier and the effective mass increases in Drude model after polaron formation. This explanation matches the observation that the parameter c drops on a sub-picosecond time scale after free carrier generation. Meanwhile, the carrier density does not change in this period of time. Therefore, we conclude that the decrease of σ_{real} in the first picosecond after photoexcitation is attributed to an increase in m^* due to polaron formation on a sub-picosecond time scale.

Comparing the rate of decrease of the parameter c for different P3HT/PCBM blends in Table 1, we notice that the higher weight ratio of PCBM in the blend has the higher rate of decrease. The result implies that the charge carriers in PCBM form stronger polarons than those in P3HT.

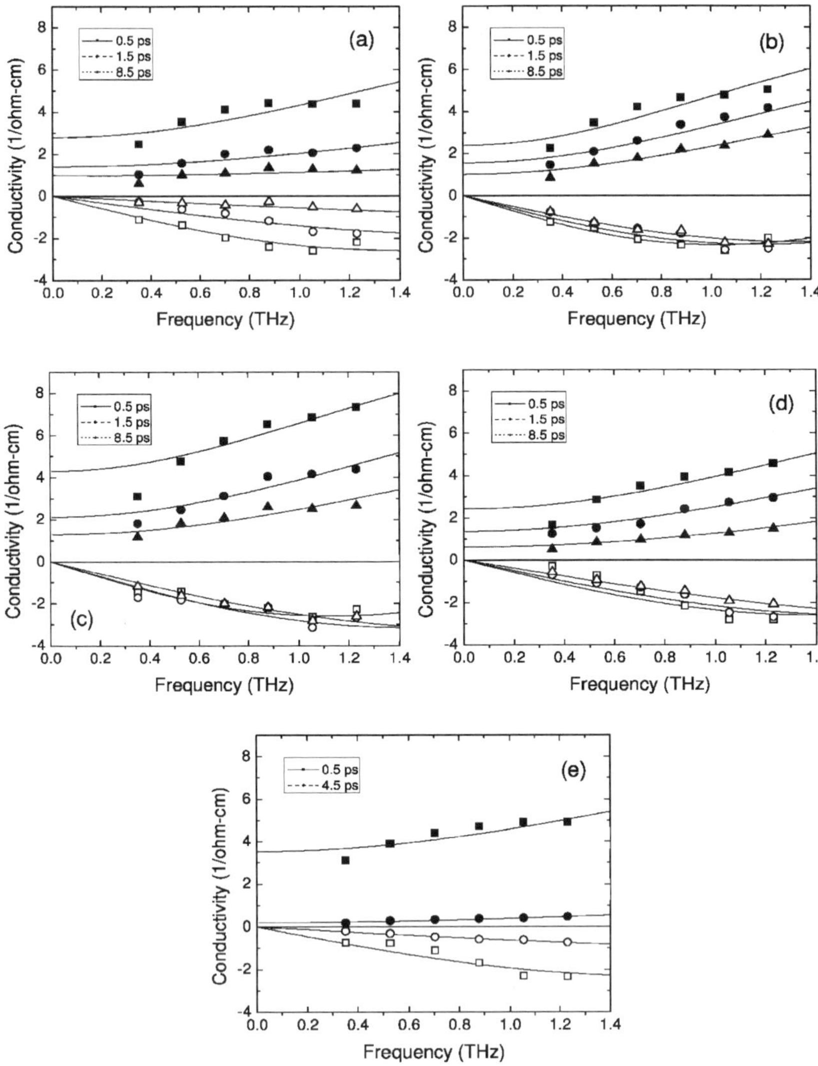

Figure 10. Complex conductivity at the probe delay times 0.5, 1.5, and 8.5 ps for (a) pure P3HT, (b) P3HT/PCBM 4:1, (c) P3HT/PCBM 1:1, (d) P3HT/PCBM 1:4, and (e) pure PCBM. Lines are the fit with Drude-Smith model.

Table 1. Results of the application of a Drude-Smith model based analysis of frequency dependent conductivity data measured for the P3HT/PCBM blends at probe delay of 0.5, 1.5, 2.5, 4.5, and 8.5 ps

Delay (ps)	τ *(fs)*	$e\tau/m^*$ *(cm^2/Vs)*	*N* *($x10^{18}$ cm^{-3})*	*Yield (%)*	*C*	μ_DC *(cm^2/Vs)*
			P3HT			
0.5	42	43	2.46	1.64	-0.84	7.1
1.5	31	32	2.11	1.41	-0.87	4.1
2.5	31	32	1.59	1.06	-0.87	4.1
4.5	30	31	1.49	0.99	-0.86	4.3
8.5	29	30	1.08	0.72	-0.83	5.1
			P3HT:PCBM 4:1			
0.5	58	60	1.56	1.04	-0.84	9.6
1.5	51	53	1.53	1.02	-0.88	6.3
2.5	48	49	1.56	1.04	-0.89	5.4
4.5	48	49	1.45	1.14	-0.9	4.9
8.5	45	46	1.5	1	-0.91	4.2
			P3HT:PCBM 1:1			
0.5	50	52	2.47	1.65	-0.79	10.9
1.5	42	44	2.55	1.7	-0.88	5.2
2.5	38	40	2.62	1.75	-0.9	4
4.5	39	40	2.47	1.65	-0.9	4
8.5	34	35	2.92	1.96	-0.92	2.8
			P3HT:PCBM 1:4			
0.5	42	44	2.33	1.55	-0.85	6.5
1.5	37	38	2.26	1.51	-0.9	3.8
2.5	36	37	2.06	1.38	-0.93	2.6
4.5	28	29	2.92	1.96	-0.94	1.7
8.5	27	28	2.84	1.9	-0.95	1.4
			PCBM			
0.5	34	36	3.09	2.06	-0.8	7.2
1.5	28	29	1.34	0.88	-0.88	3.5

Continued on next page.

Table 1. (Continued). Results of the application of a Drude-Smith model based analysis of frequency dependent conductivity data measured for the P3HT/PCBM blends at probe delay of 0.5, 1.5, 2.5, 4.5, and 8.5 ps

Delay (ps)	τ *(fs)*	$e\tau/m^*$ *(cm^2/Vs)*	*N* *($x10^{18}$ cm^{-3})*	*Yield (%)*	*C*	μ_DC *(cm^2/Vs)*
2.5	24	25	1.29	0.85	-0.94	1.5
4.5	23	23	1.36	0.9	-0.96	0.9

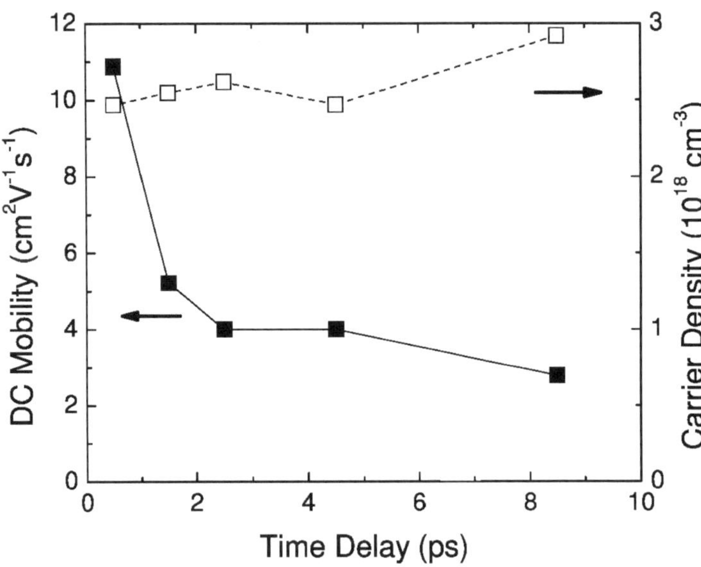

Figure 11. Comparison of time dependant DC mobility (left axis) and carrier density (right axis) for P3HT/PCBM 1:1 blend.

Summary

We have investigated the photoinduced carrier dynamics of P3HT/PCBM blends using OPTP-TDS with subpicosecond resolution. The data analysis process suggested by Nienhuys and Sundstrom is used for the first time to obtain the time resolved frequency dependent complex photoconductivity, and is verified by the GaAs OPTP-TDS measurement. The time resolved conductivity is analyzed using the Drude-Smith model to determine the photon-to-carrier yields, average carrier mobility, and carrier density. For the blends investigated, the 1:1 blend exhibits the highest conductivity. This is consistent with photovoltaic measurements. The significant drop of the real part of the conductivity in the first picosecond after photoexcitation for P3HT/PCBM blends is attributed to the time dependent mobility caused by polaron formation on a sub-picosecond time scale.

Acknowledgments

We would like to thank Dr. J. Goldhar for valuable discussions.

References

1. Nienhuys, H.; Sundstrom, V. *Phys. Rev. B* **2005**, *71*, 235110.
2. Smith, N. V. *Phys. Rev. B* **2001**, *64*, 155106.
3. Kim, J. Y.; Kim, S. H.; Lee, H.; Lee, K.; Ma, W; Gong, X.; Heeger, A. J. *Adv. Mater.* **2006**, *18*, 572.
4. Reyes-Reyes, M.; Kim, K.; Carroll, D. L. *Appl. Phys. Lett.* **2005**, *87*, 083506.
5. Kim, J. Y.; Lee, K; Coates, N. E.; Moses, D.; Nguyen, T.; Dante, M.; Heeger, A. J. *Science* **2007**, *317*, 222.
6. Bredas, J.; Beljonne, D.; Coropceanu, V.; Cornil, J. *Chem. Rev.* **2004**, *104*, 4971.
7. Exter, M.; Grischkowsky, D. *Appl. Phys. Lett.* **1990**, *56*, 1694.
8. Beard, M. C.; Turner, G. M.; Schmuttenmaer, C. A. *Phys. Rev. B* **2000**, *62*, 15764.
9. Hendry, E.; Schins, J. M.; Candeias, L. P.; Siebbeles, L. D. A.; Bonn, M. *Phys. Rev. Lett.* **2004**, *92*, 196601.
10. Ai, X.; Beard, M. C.; Knutsen, K. P.; Shaheen, S. E.; Rumbles, G.; Ellingson, R. J. *J. Phys. Chem. B* **2006**, *110*, 25462.
11. Hendry, E.; Kowberg, M.; Schins, J. M.; Siebbeles, L. D. A.; Bonn, M. *Chem. Phys. Lett.* **2006**, *432*, 441.
12. Cunningham, P. D.; Hayden, L. M. *J. Phys. Chem. C* **2008**, *112*, 7928.
13. Kindt, J. T.; Schmuttenmaer, C. A. *J. Chem. Phys.* **1999**, *110*, 8589.
14. Gallot, G.; Grischkowsky, D. *J. Opt. Soc. Am. B* **1999**, *16*, 1204.
15. Yariv, A.; Yeh, P. *Optical Waves in Crystals: Propagation and Control of Laser Radiation*; Wiley: New York, 1984
16. Sturge, M. D. *Phys. Rev.* **1962**, *127*, 768.
17. Lee, K.; Menon, R.; Yoon, C. O.; Heeger, A. J. *Phys. Rev. B* **1995**, *52*, 4779.
18. Bettignies, R. DE; Leroy, J.; Firon, M.; Sentein, C. *Proc. Org. Photovoltaics VI* **2005**, *5938*, 5938C.
19. Li, G.; Shrotriya, V.; Huang, J.; Yao, Y.; Moriarty, T.; Emery, K; Yang, Y. *Nat. Mater.* **2005**, *4*, 864.
20. Dicker, G.; Haas, M. P. D.; Warman, J. M.; Leeuw, D. M. D.; Siebbeles, L. D. A. *J. Phys. Chem. B* **2004**, *108*, 17818.
21. Prins, P.; Grozema, F. C.; Siebbeles, L. D. A. *J. Phys. Chem. B* **2006**, *110*, 14659.

Indexes

Author Index

Subject Index

E

I

L

M

N

O

P

Q

R

S

V

W

X

Z